MICROWAVE CIRCUITS

Analysis and Computer-aided Design

MICROWAVE CIRCUITS

Analysis and Computer-aided Design

VINCENT F. FUSCO
The Queen's University of Belfast

Prentice-Hall International

Englewood Cliffs, NJ   London   Mexico
New Delhi   Rio de Janeiro   Singapore
Sydney   Tokyo   Toronto

Library of Congress Cataloging-in-Publication Data

Fusco, Vincent F., 1957–
  Microwave circuits.

  Includes bibliographies and indexes.
  1. Microwave circuits – Design and construction –
Data processing. I. Title.
TK7876.F87 1987   621.381'32   86-16882
ISBN 0-13-581588-6

British Library Cataloguing in Publication Data

Fusco, Vincent F.
  Microwave circuits: analysis and computer-
  aided design.
  1. Microwave circuits – Design and
  construction 2. Digital electronics
  I. Title
  621.381'32   TK7876

  ISBN 0-13-581588-6
  ISBN 0-13-581562-2 Pbk

Prentice-Hall Inc., *Englewood Cliffs, New Jersey*
Prentice-Hall International (UK) Ltd, *London*
Prentice-Hall of Australia Pty Ltd, *Sydney*
Prentice-Hall Canada Inc., *Toronto*
Prentice-Hall Hispanoamericana S.A., *Mexico*
Prentice-Hall of India Private Ltd, *New Delhi*
Prentice-Hall of Japan Inc., *Tokyo*
Prentice-Hall of Southeast Asia Pte Ltd, *Singapore*
Editora Prentice-Hall do Brasil Ltda, *Rio de Janeiro*

Printed and bound in Great Britain for
Prentice-Hall International (UK) Ltd,
66 Wood Lane End, Hemel Hempstead, Hertfordshire, HP2 4RG
at the University Press, Cambridge.

1 2 3 4 5   90 89 88 87 86

ISBN 0-13-581588-6
ISBN 0-13-581562-2 PBK

Contents

Preface

Microwave Circuits provides an introduction to the techniques of lumped and distributed circuitry applied at very high frequencies in the microwave and UHF frequency regions. The text of this book is sufficiently detailed to allow both the analysis and synthesis of simple and complex microwave circuits. These circuits are formulated in a way that can be implemented easily in real design situations. The book is written to encourage self-study and contains the solutions to many realistic problems. These problems illustrate the use of the governing equations and provide typical numerical values for real life problems. Manual design is emphasized strongly throughout the text. This approach is complemented by a suite of thirty useful computer programs.

The organization of the book is in five chapters, these lead the reader progressively from the fundamental aspects of microwave circuitry to the design of components for subsystem use and to the creation of design data for novel circuit element types. It is hoped that the text is flexible enough to allow individuals the freedom to select topics particular to their own needs. Sufficient theoretical material is included to provide a course in the penultimate year of a three year undergraduate degree course in elementary microwave techniques or indeed a final year degree course in numerical and analytical microwave circuit design. The book should also appeal to final year higher diploma students because of its practical nature. Interested amateurs and neophyte engineers working in this area should also find the text of value. At the end of each chapter sufficient references together with comments on their content have been included. These will enable independent study of a particular subject area in greater depth and also provide background reading for those equations quoted without proof in the text.

In order to obtain a feel for problems under investigation, manual design is emphasized and is not discounted totally in favor of computer-aided design. However, it is recognized that a computer is often indispensible when solving problems where the algebra is complicated or where tedious iterative procedures are required. Computer programs are therefore given in order to augment the manual design tools developed. These programs are written in an elementary way for ease of understanding using the BASIC language. Each program is provided with a sample set of results for test purposes. The programs developed provide a working suite of powerful CAD tools that can be operated on most small microcomputer systems. The programs can, with little ingenuity, be modified to operate together, so that a large class of custom problems can be solved once the design task has been fully formulated.

ACKNOWLEDGEMENTS

The author wishes to acknowledge the following, for their assistance in furnishing permission to use material reproduced in this text.

Equation (2.22) which originally appeared in reference [5] chapter 2. Copyright © 1965 IEEE.

Figure 2.8   Characteristic Impedance Curve for Stripline, Source Howe, H., 'Stripline Circuit Design', 1974. Reproduced by permission of Artech House, 888 Washington Street, Dedham, Massachusetts, USA 02026.

Figure 4.1   A commercially available form of the Smith Chart with normalized impedance coordinates; copyrighted by the Kay Electric Company, Pine Brook, N.J. and reproduced with their permission.

Figure 4.7   A commercially available form of the Admittance Chart; copyrighted by the Kay Electric Company, Pine Brook, N.J. and reproduced with their permission.

Thanks are also due to the Kay Electric Company for the base charts used in Figs. 4.5, 4.8, 4.9, 4.11, 4.22, 4.27, 4.28, 4.34, 4.35, 4.36.

Section 4.6.3 Based on H.P. Application Note 154, April 1972, 'S-Parameter Design' and used with their kind permission.

Equation given on page 302, chapter 5, Hammerstad, E. O. and Bekkadahl, F., 'A Microstrip Handbook', ELAB Report STF44 A74169, N7034, 1976.

Equations on page 329 reference [10] chapter 5; copyright © 1965 IEEE.

Equations 5.20 to 5.23 which originally appeared in reference [5] chapter 5; Copyright © 1958 IEEE.

Equations 5.31 and 5.32 reference [15] chapter 5; copyright © 1975 IEEE.

Equations 5.33 and 5.34 reference [14] chapter 5; copyright © 1981 IEEE.

Figure 5.40 derived from reference [5] chapter 5; copyright © 1955 IEEE.

Figure 5.43 derived from reference [15] chapter 5; copyright © 1975 IEEE.

V.F.F.

1

Basic Transmission Line Properties

A transmission line can be thought of as a circuit element that transfers energy in the form of electromagnetic waves from one place to another. Transmission lines can be divided into two classes: balanced and unbalanced. A balanced transmission line is one where two signal wires are used to propagate electromagnetic waves relative to some fixed potential, usually assumed to be ground. In an unbalanced line, one conductor forms the signal side while the other is ground. Examples of balanced and unbalanced transmission lines are flat twin (balanced) and coaxial cable (unbalanced). Both of these types of line together with a variety of other types are discussed in chapter 2.

Transmission lines find a bewildering variety of applications from zero hertz up to millions of hertz in the optical frequency range. The construction of the line will vary depending on its end use. Copper wire is used for low frequency audio applications, copper dielectric mixtures for VHF, UHF and microwave use and solid dielectric such as plastic or glass for optical use. By carefully exploiting the properties of a given line configuration, useful circuit components such as filters and impedance matching networks can be designed and built.

The design of these components is complicated somewhat by the distributed nature of the transmission line. For frequencies of 10 MHz and below, wavelengths are long, greater than thirty meters for the case cited. This means that standard electronic components, capacitors, inductors, etc., that are typically several centimeters in length, appear very short when compared with long wavelengths. This, then, gives rise to the notion of lumped circuits. That is to say circuits whose dimensions are negligible when compared with the wavelength of their excitation signals. However, a point is reached when, as the signal frequency is increased, the wavelength of the excitation signal becomes comparable with that of the circuit components. At this point distributed circuit techniques must be used. These differ from the techniques used for lumped circuit design and sets of rules appropriate to this type of design must be developed. The division between lumped and distributed circuit considerations occurs when the dimension of the electronic component is not greater than about one-twentieth times the signal wavelength.

1.1   CHARACTERISTIC IMPEDANCE: THE LUMPED LINE ANALOGY

In its simplest form, a transmission line can be considered to be a pair of guiding conductors. As such certain properties are exhibited, these conductors exhibit series impedance and loss. One simple way of describing these properties is to use a lumped circuit representation employing conventional passive circuit elements (figure 1.1).

Here an infinitesimally short section of line of length Δl is represented by three lumped sections each comprising a series loss resistance R representing copper losses perhaps, a shunt loss conductance G representing losses in the dielectric supporting structure, a series inductance L and a shunt capacitance C representing energy storage within the line. All of these quantities are calculated on a per unit length basis such that the series loss element R, for example, would be specified in terms of resistance per meter. It is assumed in figure 1.1 that the lower conductor presents a zero impedance return path to signal currents in the line and that the incremental length of line is terminated at both ends by an impedance Z_0.

If an infinitely long pair of wires were considered and voltage and current were somehow measured at uniformly spaced points along the line (figure 1.2) then

$$\frac{V_1}{I_1} = \frac{V_2}{I_2} = \cdots = \frac{V_k}{I_k} = \text{constant} = Z_0 \text{ ohms}$$

This is termed the characteristic impedance of the line and is denoted Z_0. Provided the line is infinitely long then the ratio V/I will always produce the same answer, Z_0. In general Z_0 is complex and will vary with frequency. Redrawing figure 1.1 in terms of a series impedance and a shunt admittance for one lumped section, the L network in figure 1.3 results.

The input impedance Z_{in} for the L-section shown in figure 1.3 is given by simple circuit theory as

$$Z_{\text{in}} = Z_0 = \frac{(Z_0 + \Delta Z)1/\Delta Y}{Z_0 + \Delta Z + 1/\Delta Y} = \frac{Z_0 + \Delta Z}{1 + Z_0 \Delta Y} \qquad \Delta Y, \Delta Z \to 0$$

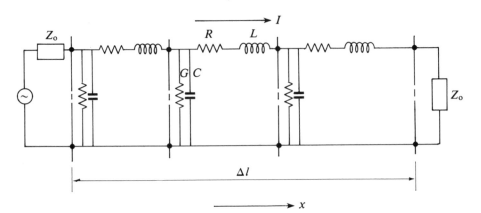

Figure 1.1   Lumped Equivalent Circuit of a Transmission Line

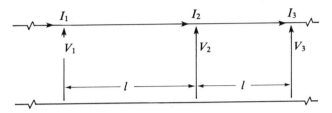

Figure 1.2   Uniform Transmission Line

Solving this expression for Z_o gives

$$Z_o = \left(\frac{\Delta Z}{\Delta Y}\right)^{1/2}$$

where $\Delta Z = R + j\omega L$ and $\Delta Y = G + j\omega C$ hence

$$Z_o = \left(\frac{R + j\omega L}{G + j\omega C}\right)^{1/2} \text{ ohms} \tag{1.1}$$

This expression is of fundamental importance since it relates the lumped circuit model for the transmission line to one of the primary line constants, characteristic impedance.

At very low frequencies ω tends to zero so that equation (1.1) becomes

$$Z_o = \left(\frac{R}{G}\right)^{1/2} \text{ ohms} \tag{1.2}$$

and for high frequencies $\omega L \gg R$ and $\omega C \gg G$ then

$$Z_o = \left(\frac{L}{C}\right)^{1/2} \text{ ohms} \tag{1.3}$$

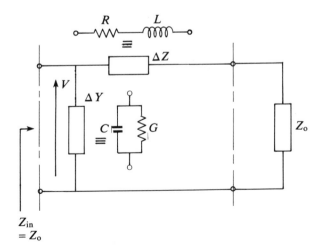

Figure 1.3   Evaluation of Characteristic Line Impedance

Since most transmission line circuit design is done at high frequencies, equation (1.3) is often used as an alternative to equation (1.1). Sometimes it is convenient to neglect line losses if they are small. In this case, the line is said to be lossless. If losses cannot be neglected, then the line is referred to as being lossy.

* * *

Example 1.1

A lossless transmission line has a characteristic impedance of 50 Ω and a self-inductance of 0.08 μH/m. Calculate the capacitance of a 4 meter length of line.

Solution

The line is lossless, therefore

$$R = 0: \ G = \infty$$

so that

$$Z_o = \left(\frac{L}{C}\right)^{1/2} \rightarrow C = \frac{L}{Z_o^2} = \frac{0.08 \times 10^{-6}}{50 \times 50} = 32 \text{ pF/m}$$

The total capacitance for a 4 meter length must then be

$$C_{TOT} = 4 \times 32 \times 10^{-12} = 128 \text{ pF}$$

This capacitance will limit the high frequency response of the cable.

* * *

1.2   FUNDAMENTAL LINE CONSTANTS

In the last section, the idea of characteristic impedance was introduced and expressions were derived from a lumped equivalent model of a uniform transmission line with incremental length (figure 1.1). Returning again to figure 1.1 a number of other important line parameters can be defined. The first is a quantity called the propagation coefficient or propagation constant ψ.

From figure 1.1 the voltage drop ΔV across one lumped section will be

$$\Delta V = - I(R + j\omega L) \, \Delta x$$

here Δx is the incremental length of line represented by the lumped section, remember R and L are defined on a per unit length basis. Dividing both sides of the above expression by Δx yields

$$\frac{\Delta V}{\Delta x} = -(R + j\omega L)I \tag{1.4}$$

As Δx tends to zero then equation (1.4) becomes

$$\frac{dV}{dx} = -(R + j\omega L)I \tag{1.5}$$

Similarly, for the shunt arm of the lumped section

$$\Delta I = -(G + j\omega C)V\,\Delta x$$

so that in the limit as Δx goes to zero

$$\frac{dI}{dx} = -(G + j\omega C)V \tag{1.6}$$

By differentiating equation (1.5) and substituting equation (1.6) into the result we get

$$\frac{d^2V}{dx^2} = (R + j\omega L)(G + j\omega C)V$$

This expression can be rewritten as

$$\frac{d^2V}{dx^2} = \psi^2 V \tag{1.7}$$

where $\psi = [(R + j\omega L)(G + j\omega C)]^{\frac{1}{2}}$ is termed the propagation constant or propagation coefficient. The propagation coefficient is normally expressed as a complex number

$$\psi = \alpha + j\beta \tag{1.8}$$

where α represents line attenuation per unit length and β the phase shift per unit length. Both of these quantities will be discussed in more detail later.

Returning for the moment to equation (1.7), it can be seen that this expression is a second order differential equation and as such will have a solution at a fixed frequency of the form

$$V(x) = A \exp(-\psi x) + B \exp(+\psi x) \tag{1.9}$$

Equation (1.9) suggests that the line will contain two waves, one traveling in the positive x direction given by $\exp(-\psi x)$ and the other propagating in the negative x direction given by $\exp(+\psi x)$. The constants A and B represent the amplitude and phase of the forward and backward waves at position $x = 0$ in the line. In order to evaluate the A and B terms consider what happens at the end of an infinite length of line when a sine wave of amplitude V_{in} is used to excite the line at position $x = 0$. If the line contains resistive elements then at $x = \infty$ any voltage across the line would have decayed to zero. In this case, equation (1.9) becomes

$$0 = A \exp(-\psi\infty) + B \exp(+\psi\infty)$$
$$= A(0) \qquad + B(\text{large quantity})$$

this means that B must equal zero.

Now at the driving end of the line, $x = 0$, the voltage is V_{in}, therefore equation

(1.9) reduces to

$$V_{\text{in}} = A \exp\left(-\psi \cdot 0\right) + 0 = A$$

Collecting the information obtained about constants A and B and substituting into equation (1.9) gives

$$V(x) = V_{\text{in}} \exp\left(-\psi x\right)$$

Consider now the propagation coefficient in more detail. Since ψ is defined as being $\alpha + j\beta$ from equation (1.8), then

$$V(x) = V_{\text{in}} \exp\left(-\alpha x\right) \exp\left(-j\beta x\right) \tag{1.10}$$

First consider the attenuation term α. From figure 1.2 it can be seen, for a line containing uniform loss per unit length of line, that

$$V_2 \quad = kV_1$$
$$V_3 \quad = kV_2$$
$$\vdots \qquad \vdots$$
$$V_{n+1} = kV_n$$

where k has a value of less than one and represents attenuation within the lossy line. For a lossless line, k is equal to one. Now, since the characteristic impedance of the line is constant, then

$$\frac{V_n}{I_n} = \frac{V_{n+1}}{I_{n+1}}$$

However, since for the lossy line $V_{n+1} = kV_n$, then

$$\frac{V_n}{I_n} = \frac{kV_n}{I_{n+1}} \rightarrow I_{n+1} = kI_n$$

This means for the lossy line that

$$\frac{V_1}{I_1} = \frac{V_2}{kI_1} = \frac{V_3}{k^2 I_1} = \cdots = \frac{V_{n+1}}{k^n I_1} = Z_o$$

So that finally

$$V_{n+1} = V_1 k^n \tag{1.11}$$

Equation (1.11) is normally represented in a compact form by taking natural logarithms. Hence

$$\log_e \left(\frac{V_{n+1}}{V_1}\right) = n \log_e k$$

Selecting $k = \exp\left(-\alpha x\right)$ in equation (1.10) yields the useful result

$$\log_e \left(\frac{V_{n+1}}{V_1}\right) = -n\alpha x$$

The term $n\alpha x$ is defined as the total line attenuation and is measured in units called *nepers*. In order to relate the unit of nepers to a more familiar unit, the *decibel*, the following development can be used.

If the power delivered by the transmission line to a load is termed P_L and the power injected into the line at the sending end is called P_{in}, then the line attenuation in decibels is defined as

$$\text{Attenuation in dB} = 10 \log_{10} \frac{P_L}{P_{in}}$$

However, in terms of voltage and current magnitudes

$$P_L = V_L I_L \quad \text{and} \quad P_{in} = V_{in} I_{in}$$

so that

$$\frac{P_L}{P_{in}} = \frac{V_L}{V_{in}} \frac{I_L}{I_{in}}$$

Referring back to the development of attenuation in terms of nepers it was shown that

$$\exp(-\alpha x) = k^n = \frac{V_{n+1}}{V_1} = \frac{V_L}{V_{in}}$$

where the $(n+1)$th section of line is assumed to be connected to the load and the first section to the input.

A similar expression to the one above can be written for current

$$\exp(-\alpha x) = k^n = \frac{I_{n+1}}{I_1} = \frac{I_L}{I_{in}}$$

Gathering these terms together, it can be seen that

$$\frac{P_L}{P_{in}} = \frac{V_L I_L}{V_{in} I_{in}} = \exp(-2\alpha x)$$

This means that

$$\begin{aligned}
\text{Attenuation in dB} &= 10 \log_{10}[\exp(-2\alpha x)] \\
&= -20\,\alpha x \log_{10}[\exp(1)] \\
&= -8.686\,\alpha x
\end{aligned}$$

This expression allows the connection between nepers and dB units of measurement for attenuation to be made.

From this it is evident that

1 neper must equal 8.686 dB

Going back to equation (1.10), it can be seen that

$$V(x) = V_{in} \text{ (attenuation as a function of length) } \exp(-j\beta x)$$

Consider now the term $\exp(-j\beta x)$. For a moment refer to figure 1.1 and

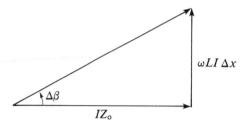

Figure 1.4   Phasor diagram for Ideal Lumped Transmision Line Section

neglect any line losses. Taking current I as a reference, then the voltage drop across a single incremental inductance element L is $j\omega LI\Delta x$ while the voltage across the element Z_0 is IZ_0. Between these two voltages exists a phase angle $\Delta\beta$ which can be determined by simple geometry (figure 1.4) as

$$\Delta\beta = \tan^{-1}\left(\frac{\omega LI\Delta x}{IZ_0}\right)$$

For small angles $\tan\theta \approx \theta$ so that for an incremental line length

$$\Delta\beta = \frac{\omega L\Delta x}{Z_0} \tag{1.12}$$

For a lossless line, equation (1.3) can be substituted into equation (1.12). This gives

$$\Delta\beta = \frac{\omega L\Delta x}{(L/C)^{1/2}} = \omega(LC)^{1/2}\,\Delta x$$

The quantity $\Delta\beta/\Delta x$ is called the phase shift, change per unit length or the wave number and is usually designated β,

$$\beta = \omega(LC)^{1/2} \tag{1.13}$$

Since the transmission line is assumed to be operated at a fixed frequency then

$$V(x) = V_{in}(\text{attenuation as a function of length}) \sin(\omega t - \beta x)$$

Rewriting the last term in this expression gives

$$\sin(\omega t - \beta x) = \sin\omega\left(t - \frac{\beta x}{\omega}\right)$$

From the above expression it can be seen that the voltage at any point x in a lossy transmission line is the original voltage attenuated, and delayed by an amount $\beta x/\omega$ seconds (figure 1.5).

Armed now with some basic equations governing the behavior of a uniform transmission line, further useful expressions can be derived.

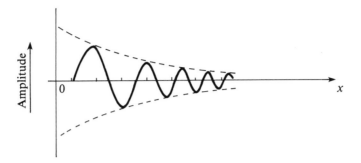

Figure 1.5   Current or Voltage Distribution in a Lossy Transmission Line

The velocity of propagation V_p of the phase front of a wave can be determined from the product of its wavelength λ and frequency f.

$$V_p = f\lambda \quad \text{or} \quad V_p = \frac{\omega}{2\pi}\lambda \quad \text{where } \omega = 2\pi f$$

Here λ is one wavelength or the physical distance required for the phase to change by 2π radians. On the other hand β is phase change per unit length, this means that

$$\lambda = \frac{2\pi}{\beta} \quad \text{or} \quad \beta = \frac{2\pi}{\lambda} \tag{1.14}$$

so that

$$V_p = \frac{\omega}{\beta} \text{ m/s}$$

Applying equation (1.13) to this result gives

$$V_p = \frac{1}{(LC)^{\frac{1}{2}}} \text{ m/s} \tag{1.15}$$

Equation (1.15) is rather useful since it allows characteristic impedance to be expressed in several useful forms. It will be recollected from equation (1.3) that

$$Z_0 = \left(\frac{L}{C}\right)^{\frac{1}{2}}$$

Rewriting this expression with equation (1.15) in mind gives

$$Z_0 = V_p L \tag{1.16}$$

or

$$Z_0 = \frac{1}{V_p C} \tag{1.17}$$

Equations (1.16) and (1.17) are especially useful when numerical techniques are employed for the design of transmission lines. See, for example, chapter 3.

* * *

Example 1.2

A section of lossy transmission line has characteristic impedance of 50 ohms and capacitance per unit length of 100 pF/m. When operated at 100 MHz, the time delay and phase shift of the output signal relative to the input signal over a 25 meter length of cable are required. Measurement on the cable shows that after a 15 meter length the magnitude of the input voltage has been reduced to 80 percent of its original value. Find the attenuation per unit length together with the total attenuation for a 25 meter section of cable.

Solution

$$\beta = \omega(LC)^{\frac{1}{2}} \quad \text{also} \quad Z_o \approx \left(\frac{L}{C}\right)^{\frac{1}{2}}$$

$$\therefore \quad L \approx Z_0^2 C = 50 \times 50 \times 100 \times 10^{-12}$$
$$= 250 \text{ n/Hm}$$
$$\beta = 2\pi \ 100 \times 10^6 (250 \times 100 \times 10^{-21})^{\frac{1}{2}}$$
$$= 3.14159 \text{ rad/m}$$

after a 25 meter length

$$\beta = 78.54 \text{ rad}$$
$$= 4500 \text{ degrees}$$
$$= 12 \times 360° + 180°$$

hence the received voltage will lag the transmitted voltage by 180°.

The time delay between transmitted and received voltages is given by

$$t = \frac{\beta l}{\omega} = \frac{3.14159 \times 25}{2\pi \times 100 \times 10^6} = 0.125 \ \mu s$$

After a 15 m length of cable, the input voltage to the line has reduced to 0.8 times its original value

$$\therefore \quad 20 \log_{10}\left(\frac{V_{in}}{0.8 \ V_{in}}\right) = 1.94 \text{ dB or } 0.22 \text{ nepers}$$

Hence, attenuation per unit length is

$$\frac{1.94}{15} = 0.13 \text{ dB/m or } 0.015 \text{ nepers/m}$$

Over the complete length of the cable, 25 m, the attenuation x is

$$20 \log_{10}\left(\frac{xV_{in}}{V_{in}}\right) = -0.13 \times 25$$

$$\therefore \quad x = 10^{-0.163} = 0.687 \text{ dB or } 0.0775 \text{ nepers}$$

Therefore, after 25 meters, the input voltage has fallen to about 69 percent of its original value.

$$* \quad * \quad *$$

For routine computation, a variety of calculations similar to those described in the above examples can be carried out using computer program 1.1 LUMPT. LUMPT is given complete with a number of sample runs so that the user is provided with a self-check until familiarity with the program is gained.

```
][ FORMATTED LISTING
FILE: PROGRAM 1.1 LUMPT
PAGE-1


 10   REM
 20   REM   **** LUMPT ****
 30   REM
 40   REM   THIS PROGRAM COMPUTES
 50   REM   CHARAC. IMP., WAVE NO.
 60   REM   TIME DELAY, PHASE VEL.
 70   REM   GIVEN FREQ., L, C PER
 80   REM   UNIT LENGTH.
 90   REM   IT ALSO CALCULATES
100   REM   ATTN. FROM MEASURED
110   REM   DATA
120   REM
130   REM   ZO=CHARAC. IMP.(OHMS)
140   REM   B=WAVE NUMBER
150   REM   T=TIME DELAY/METER
160   REM   VP=PHASE VELOCITY(M/S)
170   REM   AL=ATTENUATION(DB/M)
180   REM   L=INDUCTANCE/M
190   REM   C=CAPACITANCE/M
200   REM   F=FREQUENCY(GHZ)
210   REM
220   HOME
230   PRINT
240   PRINT "DO YOU WANT ATTENUATION ?"
250   PRINT "ENTER 1 IF YES"
260   INPUT I
270   IF I = 1 THEN
          540ELSE220
280   PRINT
290   PRINT "I/P INDUCTANCE (nH/M)"
300   INPUT L
310   PRINT "INPUT CAPACITANCE (pF/M)"
320   INPUT C
330   LET L = L * 1E - 9
340   LET C = C * 1E - 12
350   PRINT "INPUT FREQUENCY (GHZ)"
360   INPUT F
365   LET F = F * 1E9
370   LET ZO =   SQR (L / C)
380   LET T =   SQR (L * C)
390   LET B = 2 * 3.1415927 * F * T
400   LET VP = 1 / T
410   PRINT
420   PRINT "***********************"
```

```
430   PRINT
440   PRINT "LINE CAPACITANCE PER METER = "C / 1E - 12" pF/M"
450   PRINT "LINE INDUCTANCE PER METER = "L / 1E - 9" nH/M"
460   PRINT "CHARACTERISTIC IMP. = " INT (ZO * 1000 / 1000 + .5)" OHMS"
470   PRINT "WAVE NUMBER = "B" RADS/M"
480   PRINT "TIME DELAY PER METER = "T * 1E12" pSEC/M"
490   PRINT "PHASE VELOCITY = " INT (VP * 1000 / 1000 + .5)" M/S"
500   PRINT "FOR A SIGNAL FREQ. OF "F / 1E9" GHZ"
510   PRINT
520   PRINT "***********************"
530   GOTO 680
540   PRINT
550   PRINT "REDUCTION IN SIGNAL AS A DECIMAL"
560   PRINT "AFTER A GIVEN LTH. IN METERS"
570   INPUT FR
580   PRINT "I/P THE LENGTH"
590   INPUT L
600   LET AL = 20 * LOG (1 / FR) / LOG (10)
610   LET AL = AL / L
620   PRINT
630   PRINT "******************"
635   PRINT
640   PRINT "ATTENUATION = "AL" DB/M"
650   PRINT "OR "AL / 8.686" NEPERS/M"
660   PRINT
670   PRINT "******************"
680   PRINT "DO YOU WANT ANOTHER GO ?"
690   PRINT "ENTER 1 IF YES"
700   INPUT I
710   IF I = 1 THEN
               220
720   PRINT
730   PRINT "**** END OF PROGRAM ****"
740   END


END-OF-LISTING

]RUN

DO YOU WANT ATTENUATION ?
ENTER 1 IF YES
?1

REDUCTION IN SIGNAL AS A DECIMAL
AFTER A GIVEN LTH. IN METERS
?.5
I/P THE LENGTH
?10

*****************

ATTENUATION = .602059991 DB/M
OR .0693138374  NEPERS/M

******************
DO YOU WANT ANOTHER GO ?
ENTER 1 IF YES
?1

DO YOU WANT ATTENUATION ?
ENTER 1 IF YES
?0

I/P INDUCTANCE (nH/M)
?2
INPUT CAPACITANCE (pF/M)
?4
INPUT FREQUENCY (GHZ)
?10

*********************

LINE CAPACITANCE PER METER = 4 pF/M
```

```
LINE INDUCTANCE PER METER = 2 nH/M
CHARACTERISTIC IMP. = 22 OHMS
WAVE NUMBER = 5.61985186 RADS/M
TIME DELAY PER METER = 89.442719 pSEC/M
PHASE VELOCITY = 1.11803399E+10 M/S
FOR A SIGNAL FREQ. OF 10 GHZ

*************************
DO YOU WANT ANOTHER GO ?
ENTER 1 IF YES
?0

**** END OF PROGRAM ****
```

1.3 EQUIVALENT MODEL FOR LUMPED TRANSMISSION LINE

The lumped circuit discussed in section 1.1 can be recast into a form that is useful in various applications such as the design of attenuator sections or for crude estimation of discontinuities within cascaded transmission line sections. In this model small sections of transmission line, having uniform characteristic impedance along their length, are replaced by PI or TEE equivalent circuits. As was the case with the lumped circuit analogy discussed in section 1.1, these PI and TEE equivalent circuits represent the line exactly only at one specific frequency (the frequency at which the elements of the equivalent circuit have been calculated).

The equivalent representation of a transmission line segment using a TEE section is shown in figure 1.6. Here the line section assumed to be represented by figure 1.6 has length l and propagaton coefficient ψ. Simple circuit theory applied to the TEE section gives

$$Z_S = Z_1 + \frac{Z_2(Z_1 + Z_o)}{Z_o + Z_1 + Z_2} \tag{1.18}$$

If the TEE section is to be equivalent to a transmission line with characteristic impedance Z_o then $Z_s = Z_o$. But

$$I_{out} = \left(\frac{Z_2}{Z_1 + Z_2 + Z_o}\right) I_{in}$$

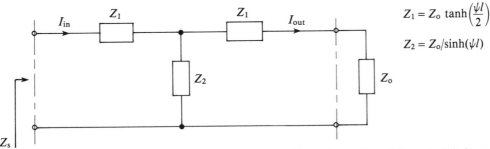

$$Z_1 = Z_o \tanh\left(\frac{\psi l}{2}\right)$$

$$Z_2 = Z_o/\sinh(\psi l)$$

Figure 1.6   Equivalent TEE Representation for Transmission Line of Characteristic Imped-ance Z_o, Propagation Constant ψ and Length l

Also, it is known from section 1.2 that if the propagation coefficient for a uniform line is ψ then

$$\frac{I_{\text{out}}}{I_{\text{in}}} = \exp(-\psi l)$$

Hence

$$\frac{Z_2}{Z_{\text{o}} + Z_1 + Z_2} = \exp(-\psi l) \tag{1.19}$$

From equation (1.18) with the equivalent $Z_s = Z_o$ then

$$Z_o = Z_1 + \exp(-\psi l)(Z_1 + Z_o) \tag{1.20}$$

Solving for Z_1

$$Z_1 = Z_o \frac{1 - \exp(-\psi l)}{1 + \exp(-\psi l)}$$

Rewriting the right-hand side of this expression gives

$$Z_1 = Z_o \frac{\exp\left(\frac{\psi l}{2}\right) - \exp\left(-\frac{\psi l}{2}\right)}{\exp\left(\frac{\psi l}{2}\right) + \exp\left(-\frac{\psi l}{2}\right)}$$

The complicated expression on the right-hand side of this equation is the hyperbolic tangent, tanh. Substituting this piece of information into the above expression gives

$$Z_1 = Z_o \tanh\left(\frac{\psi l}{2}\right) \tag{1.21}$$

Next, find Z_2 by using equation (1.19) and (1.20) so that

$$Z_0 = Z_0 \frac{1 - \exp(-\psi l)}{1 + \exp(-\psi l)} + Z_2[1 - \exp(-\psi l)]$$

Rearranging this expression in terms of Z_2 gives

$$Z_2 = Z_0 \left[\frac{2 \exp(-\psi l)}{1 - \exp(-2\psi l)}\right]$$

which simplifies on using the hyperbolic sine function to become

$$Z_2 = \frac{Z_o}{\sinh(\psi l)} \tag{1.22}$$

The values obtained for Z_1 and Z_2 in equations (1.21) and (1.22) can now be substituted into figure 1.6. These equations represent the desired design tools necessary to evaluate the TEE model.

Figure 1.7 shows the PI equivalent to the same transmission line that was modeled using the TEE circuit. The PI and TEE circuits are complementary and

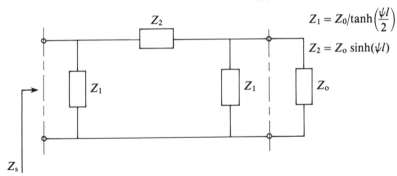

Figure 1.7   Equivalent PI Representation for Transmission Line of Characteristic Impedance Z_o, Propagation Constant ψ and Length l

either can be used for line representation with equal facility. The circuit elements for the PI representation can be derived using a procedure similar to that outlined for the TEE circuit described above.

For both the PI and TEE circuits, if length l is made very small then $\sinh(\psi l)$ approaches (ψl) and $\tanh(\psi l/2)$ tends to $(\psi l/2)$. Therefore, the transmission line representations shown in figures 1.6 and 1.7 reduce to the lumped transmission line representation discussed in section 1.1.

1.4   SENDING END IMPEDANCE

When a load is placed at the end of a section of transmission line the line will produce a transformation effect whereby the load impedance at the end of the line will appear different when viewed from the sending end of the line. This introduces a complication when calculations on transmission lines are to be performed but is advantageous to a circuit designer if exploited correctly. To illustrate one advantage and one disadvantage associated with transmission line impedance transformation, consider the following examples. The impedance transformation caused by inserting a section of transmission line between a given load impedance and a measurement point can be carefully controlled to allow a variety of useful impedance matching circuits to be designed. Conversely, if measurements are to be made on an unknown load impedance, the impedance transformation caused by inserting a connecting section of transmission line between the unknown load and the measurement equipment would have to be taken into account before the measurement results could be used. In either case, what is needed is a method for predicting this transformation.

Rewriting equation (1.9) for convenience and considering figure 1.8

$$V(x) = A \exp(-\psi x) + B \exp(+\psi x) \tag{1.23}$$

the total current flowing in the line I_T is the sum of the current traveling in the

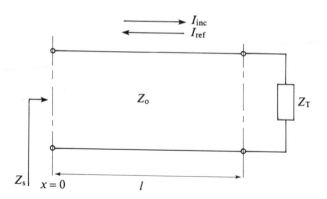

Figure 1.8   Finite Length of Transmission Line Terminated in Load Impedance Z_T

forward direction from $x = 0$ to $x = l$, I_{inc}, and the amount of current that is reflected from the load termination Z_T, I_{ref}.

This reflected current is the portion of the forward, incident, current that is not absorbed by the load.

$$I_T = I_{inc} - I_{ref}$$

$$I_T = \frac{A}{Z_o} \exp(-\psi x) - \frac{B}{Z_o} \exp(+\psi x) \tag{1.24}$$

Consider what happens at position $x = l$, here, on taking the ratio of equation (1.23) to (1.24), equation (1.25) results

$$Z_T = \frac{V_T}{I_T} = \frac{A \exp(-\psi l) + B \exp(+\psi l)}{A \exp(-\psi l) - B \exp(+\psi l)} Z_o \tag{1.25}$$

Therefore

$$\frac{B}{A} = \exp(-2\psi l) \frac{Z_T - Z_o}{Z_T + Z_o} \tag{1.26}$$

The term $\exp(-2\psi l)$ on the right-hand side of the above expression represents the loss encountered by a signal traversing from source to termination and back again.

Having looked at what happens at $x = l$, consider position $x = 0$. Substituting $x = 0$ into the ratio of equations (1.23) and (1.24) gives

$$Z_s = \frac{V_s}{I_s} = \frac{A + B}{A - B} Z_o$$

resulting in

$$\frac{Z_s}{Z_o} = (1 + B/A)/(1 - B/A) \tag{1.27}$$

Here Z_s is the sending end impedance, i.e. the impedance seen looking into the line toward the load (see figure 1.8). Substitution of equation 1.26 into 1.27 will eventually lead to the desired result.

$$\frac{Z_0}{Z_s} = \frac{1 + \exp(-2\psi l)}{1 - \exp(-2\psi l)} \left[\left(\frac{Z_T - Z_0}{Z_T + Z_0} \right) \middle/ \left(\frac{Z_T - Z_0}{Z_T + Z_0} \right) \right]$$

$$= \frac{Z_T\{1 + \exp(-2\psi l) + Z_0[1 - \exp(-2\psi l)]\}}{Z_T\{1 - \exp(-2\psi l) + Z_0[1 + \exp(-2\psi l)]\}} \qquad (1.28)$$

Stepping aside for a moment, it is known that

$$\sinh \psi l = \tfrac{1}{2} [\exp(\psi l) - \exp(-\psi l)]$$

$$\therefore \quad 1 - \exp(-2\psi l) = \exp(-\psi l)[\exp(\psi l) - \exp(-\psi l)]$$

$$= 2 \sinh(\psi l) \exp(-\psi l)$$

Similarly $1 + \exp(-2\psi l)$ is equal to $2 \cosh(\psi l) \exp(-\psi l)$

Coming back to the problem under consideration and substituting the two expressions just derived into equation (1.28) gives, on cancellation of a $2 \exp(-\psi l)$ term,

$$\frac{Z_s}{Z_0} = \frac{Z_T \cosh(\psi l) + Z_0 \sinh(\psi l)}{Z_T \sinh(\psi l) + Z_0 \cosh(\psi l)} \qquad (1.29)$$

Equation (1.29) represents the desired result since it enables the sending end impedance at the end of a length l of lossy cable to be evaluated.

Equation (1.29) can be stated in a variety of ways, for example, by division above and below by $Z_0 \cosh(\psi l)$ on the right-hand side of equation (1.29), equation (1.30) results:

$$\frac{Z_s}{Z_0} = \left[\frac{Z_T}{Z_0} + \tanh(\psi l) \right] \middle/ \left[1 + \frac{Z_T}{Z_0} \tanh(\psi l) \right] \qquad (1.30)$$

Equation (1.30) makes clear the fact that if the load termination Z_T is made equal to the characteristic impedance Z_0 of the line, then the sending end impedance Z_s will equal the characteristic impedance of the line. In this case no reflection will occur from the load, the line is said to be matched.

Equation (1.30) can be evaluated either graphically or by using tabulated data for $\tanh(\psi l)$. It should be recalled that, since ψ is a complex quantity equal to $\alpha + j\beta$, the hyperbolic tangent is also complex. In order to see how the complex hyperbolic tangent function can be computed manually, it is necessary to expand the tanh function

$$\tanh(\psi l) = \tanh(\alpha l + j\beta l)$$

$$= \frac{\tanh(\alpha l) + j \tan(\beta l)}{1 - j \tanh(\alpha l) \tan(\beta l)}$$

Here $j \tan(\beta l) = \tanh(j\beta l)$ is used. Tanh (αl) and $\tan(\beta l)$ can be evaluated in the normal manner.

Alternatively, the expansion

$$\tanh (\alpha \pm j\beta l) = \frac{\sinh (2\alpha l) \pm j \sin (2\beta l)}{\cosh (2\alpha l) + \cos (2\beta l)}$$

may prove useful.

*   *   *

Example 1.3

A transmission line of length 100 meters is operated at a frequency of 10.0 MHz and has an attenuation per unit length of 0.002 nepers/m. If the phase velocity of the line is 2.7×10^8 m/s and the line has a characteristic impedance of 50 ohms what is the value of the terminal load impedance if the input impedance looking into the line is measured to be $30 - j10$ ohms.

Solution

This problem is posed as a de-embedding exercise of the type associated with a measurement problem, i.e. given Z_s and knowing the line parameters, find Z_T.
Rearranging equation (1.30) results in

$$\frac{Z_T}{Z_s} = \left[\frac{Z_s}{Z_o} - \tanh (\psi l) \right] \Big/ \left[1 - \frac{Z_s}{Z_o} \tanh (\psi l) \right]$$

It will be noticed that this expression is almost identical in form to equation (1.30), so that all that is necessary is to note from the question that a movement along the line toward the load is needed. This is the opposite direction to that for which equation (1.30) was derived, so that it is necessary to use $-l$ instead of $+l$ in equation (1.30) to indicate a movement towards the load and let the tanh function take care of itself ($\tanh (-\psi l) \equiv -\tanh (\psi l)$).

Z_T = quantity to be found
$Z_S = 30 - j10$ ohms
$Z_o = 50 + j0$ ohms
$l = 100$ meters
$\alpha = 0.002$ nepers/m
$V_p = 2.7 \times 10^8$ m/s

so that

$$\frac{Z_s}{Z_o} = 0.6 - 0.2\, j$$

$$\alpha l = 2 \times 10^{-3} \times 100 = 0.2 \text{ nepers}$$

$\therefore \quad 2\alpha l = 0.4 \text{ nepers}$

$$\beta l = \frac{\omega l}{v_p} = \frac{2\pi \times 10 \times 10^6 \times 100}{2.7 \times 10^8} = 23.27 \text{ radians}$$

$\therefore \quad 2\beta l = 46.54 \text{ radians}$

This makes

$$\tanh(\psi l) = \tanh(\alpha l + j\beta l)$$

$$= \tanh(0.2 + j23.27)$$

$$= \frac{\sinh(0.4) + j\sin(46.54)}{\cosh(0.4) + \cos(46.54)}$$

here

$$\sinh(0.4) = \frac{\exp(0.4) - \exp(-0.4)}{2} = 0.411$$

$$\cosh(0.4) = 1.081$$

$$\sin(46.54) = \sin(46.54 - 14\pi) = \sin(2.578) = 0.5513$$

$$\cos(46.54)\cos(2.578) = -0.8343$$

so that

$$\tanh(\psi l) = \frac{0.4111 + j0.5513}{1.081 - 0.8343}$$

$$= 1.666 + j2.2347$$

Substituting this into the governing equation gives

$$\frac{Z_T}{Z_o} = \frac{0.6 - j0.2 - 1.666 - j2.2347}{1 - (0.6 - j0.2)(1.666 + j2.2347)}$$

$$= \frac{4.55 - j0.7063}{3.0}$$

and finally the desired result

$$Z_T = 76 - j12 \text{ ohms}$$

*   *   *

This calculation is fairly long due to the complexities of the algebraic manipulations required. Program 1.2 ZSEND allows sending end calculations to be carried out rapidly. A sample set of results together with input data is presented with a BASIC listing of the program. This program has been written in such a way that equation (1.30) is not evaluated directly, instead the problem is framed in terms of the reflection coefficient. This approach is more in keeping with the graphical solution methods to be discussed in chapter 4. Section 1.7 should be consulted for further details.

```
][ FORMATTED LISTING
FILE: PROGRAM 1.2 ZSEND
PAGE-1


    10    REM
    20    REM    **** ZSEND ****
    30    REM
    40    REM    CALCULATES SENDING END
    50    REM    IMPEDANCE FOR A
    60    REM    UNIFORM TRANSMISSION
    70    REM    LINE, WITH OR WITHOUT
    80    REM    LOSS.
    90    REM    -L GIVES MOVEMENT
   100    REM    TOWARDS LOAD
   110    REM    +L GIVES MOVEMENT
   120    REM    TOWARDS GENERATOR
   130    REM
   140    REM    F=FREQ.(GHZ)
   150    REM    L=LINE LTH.(CM)
   160    REM    A2=ATTENUATION (DB/CM)
   170    REM    ZO=CHARC. IMP.(OHMS)
   180    REM    E=REL./EFF. PERM.
   190    HOME
   200    PRINT "INPUT CHARC. IMP.(OHMS)"
   210    INPUT ZO
   220    PRINT "INPUT FREQ.(GHZ)"
   230    INPUT F
   240    PRINT "INPUT ATTN.(DB/CM)"
   250    INPUT A2
   260    PRINT "INPUT REL.OR EFF. PERM."
   270    INPUT E
   280    PRINT "INPUT LINE LTH.(CM)"
   290    INPUT L
   300    PRINT "I/P LOAD RES.(OHMS)"
   310    INPUT R1
   320    PRINT "INPUT LOAD REACTANCE (OHMS)"
   330    INPUT X1
   340    LET R = R1
   350    LET X = X1
   360    LET W1 = 30 / (F *  SQR (E))
   370    LET R2 = (R1 * R1 - ZO * ZO + X1 * X1) / ((R1 + ZO) * (R1 + ZO) + X1 * X1
          )
   380    LET X2 = 2 * X1 * ZO / ((R1 + ZO) * (R1 + ZO) + X1 * X1)
   390    IF X2 <  > 0 THEN
          430
   400    IF R2 >  = 0 THEN
          420
   410    LET G =  - 3.1415927:
          GOTO 53
   420    LET G = 1E - 20:
          GOTO 530
   430    IF R2 <  = 0 THEN
          460
   440    IF X2 = 0 THEN
          390
   450    LET G =  ATN (X2 / R2):
          GOTO 530
   460    IF R2 <  > 0 THEN
          500
   470    IF X2 >  = 0 THEN
          490
   480    LET G =  - 1.5707963:
          GOTO 530
   490    LET G = 1.5707963:
          GOTO 530
   500    IF R2 >  = 0 THEN
          530
   510    IF X2 = 0 THEN
          390
   520    LET G = 3.1415927 +  ATN (X2 / R2)
   530    LET T1 = G
   540    LET M1 =  SQR (R2 * R2 + X2 * X2)
   550    LET A2 = A2 / 8.686
```

```
560   LET T2 = T1 - 4 * 3.1415927 * L / W1
570   LET M2 = M1 *  EXP ( - (2 * A2 * L))
580   LET D = 1 - 2 * M2 *  COS (T2) + M2 * M2
590   IF D = 0 THEN
          LET D = 1E - 20
600   LET R1 = ZO * (1 - M2 * M2) / D
610   LET X1 = ZO * 2 * M2 *  SIN (T2) / D
620   PRINT
630   PRINT
640   PRINT "********RESULTS********"
650   PRINT
660   PRINT "LINE LTH.(CMS) "L
670   PRINT "REL. OR EFF. PERM. "E
680   PRINT "OPERATING FREQ.(GHZ) "F
690   PRINT "ATT.(DB/CM) " INT (A2 * 1000 * 8.686 + .5) / 1000
700   PRINT "CHARAC. IMP.(OHMS) "ZO
710   PRINT
720   PRINT "LOAD IMPEDANCE"
730   PRINT "   " INT (1000 * R / 1000 + .5)"  " INT (1000 * X / 1000 + .5)" j O
      HMS"
740   PRINT
750   PRINT "SENDING END IMPEDANCE"
760   PRINT "   " INT (1000 * R1 / 1000 + .5)"  " INT (1000 * X1 / 1000 + .5)" j
       OHMS"
770   PRINT
780   PRINT
790   PRINT "**********************"
800   PRINT
810   PRINT "FINISHED? IF NO ENTER 1"
820   INPUT T
830   IF T = 1 THEN
          190
840   PRINT
850   PRINT "££££££ END OF PROGRAM ££££££"
860   END

END-OF-LISTING

]RUN
INPUT CHARC. IMP.(OHMS)
?50
INPUT FREQ.(GHZ)
?10
INPUT ATTN.(DB/CM)
?0
INPUT REL.OR EFF. PERM.
?1
INPUT LINE LTH.(CM)
?.75
I/P LOAD RES.(OHMS)
?50
INPUT LOAD REACTANCE (OHMS)
?50

********RESULTS********

LINE LTH.(CMS) .75
REL. OR EFF. PERM. 1
OPERATING FREQ.(GHZ) 10
ATT.(DB/CM) 0
CHARAC. IMP.(OHMS) 50

LOAD IMPEDANCE
   50   50 j OHMS

SENDING END IMPEDANCE
   25  -25 j OHMS

**********************

FINISHED? IF NO ENTER 1
?0

££££££ END OF PROGRAM ££££££
```

1.5   SPECIAL CASE TERMINATIONS

If the load impedance terminating a transmission line is either an open circuit $Z_T = \infty$ or a short circuit $Z_T = 0$ then a number of useful circuit elements can be constructed. From equation (1.30) governing the sending end impedance of a section of terminated transmission line or alternatively from the TEE or PI transmission line representations, if a short circuit is used as the load termination then the sending end impedance becomes

$$Z_s \big|_{SC} = Z_o \tanh (\psi l)$$

for short lengths of line loss can be neglected so that

$$Z_s \big|_{SC} = Z_o \tanh (j\beta l) = jZ_o \tan (\beta l) \tag{1.31}$$

When the line is terminated in an open circuit, i.e. no load connected then

$$Z_s \big|_{OC} = \frac{Z_o}{\tanh (\psi l)}$$

Again, for short lengths $\alpha \to 0$ so that

$$Z_s \big|_{OC} = \frac{-jZ_o}{\tan (\beta l)} = -jZ_o \cot (\beta l) \tag{1.32}$$

Examination of equations (1.31) and (1.32) for line segments with short circuit and open circuit loads indicates that by correctly choosing length l, then the open or short circuit line section, or stub as it is called, can be made to have a sending end impedance that is equivalent to that obtained for a lumped capacitor or inductor. Equations (1.31) and (1.32) are plotted for $l = 0$ to $l = \lambda/4$ in figure 1.9. The points raised so far in this section will be discussed further in chapter 4.

When a transmission line is terminated in a perfect open or short circuit all

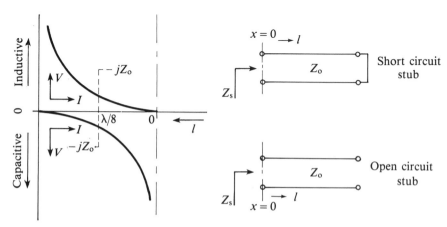

Figure 1.9   Reactance of Open and Short Circuited Line Sections as a Function of Frequency

the energy incident on the termination is reflected back into the line. This is called total reflection and will be dealt with in more detail in the next section.

Taking the product of equations (1.31) and (1.32) gives

$$Z_s|_{OC} \times Z_s|_{SC} = Z_o{}^2 \tag{1.33}$$

This happens to provide a useful method by which the characteristic impedance of a transmission line can be measured in terms of open and short circuit sending impedances. One practical point worth noting is that the line lengths selected should have some value close to an odd number of $\lambda/8$. This will ensure that $Z_s|_{OC}$, $Z_s|_{SC}$ and Z_o all have magnitudes that are roughly equal (see figure 1.9). For cases when the guide wavelength for the transmission line under test is not accurately known, trial and error on several test lines tuned to meet the condition specified by equation (1.33) is normally sufficient in order to achieve good results.

1.6   STANDING WAVES

For any transmission line, a sinusoidal signal of appropriate frequency introduced at the sending end by a generator will propagate along the length of the line. If the line has infinite length, then the signal never reaches the end of the line. If the signal is viewed at some distance down the line far away from the generator, then it will appear to the viewer to have the same frequency but will exhibit a smaller peak to peak voltage swing than it had at the generator end. The signal has been attenuated by the line losses due to conductor resistance and dielectric imperfections. If the line can be considered as having zero loss, then the signal viewed at some remote measurement point will be identical to that at the generator but time delayed by an amount dependent on the position at which the signal is sampled.

The real world consists of finite length lines that exhibit attenuation to some degree. For the purpose of this explanation, attenuation can be neglected. So then, what happens when a sinusoidal signal reaches the open end of a section of lossless transmission line? Since the line is lossless, it can dissipate no energy. In an ideal open circuit no current can flow, therefore no energy is absorbed by the open circuit. This means that all the energy propagating along the line in the forward direction (the incident signal) will be reflected completely on reaching the open circuit termination. The reflected wave (backward wave) must be such that the total current at the open circuit is zero. As the reflected signal travels back along the line toward the generator, it can reinforce the incident waveform at certain points along the line thereby forming maxima (nodes). Similarly, it can cancel the incident waveform at certain other points producing minima (antinodes). In an open-circuited line, node voltage points will occur at the same position as antinode current points. The converse is also true. Figure 1.10 illustrates the formation of standing waves for signals incident on perfectly reflecting terminations. This diagram shows, by way of

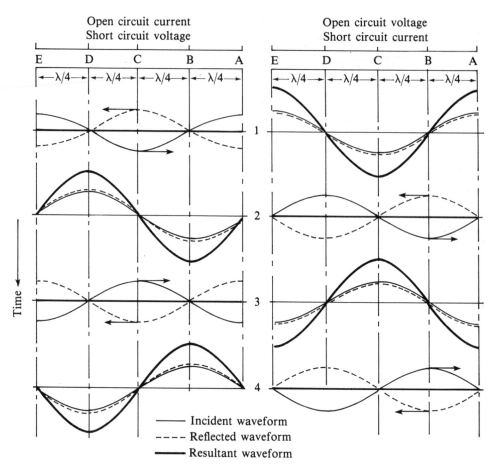

Figure 1.10   Standing Wave Formation on a Perfectly Reflecting Lossless Transmission Line

a sequence of snapshots, the formation of standing waves. The incident waveform is depicted as a thin solid line while the reflected wave is given by the dashed line. The sum of the incident and reflected waves known as the resultant is described by the heavy solid line.

In figure 1.10 it is evident from snapshot 1 that the incident and reflected current wavefronts cancel each other since they are in antiphase. The voltage waveforms at the same instant in time reinforce each other to produce a peak voltage of twice that which would normally flow in an infinite length or perfectly matched line. Here a perfectly matched line is defined as being a transmission line having finite length and terminated with a load impedance of value equal to the characteristic impedance of the line. At the next instant in time denoted by snapshot 2, the backward wave has moved a physical distance $\lambda/4$ from right to left while the incident wave has traversed the same distance but in the opposite direction. The end result is that the two waveforms are in phase and will add to produce a peak current that is double the magnitude of either the incident or reflected waveforms.

Examination of the voltage waveforms at this instant in time indicates destructive interference (cancellation) has occurred. The process of sliding the incident and reflected waveforms along the horizontal axis can be continued to produce snapshots 3 and 4 followed by 1 and 2 once again. When a fixed point on the line is viewed over a period of time, the voltage at this point will be observed to vary in a cyclic fashion. The incident and reflected waveforms in the transmission line are said to form a standing wave. When the line is terminated with a perfectly conducting short circuit, the discussion above for a perfect open circuit is valid. However, this time the current and voltage waveforms in figure 1.10 should be interchanged.

A simple demonstration program, program 1.3 SWAVE, has been developed specifically for use on a BBC B microcomputer. This program demonstrates by means of simple animation the formation of standing waves in a perfectly terminated transmission line. In the animation, forward and reflected waves are seen to slide past each other while nodes and antinodes formed in the resultant wave appear at fixed positions along the line. With program SWAVE as a model, suitable modifications to the graphical commands used will allow the program to run on a variety of microcomputers.

```
][ FORMATTED LISTING
FILE: PROGRAM 1.3   SWAVE
PAGE-1
 10 REM
 20 REM **** STANDING WAVE DEMO ****
 30 REM       **** FOR BBC B ****
 40 REM
 50 DIMA(200),C(200),B(200)
 60 Z=5
 70 MODE1
 80 PRINTTAB(8,6)"STANDING WAVE DEMONSTRATION"
 90 PRINTTAB(8,7)"============================"
100 FOR I = 0 TO 144
110    A(I)=INT(50*SIN(0.043633*I))+127
120    PRINTTAB(2,15)"I"I,"SIN(I)"INT(100*SIN(I))/100
130    NEXT I
140 CLS
150 PRINTTAB(8,4)"STANDING WAVE DEMONSTRATION"
160 PRINTTAB(8,6)"============================"
170 MOVE98,90
180 PLOT5,685,90
190 PLOT5,685,730
200 PLOT5,98,730
210 PLOT5,98,90
220 MOVE459,90
230 PLOT21,459,730
240 MOVE315,90
250 PLOT21,315,730
260 MOVE603,90
270 PLOT21,603,730
280 MOVE171,90
290 PLOT21,171,730
300 PRINTTAB(24,25)" FORWARD WAVE"
310 PRINTTAB(24,27)"    ---->"
```

```
320 PRINTTAB(24,21)"REFLECTED WAVE"
330 PRINTTAB(24,23)"     <----"
340 PRINTTAB(24,14)"RESULTANT WAVE"
350 VDU 23;8202;0;0;0;
360 PRINTTAB(3,30)"N A N A  N AN  A N"
370 K=144
380 G=0
390 MOVE100,100+126
400 FORL=0TO2*144STEP288
410    M=0:X=0
420    FOR I = K TO 144
430       PLOTZ,(100+L+M),A(I)+100
440       B(X)=A(I)
450       M=M+2:X=X+1
460       NEXTI
470    FOR F = 0 TO K
480       PLOTZ,(100+L+M),A(F)+100
490       B(X)=A(F)
500       M=M+2:X=X+1
510       NEXTF
520    NEXTL
530 MOVE459,90
540 PLOT21,459,730
550 MOVE315,90
560 PLOT21,315,730
570 MOVE603,90
580 PLOT21,603,730
590 MOVE171,90
600 PLOT21,171,730
610 MOVE100,90
620 PLOT5,100,730
630 MOVE100,220+126
640 FORL=0TO2*144STEP288
650    M=0:X=0
660    FOR I = G TO 144
670       PLOTZ,(100+L+M),A(I)+220
680       C(X)=A(I)
690       M=M+2:X=X+1
700       NEXTI
710    FOR F = 0 TO G
720       PLOTZ,(100+L+M),A(F)+220
730       C(X)=A(F)
740       M=M+2:X=X+1
750       NEXTF
760    NEXTL
770 MOVE459,90
780 PLOT21,459,730
790 MOVE315,90
800 PLOT21,315,730
810 MOVE603,90
820 PLOT21,603,730
830 MOVE171,90
840 PLOT21,171,730
850 MOVE100,90
860 PLOT5,100,730
```

```
 870 MOVE100,420+126
 880 FORL=0TO2*144STEP288
 890    M=0:X=0
 900    FORI=0TO144
 910      PLOTZ,(100+L+M),B(X)+C(X)+300
 920      M=M+2:X=X+1
 930     NEXTI
 940    NEXTL
 950 MOVE459,90
 960 PLOT21,459,730
 970 MOVE315,90
 980 PLOT21,315,730
 990 MOVE603,90
1000 PLOT21,603,730
1010 MOVE171,90
1020 PLOT21,171,730
1030 MOVE100,90
1040 PLOT5,100,730
1050 IFZ=5THENZ=7:GOTO390
1060 IFZ=5THENZ=7:GOTO630
1070 Z=5
1080 K=K-12
1090 G=G+12
1100 IFK=0THENGOTO370
1110 GOTO390
>
```

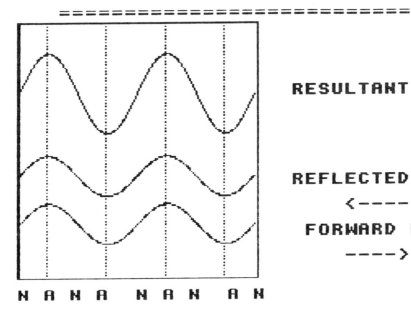

1.7   RESISTIVE AND REACTIVE TERMINATIONS

The standing wave representation shown in figure 1.10 is seldom used in practice due to the way by which voltage and current maxima are assessed experimentally. In an experimental set-up a metallic probe is placed in electrical proximity (coupled) to the magnetic fields or the electrical fields present within or close to the transmission line. The current or voltage induced in the probe by the time-varying electromagnetic fields associated with the transmission line is usually averaged by sensing equipment, so that an r.m.s. or d.c. quantity related to the magnitude of the field component of interest is normally the only information available, in other words, relative phase information is lost. For this reason, it is customary to represent the voltage and current distributions along the line in the manner shown in figure 1.11.

In figure 1.11 it is evident that a point one-quarter wavelength away from a short circuit will have voltage and current magnitude equivalent to those obtained for an open circuit. This is of considerable practical significance since a quarter wavelength section of line could be removed from the end of an open-circuited lossless transmission line without disturbing the existing current and voltage relationship already present in the line. Further reference to figure 1.11 shows the period of the standing wave patterns to be one-half wavelength. Once again this turns out to be a point of practical significance. Both half and quarter wavelength line sections will be discussed in more detail in chapter 4.

In the last section it was shown how, for a lossless line with a perfectly reflecting termination, the peak value of the standing wave envelope is twice that of the incident wave. Further, it was demonstrated how complete cancellation of the incident and reflected waves would cause a series of nulls to appear along the line length. Practical lines unfortunately have loss associated with them. These losses will prevent peaks from building up to the maximum theoretical value of twice the peak incident voltage or current. Similarly, complete cancellation of the forward and backward waves will rarely result. This produces a 'filling in' of the ideal lossless standing wave pattern nulls as shown in figure 1.12.

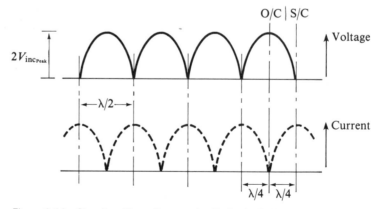

Figure 1.11   Standing Wave Patterns for Perfectly Reflecting Loads

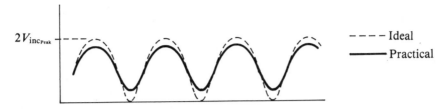

$2V_{\text{inc}_{\text{Peak}}}$

---- Ideal
——— Practical

Figure 1.12   Standing Wave Ratio for Partially Reflecting Loads

In figure 1.12 the ideal situation exhibits a peak to node voltage ratio of infinity. This peak to node ratio is known as the voltage standing wave ratio (VSWR) or simply as standing wave ratio (SWR). When line losses are considered, as they would be of necessity in a real line, then VSWR values will always be less than infinity. The discussion above for VSWR applies equally well to the standing wave representation for current.

For real terminations, partial reflection of the incident waveform will result. The amount of reflection will depend upon the termination impedance. If the termination contains a resistive component then energy in the incident wave will be absorbed. Figure 1.13 shows a modified version of figure 1.10 that is applicable to the case of partially reflecting terminations.

From this diagram, the voltage standing wave ratio is by definition given as

$$\text{VSWR} = \frac{|\,V_{\text{inc}}\,| + |\,V_{\text{ref}}\,|}{|\,V_{\text{inc}}\,| - |\,V_{\text{ref}}\,|} \tag{1.34}$$

Here, peak values have been used, but r.m.s. values ($1/\sqrt{2}$ peak) could also have been used. Load terminations can be formed by lumped elements or by sections of transmission line of different characteristic impedance to the main line, figure 1.14.

In either case, some of the energy present in the incident wave will be reflected at the discontinuity formed at the load. The discontinuity interface is assumed to

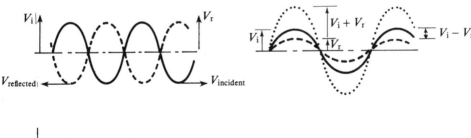

Figure 1.13   Standing Waves for a Partially Reflecting Load

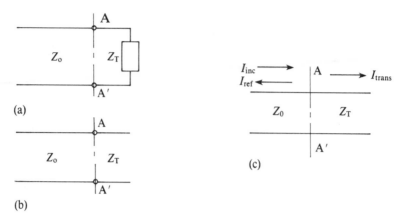

(a)

(b)

(c)

Figure 1.14   Line Discontinuities Due to Terminating Impedances (a) Lumped Termination (b) Distributed Termination (c) Current Convention

lie along line A–A′ in figure 1.14. The energy in the incident wave that is not reflected at this discontinuity will be transmitted.

Applying Kirchoff's laws for current continuity across the boundary A–A′ gives

$$\frac{V_{inc}}{Z_o} - \frac{V_{ref}}{Z_o} = \frac{V_{trans}}{Z_T} \tag{1.35}$$

and for voltage continuity across the same boundary

$$V_{inc} + V_{ref} = V_{trans} \tag{1.36}$$

Substituting for V_{trans} from equation (1.36) into equation (1.35) gives the reflection coefficient Γ as

$$\Gamma = \frac{V_{inc}}{V_{ref}} = \frac{V_T - Z_o}{Z_T + Z_o} \tag{1.37}$$

Since Z_T and Z_o are, in general, complex quantities then equation (1.37) can be written in polar form as

$$\Gamma = \rho \angle \theta \tag{1.38}$$

where

$$\rho = \left| \frac{Z_T - Z_o}{Z_T + Z_o} \right| \quad \text{and} \quad \theta = \tan^{-1} \left(\frac{\text{imaginary part of } \Gamma}{\text{real part of } \Gamma} \right)$$

From equation (1.34) and equation (1.37) VSWR can be expressed as

$$\text{VSWR} = \left(1 + \left| \frac{V_{inc}}{V_{ref}} \right| \right) \Big/ \left(1 - \left| \frac{V_{inc}}{V_{ref}} \right| \right) \tag{1.39}$$

From equation (1.39), equation (1.37) can be restated as

$$\Gamma = \left| \frac{V_{inc}}{V_{ref}} \right| = \frac{VSWR - 1}{VSWR + 1} \tag{1.40}$$

It should be noted that the reflection coefficient can be complex but VSWR is always a real quantity whose value varies between unity for a perfectly matched line $Z_T = Z_o$ and infinity for an open or short circuit, i.e. $Z_T = \infty$ or 0 respectively.

To evaluate the effect of VSWR on incident power passed to the load, consider the simple argument given below

Incident power = power dissipated in the load + reflected power

or

$$P_{inc} = P_L + P_{ref}$$

Here, a lossless line is assumed, this means that the power in the incident and reflected waves can be treated individually, (for a lossy line this is not the case). Hence

$$\frac{P_L}{P_{inc}} = \frac{P_{inc} - P_{ref}}{P_{inc}} = 1 - \frac{P_{ref}}{P_{inc}} \tag{1.41}$$

The second term on the right-hand side of this expression can be expressed as $(V_{ref}/V_{inc})^2$ since V^2/Z is a measure of power. Therefore, the ratio of power passed to the load relative to the power in the incident wave is related to VSWR by equation (1.42):

$$\frac{P_L}{P_{inc}} = 1 - \left(\frac{VSWR - 1}{VSWR + 1} \right)^2 = \frac{4VSWR}{(VSWR + 1)^2} \tag{1.42}$$

The squared term of the bottom of equation (1.42) indicates that for VSWR values much greater than 1, the power delivered to the load will fall off quickly. For example, for a VSWR of 3, only 75 percent of the power contained in the incident wave is transferred to the load. For a VSWR of 20, only 18 percent energy transfer occurs.

* * *

Example 1.4

A certain transmission line has a characteristic impedance of 50 ohms and is required to deliver 1 kilowatt of power over a short distance to a load. The cable manufacturers state that the r.m.s. voltage in the line must be less than 250 volts. Determine:

(a) the maximum VSWR that may exist in the line without damage occurring, and
(b) the amount of power that must be provided by the generator.

Solution

In the question, a short run of cable has been specified and since a value of attenuation has not been given, it is a valid assumption to neglect line loss. This means incident and reflected powers can be considered to be independent.

$$P_{\text{L}} = P_{\text{inc}} - P_{\text{ref}}$$

$$\therefore \quad P_{\text{L}} = \frac{V_{\text{inc}}{}^2}{Z_{\text{o}}} - \frac{V_{\text{ref}}^2}{Z_{\text{o}}}$$

$$= \frac{1}{Z_{\text{o}}}(V_{\text{inc}} + V_{\text{ref}})(V_{\text{inc}} - V_{\text{ref}})$$

$$= 1000 \text{ watts}$$

hence

$$(V_{\text{inc}} + V_{\text{ref}})(V_{\text{inc}} - V_{\text{ref}}) = 50\,000$$

but since the r.m.s. voltage in the line must be less than 250 volts then

$$(V_{\text{inc}} + V_{\text{ref}}) < 250$$

this means

$$(V_{\text{inc}} - V_{\text{ref}}) > \frac{50\,000}{250} = 200$$

now

$$\text{VSWR} = \frac{|V_{\text{inc}}| + |V_{\text{ref}}|}{|V_{\text{inc}}| - |V_{\text{ref}}|}$$

then

$$\text{VSWR} < \frac{250}{200} = 1.25$$

For this value of VSWR 98.8 percent of the power in the incident wave is delivered to the load. Equating percentage power delivered to the load and the power launched into the cable results in

$$0.988 = 1000 \text{ watts}$$

therefore, the generator must be capable of supplying at least 1012 watts and must be properly matched to the line. This gives maximum power transfer and reduces secondary reflections of the reflected wave.

* * *

In normal design VSWR is kept below 1.1 giving 99.8 percent power delivery to the load. A computer program, program 1.4 MISS, can be used to evaluate

```
][ FORMATTED LISTING
FILE: PROGRAM 1.4 MISS
PAGE-1


10   REM
20   REM   **** MISS ****
30   REM
40   REM   THIS PROGRAM COMPUTES
50   REM   THE REFLECTION COEFF. OF
60   REM   A UNIFORM TRANS. LINE
70   REM   GIVEN THE LOAD IMP.
80   REM   AND CHARAC. IMP.
90   REM   IT ALSO COMPUTES
100  REM   VSWR, AND THE
110  REM   PROPORTION OF POWER
120  REM   IN THE INCIDENT WAVE
130  REM   DELIVERED TO THE LOAD
140  REM
150  REM   VSWR=VOLTAGE S.W. RATIO
160  REM   FAC=REFLECTION COEFF.
170  REM   PR=% POWER TO LOAD
180  REM   ZO=CHARAC. IMP.(OHMS)
190  REM   ZL=LOAD IMP.(OHMS)
200  REM
210  HOME
220  PRINT
230  PRINT "I/P CHARAC. IMP(OHMS)"
240  PRINT "REAL PART FIRST"
250  INPUT RO
260  PRINT "NOW IMG. PART"
270  INPUT IO
280  PRINT
290  PRINT "INPUT LOAD IMP.(OHMS)"
300  PRINT "REAL PART FIRST"
310  INPUT RL
320  PRINT "NOW IMG. PART"
330  INPUT IL
340  REM
350  REM   REFLECTION COEFF
360  LET A = RL - RO
370  LET B = IL - IO
380  LET C = RL + RO
390  LET D = IL + IO
400  LET E = C * C + D * D
410  LET F = A * C + B * D
420  LET G = B * C - A * D
430  LET F = F / E: F1 = F
440  LET G = G / E: G1 = G
450  PRINT
460  PRINT "****** RESULTS ******"
470  PRINT
480  PRINT "FOR A UNIFORM LINE"
490  PRINT "OF CHARAC. IMP. ( "RO","IO" j) OHMS"
500  PRINT "AND LOAD IMP.  ( "RL","IL" j) OHMS"
510  PRINT "THE CALCULATED VALUE FOR"
520  PRINT "LOAD REFLECTION COEFF IS ( " INT (1000 * F / 1000 + .5)"," INT (10
     00 * G / 1000 + .5)" j) OHMS"
530  IF G < > 0 THEN
        570
540  IF F > = 0 THEN
        560
550  LET Z =  - 3.1415927:
     GOTO 670
560  LET Z = 1E - 20:
     GOTO 670
570  IF F < = 0 THEN
        600
580  IF G = 0 THEN
        530
590  LET Z =  ATN (G / F):
     GOTO 670
600  LET G < > 0 THEN
        640
```

```
610   IF F >  = O THEN
             630
620   LET Z =  - 1.5707963:
      GOTO 670
630   LET Z = 1.5707963:
      GOTO 670
640   IF F >  = O THEN
             670
650   IF G = O THEN
             530
660   LET G = 3.1415927 +  ATN (G / F)
670   LET T1 = Z
680   PRINT "OR IN POLAR FORM " INT ((F * F + G * G) * 1000 / 1000 + .5)" AT "
      INT (T1 * 360 / 2 / 3.141597)" DEGREES"
690   REM
700   REM   CALCULATE VSWR
710   LET H =   ABS (F1 * F1 + G1 * G1)
720   IF H = 1 THEN
             LET VSWR = 10000:
             GOTO 740
730   LET VSWR = (1 + H) / (1 - H)
740   PRINT
750   PRINT "VSWR = " INT (VSWR * 1000 / 1000 + .5)
760   REM
770   REM   CALCULATE POWER RATIO
780   REM
790   LET PR = 4 * VSWR / ((VSWR + 1) * (VSWR + 1)) * 100
800   PRINT
810   PRINT "% OF INCIDENT POWER"
820   PRINT "TRANSFERRED TO THE LOAD IS   " INT (PR * 1000 / 1000 + 0.5)
830   PRINT
840   PRINT "*********************"
850   PRINT
860   PRINT "FINISHED? ENTER 1 IF NO"
870   INPUT B
880   IF B = 1 THEN
             210
890   PRINT
900   PRINT "******* END OF PROGRAM *******"
910   END


END-OF-LISTING


]RUN

I/P CHARAC. IMP(OHMS)
REAL PART FIRST
?50
NOW IMG. PART
?0

INPUT LOAD IMP.(OHMS)
REAL PART FIRST
?50
NOW IMG. PART
?0


****** RESULTS ******


FOR A UNIFORM LINE
OF CHARAC. IMP. ( 50,0 j) OHMS
AND LOAD IMP.  ( 50,0 j) OHMS
THE CALCULATED VALUE FOR
LOAD REFLECTION COEFF IS ( 0,0 j) OHMS
OR IN POLAR FORM 0 AT 0 DEGREES

VSWR = 1
```

```
% OF INCIDENT POWER
TRANSFERRED TO THE LOAD IS   100

**********************

FINISHED? ENTER 1 IF NO
?0

******* END OF PROGRAM *******
```

VSWR for a transmission line with known characteristic impedance and load termination. The proportion of incident power delivered to the load for the value of VSWR found in the first part of the program is also computed.

Sometimes it is useful to compute sending end impedance on the basis of reflection coefficient. Figure 1.15 gives the notation necessary for an appreciation of how this calculation can be performed. Hence by definition

$$\Gamma_1 = \frac{Z_T - Z_o}{Z_T + Z_o} \tag{1.43}$$

rearranging equation (1.28) in terms of $\Gamma_{in} = (Z_s - Z_o)/(Z_s + Z_o)$ gives

$$\Gamma_{in} = \Gamma_1 \exp(-2\psi l) \tag{1.44}$$

These expressions provide the mechanism for the desired computation. For a line exhibiting zero loss, equation (1.44) can be written as

$$\Gamma_{in} = \Gamma_1 \exp(-2\beta l) \tag{1.45}$$

Calculation of sending end impedance in terms of reflection coefficients corresponds exactly to the most popular graphical technique used for the manual solution of transmission line problems. The graphical technique involves the use of a diagram called the Smith Chart. The derivation and application of the Smith Chart are fully detailed in chapter 4.

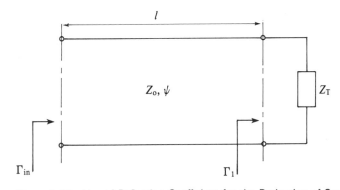

Figure 1.15   Use of Reflection Coefficient for the Derivation of Sending End Impedance

1.8   FURTHER READING

1   Slater, J. C., *Microwave Transmission,* McGraw-Hill, 1942.
Chapter 1 of this book discusses transmission line theory based on lumped two port networks. Maxwell's equations are discussed in chapter 2 and applied to the analysis of rectangular waveguide and other transmission line problems. A good blend of prose and mathematics.

2   Chipman, R.A., *Theory and Problems of Transmission Lines,* Schaum's Outline Series, McGraw-Hill, 1968.
In this book, a lumped line approach is used to set up transmission line partial differential equations. Traveling waves are examined as are standing waves. Standing waves are treated in detail with the effects of lossy lines being catered for. This title would prove a useful addition to a transmission line designer's library. It contains over 150 solved problems.

3   Stroud, K. A., *Engineering Mathematics Programs and Problems,* Macmillan, 1970.
This is a programmed student's text for self-assessed learning. Treats hyperbolic functions and complex arithmetic necessary for the handling of transmission line equations in a simple and enlightening way. The solution of simple differential equations are also considered.

4   Collin, R. E., *Foundations for microwave engineering,* McGraw-Hill, 1966.
Well-presented and useful text for a review of some of the techniques and analytical devices used to solve electromagnetic problems.

5   Cross, A. W., *Experimental Microwaves Issue 3*, Marconi Instruments Limited. Printed by Focus (Technical Services) Limited, 1975.
This practical handbook shows how the SWR of a transmission line can be measured using a standard microwave test bench. A discussion on modes within waveguide circuitry is included together with a number of useful illustrations.

2

Common Transmission Line Types

2.1 INTRODUCTORY REMARKS

This chapter provides an introduction to the main types of transmission lines currently in use for communication applications. Each line type has advantages and disadvantages when compared with other line types. For a particular application the selection of the precise transmission line to be used will depend on a number of different and often conflicting criteria. Selection of a particular line will depend on the designer's ability to assess a given electrical or mechanical specification. Two basic transmission line parameters of interest are characteristic impedance and guide wavelength. These parameters will be examined and the governing equations used to form the basis for a suite of computer-aided design routines.

2.2 COAXIAL LINE

Coaxial line (COAX as it is often called) is one of the most common transmission lines in use in industry today. Coaxial line finds a variety of applications, being used to supply high frequency signals to test equipment such as oscilloscopes or signal generators where spurious radiation of signals may cause problems. In the home, coaxial cable can be seen in the form of the down lead from a TV antenna to the television receiver. Here cable with a 75 Ω characteristic impedance is accepted as standard while for most other types of radio frequency work 50 Ω is considered the norm. Coaxial cable can be operated successfully from zero hertz up to tens of gigahertz.

To derive an expression for the characteristic impedance of coaxial line first consider how a coaxial cable is constructed. The geometry is shown in figure 2.1. Coaxial cable in its most usual form consists of a solid inner conductor, usually copper, of radius a and an outer shield with inner radius b. This shield would normally be made of a copper braid for flexible cable or by a solid outer conductor for rigid cable (rigid cable is normally associated with extremely high frequency work in the microwave frequency region). The cable is completed by introducing a

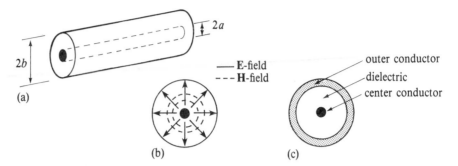

Figure 2.1   Coaxial Transmission Line (a) Geometry (b) Field Profile (c) Construction

dielectric with relative permittivity ε_r. This dielectric adds strength to the cable and acts as an insulator between the inner conductor and the shield. A typical dielectric material for VHF and UHF work is teflon, while for microwave work air-spaced cable is often used.

Before characteristic impedance can be calculated electric flux density and magnetic flux density need to be introduced. Electric flux density **D** is defined as being equal to $\varepsilon \mathbf{E}$ where **E** is the electric field intensity and ε is permittivity.

$$\mathbf{D} = \varepsilon \mathbf{E} \tag{2.1}$$

The analogous expression for magnetic fields is given as

$$\mathbf{B} = \mu \mathbf{H} \tag{2.2}$$

where **B** is magnetic flux density, μ is the permeability and H the magnetic field intensity. Equations (2.1) and (2.2) are discussed fully by Ramo *et al* [1].

Let us now calculate line capacitance per unit length for the coaxial cable given in figure 2.1. Suppose a voltage V is maintained between the inner and outer conductors and that there is a charge per unit length of $+q$ and $-q$ respectively on these conductors. The electric field intensity at some radius r lying between a and b is

$$\mathbf{D} = \frac{q}{2\pi r} \, \text{C/m}^2 \tag{2.3}$$

Since voltage V is given as $\int E \, dr$, equations (2.3) and (2.1) can be used to find voltage. This is done by integrating across the dielectric region of the coaxial line.

$$V = \int_a^b \frac{q}{2\pi \varepsilon r} \, dr$$

$$= \frac{q}{2\pi \varepsilon} (\ln b - \ln a)$$

here

$$\int \frac{1}{r} \, dr = \text{natural logarithm of } r = \ln r$$

The above expression reduces to

$$V = \frac{q}{2\pi\varepsilon} \ln\left(\frac{b}{a}\right)$$

from this expression the capacitance per unit length of the cable C can be found.

$$C = \frac{q}{V} = \frac{2\pi\varepsilon}{\ln(b/a)} \text{ F/m} \tag{2.4}$$

The easiest way to find line inductance per unit length is to follow the approach detailed above for line capacitance.

The magnetic field intensity $\mathbf{H}$ for the line is given as

$$\mathbf{H} = \frac{I}{2\pi r} \tag{2.5}$$

for a current I flowing in the line at some radius r. From equations (2.2) and (2.5)

$$\mathbf{B} = \frac{\mu I}{2\pi r} \text{ Wb/m}^2 \text{ or Tesla}$$

multiplying this expression by an area dr by one meter, measured along the line, gives the total flux ϕ linking the defined area (see figure 2.1(b)):

$$\phi = \frac{\mu I}{2\pi r} \cdot 1 \cdot dr \text{ Wb}$$

Inductance per unit length for the cable L can now be found

$$L = \frac{\phi}{I} = \frac{1}{I} \int_a^b \frac{\mu I}{2\pi r} \, dr = \frac{\mu}{2\pi} \ln\left(\frac{b}{a}\right) \text{ H/m} \tag{2.6}$$

To find the characteristic impedance of the line Z_o is now a simple task. The characteristic impedance of the line is defined in chapter 1 as

$$Z_\text{o} = \left(\frac{L}{C}\right)^{\frac{1}{2}}$$

Using equations (2.4) and (2.6)

$$Z_\text{o} = \left[\frac{\mu}{2\pi} \ln\left(\frac{b}{a}\right) \ln\left(\frac{b}{a}\right) \frac{1}{2\pi\varepsilon}\right]^{\frac{1}{2}}$$

$$= \frac{1}{2\pi} \left(\frac{\mu_\text{o}}{\varepsilon_\text{o}}\right)^{\frac{1}{2}} \left(\frac{\mu_\text{r}}{\varepsilon_\text{r}}\right)^{\frac{1}{2}} \ln\left(\frac{b}{a}\right) \tag{2.7}$$

where $\mu = \mu_\text{r}\mu_\text{o}$ and $\varepsilon = \varepsilon_\text{r}\varepsilon_\text{o}$. The suffix r denotes relative permeability or permittivity.

$$\varepsilon_\text{o} = 8.854 \times 10^{-12} \text{ F/m} \quad \text{and} \quad \mu_\text{o} = 4\pi \times 10^{-7} \text{ H/m}$$

normally μ_r will be unity for the dielectrics used in coaxial line constructions. The usual form of equation (2.7) is obtained by substituting ε_0, μ_0 and μ_r into equation (2.7)

$$Z_0 = \frac{60}{(\varepsilon_r)^{1/2}} \ln \left(\frac{b}{a}\right) \ \Omega \tag{2.8}$$

This expression is more often given in terms of logarithms to the base ten. Here the conversion $\ln x = 2.303 \log x$ is used, equation (2.8) then becomes

$$Z_0 = \frac{138}{(\varepsilon_r)^{1/2}} \log \left(\frac{b}{a}\right) \ \Omega \tag{2.9}$$

Equations (2.8) and (2.9) will of course return identical numerical results.

Equations 2.8 and 2.9 are known as analysis equations. That is, given the physical line dimensions, the electrical parameters of the line, in this case characteristic impedance, can be determined. For synthesis, however, (i.e. given the electrical parameters of the line determine the required physical dimensions) equations (2.8) or (2.9) should be rearranged. The synthesis problem usually takes the form: find the radius of the inner conductor with a known sheath radius and dielectric material for a specified characteristic impedance.

Equation 2.9 then becomes after rearranging

$$a = b \cdot 10^{-[Z_0(\varepsilon_r)^{1/2}/138]} \tag{2.10}$$

* * *

Example 2.1

Given a coaxial line with an inner diameter of 0.29 cm and sheath diameter of 1 cm calculate the characteristic impedance of the cable if the dielectric constant is 2.3.

Solution

$$Z_0 = \frac{138}{(\varepsilon_r)^{1/2}} \log \left(\frac{b}{a}\right)$$

$\varepsilon_r = 2.3$, $2b = 1$ cm, $2a = 0.29$ cm

$$\therefore \quad Z_0 = 92.4 \log \left(\frac{1}{0.29}\right) = 92.4 \times 0.54 \approx 50 \ \Omega$$

characteristic impedance $= 50 \ \Omega$

Example 2.2

A coaxial cable is required with a characteristic impedance of 25 Ω. The cable is to be constructed from outer conductor having a mean inner diameter of 1 cm and dielectric spacing material with dielectric constant of 4. What radius must the inner conductor have in order to give the desired characteristic impedance?

Solution

$$Z_o = \frac{138}{(\varepsilon_r)^{\frac{1}{2}}} \log \left(\frac{b}{a}\right)$$

rearranging

$$a = b \; 10^{-[Z_o(\varepsilon_r)^{1/2}/138]} = 1 \cdot 10^{-(25\sqrt{4}/138)} = 10^{-0.362} = 0.43 \text{ cm}$$

radius of inner conductor $= 0.43$ cm

* * *

Inspection of equation (2.8) reveals an inner to center conductor ratio b/a, this ratio is directly related to the power handling capability of the cable. The inner conductor diameter limits the current carrying capacity, the dielectric quality and thickness determines the point at which voltage breakdown occurs.

Coaxial cable is available from manufacturers with various characteristic impedance, loss and capacitance per unit length. Coaxial cable is usually specified in Britain using a UR prefix and in the USA using an RG prefix. For example 50 Ω coaxial cable with 100 pF/m capacitance could be specified as UR67 or as RG-5B/U provided other parameters such as attenuation are not considered. Some typical specifications for semi-rigid coaxial cable are shown in table 2.1.

Since coaxial cable normally contains a dielectric material the electromagnetic waves within the cable will propagate at a velocity lower than the speed of light. For this reason the guide wavelength λ_g for dielectrically loaded coaxial line will be less than the free space wavelength λ_o. A simple expression which can be used to determine guide wavelength is given in equation (2.11) (see equation 2.24)

$$\lambda_g = \frac{\lambda_o}{(\varepsilon_r)^{\frac{1}{2}}} \qquad\qquad (2.11)$$

Table 2.1   Typical semi-rigid coaxial cable data

RG No.	405/u		402/u		401/u	
Inner Conductor O.D(in)	0.0201		0.0359		0.0641	
Dielectric O.D(in)	0.066		0.1175		0.210	
Outer Conductor O.D(in)	0.085		0.141		0.250	
Frequency GHz	α	p	α	p	α	p
0.1	5.6	0.785	3.6	2.20	1.9	4.90
1.0	18.7	0.222	11.6	0.60	7.3	1.20
5.0	46.0	0.082	28.5	0.23	18.9	0.44
10.0	68.5	0.048	44.5	0.16	29.0	0.28
18.0	115.0	0.028	68.0	0.10	—	—

α = attenuation, dB/100 ft
p = average power rating, KW at VSWR = 1

A computer program, program 2.1, COAX, based on equations (2.4), (2.6) and (2.10) enables rapid calculation on coaxial line to be carried out. This program allows line inductance, capacitance per unit length and characteristic impedance to be calculated given ε_r, b and a. The program also allows an appropriate inner conductor diameter to be found given Z_o, d and ε_r. A simple check is made on the physical realizability of the geometry selected by synthesis equation (2.10). A sample run is included to test the program for both analysis and synthesis.

```
][ FORMATTED LISTING
FILE: PROGRAM 2.1 COAX
PAGE-1


 10   REM
 20   REM   -----PROGRAM COAX-----
 30   REM
 40   REM      THIS PROGRAM COMPUTES
 50   REM      LINE INDUCTANCE, CAP.
 60   REM      AND CHARAC. IMP FOR
 70   REM      COAXIAL LINE.
 80   REM      IT CAN ALSO BE USED
 90   REM      TO FIND THE REQUIRED
100   REM      INNER CONDUCTOR SIZE
110   REM      FOR A GIVEN CHARAC.
120   REM      IMPEDANCE.
130   REM
140   REM.  IDUCT=LINE INDUCTANCE
150   REM   CAP=LINE CAPACITANCE
160   REM   OCND=OUTER COND.(CMS)
170   REM   ICND=INNER COND.(CMS)
180   REM   ZO=CHARAC. LINE IMP.
190   REM   RDIE=REL. DIELECTRIC
200   REM
210   REM   INPUT DATA
220   HOME
230   PRINT "IF YOU WISH TO FIND CHARAC. IMP. ENTER 1"
240   PRINT "TO FIND INNER CONDUCTOR DIAMETER ENTER 2"
250   PRINT "TO ABORT ENTER 3"
260   INPUT K
270   IF K = 1 THEN
           310
280   IF K = 2 THEN
           350
290   IF K = 3 THEN
           890
300   GOTO 220
310   PRINT "ENTER DIA. OF INNER CONDUCTOR"
320   PRINT "IN CMS."
330   INPUT ICND
340   GOTO 370
350   PRINT "ENTER CHARAC. IMP."
360   INPUT ZO
370   PRINT "ENTER REL. DIELECTRIC"
380   INPUT RDIE
390   PRINT "ENTER DIA. OF OUTER CONDUCTOR"
400   PRINT "IN CMS."
410   INPUT OCND
420   IF K = 2 GOTO 730
430   REM
440   REM   FIND LINE INDUCTANCE
450   LET IDUCT = 2E - 7 *  LOG (OCND / ICND)
460   LET I =  INT (IDUCT * 1E9) / 1E9
470   PRINT
480   PRINT "********** RESULTS **********"
490   PRINT
500   PRINT "LINE INDUCTANCE (H/M) "I
510   REM
```

```
520   REM   FIND LINE CAPACITANCE
530   LET CAP = 5.563252E - 11 * RDIE
540   LET CAP = CAP /  LOG (OCND / ICND)
550   LET I =  INT (CAP * 1E13) / 1E13
560   PRINT
570   PRINT "LINE CAPACITANCE (F/M) "I
580   REM
590   REM   FIND CHARAC. IMPEDANCE
600   PRINT
610   LET ZO =  SQR (IDUCT / CAP)
620   LET I =  INT (ZO * 100 + 0.5) / 100
630   PRINT "CHARAC. IMPEDANCE (OHMS) "I
640   PRINT
650   PRINT "****************************"
660   PRINT
670   PRINT "DO YOU WANT ANOTHER GO ?"
680   PRINT "ENTER 1IF YES ; 0 IF NO"
690   INPUT K
700   IF K = 1 GOTO 220
710   IF K = 0 GOTO 890
720   GOTO 670
730   REM
740   REM   TO CHOOSE INNER
750   REM   CONDUCTOR
760   LET ICND = OCND * 10 ^ ( - ZO *  SQR (RDIE) / 138)
770   IF ICND > = OCND THEN
          860
780   LET I =  INT (ICND * 1000 + 0.5) / 1000
790   PRINT
800   PRINT "********** RESULTS **********"
810   PRINT
820   PRINT "DIA. OF INNER CONDUCTOR (CMS) "I
830   PRINT
840   PRINT "****************************"
850   GOTO 660
860   PRINT
870   PRINT "REALIZABILITY ERROR"
880   GOTO 660
890   PRINT
900   PRINT "+++++ END OF RUN +++++"
910   END


END-OF-LISTING


]RUN
IF YOU WISH TO FIND CHARAC. IMP. ENTER 1
TO FIND INNER CONDUCTOR DIAMETER ENTER 2
TO ABORT ENTER 3
?1
ENTER DIA. OF INNER CONDUCTOR
IN CMS.
?0.25
ENTER REL. DIELECTRIC
?1
ENTER DIA. OF OUTER CONDUCTOR
IN CMS.
?1

********** RESULTS **********

LINE INDUCTANCE (H/M) 2.77E-07

LINE CAPACITANCE (F/M) 4.01E-11

CHARAC. IMPEDANCE (OHMS) 83.12

****************************

DO YOU WANT ANOTHER GO ?
ENTER 1IF YES ; 0 IF NO
?1
IF YOU WISH TO FIND CHARAC. IMP. ENTER 1
```

```
TO FIND INNER CONDUCTOR DIAMETER ENTER 2
TO ABORT ENTER 3
?2
ENTER CHARAC. IMP.
?50
ENTER REL. DIELECTRIC
?2
ENTER DIA. OF OUTER CONDUCTOR
IN CMS.
?1

********* RESULTS *********

DIA. OF INNER CONDUCTOR (CMS) .307

***************************

DO YOU WANT ANOTHER GO ?
ENTER 1IF YES ; 0 IF NO
?0

+++++ END OF RUN +++++
```

2.2.1 Flat Twin

Flat twin, twinex or twin lead transmission line is a parallel wire system, figure 2.2. Here two cylindrical conductors with radius a are placed in parallel, a distance d between centers. It is assumed that the effect of ground planes on the field profiles around the conductors is negligible. This type of cable is commonly used to connect a dipole antenna to a TV receiver. In practice the geometry given in figure 2.2 is augmented with a thin dielectric spacer in order to maintain conductor separation. Flat twin cable is available commercially with characteristic impedance ranging from 75 to 600 Ω. Applying the procedures given in section 2.1 to the case of flat twin line an expression for characteristic impedance can be derived.

As before, the capacitance per unit length of the line has to be calculated. To do this let a charge of $+q$ and $-q$ coulomb be placed on each meter of wire. Since the structure is symmetrical about plane $X-X'$, shown in figure 2.2, the electric flux

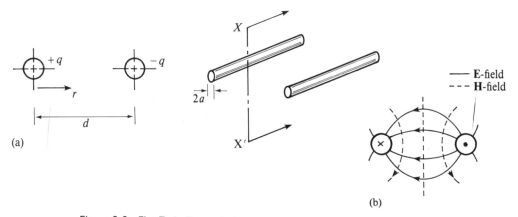

(a)

(b)

Figure 2.2   Flat Twin Transmission Line (a) Geometry (b) Field Profile

density at radius r is the sum of that due to each conductor

$$\mathbf{D} = \frac{q}{2\pi r} - \frac{-q}{2\pi(d-r)} \tag{2.12}$$

The voltage between conductors is twice that found between one conductor and plane $X - X'$. Therefore

$$V = \int_a^{d-a} E\, dr$$

from equation (2.1) and equation (2.12)

$$V = \int_a^{d-a} \frac{1}{\varepsilon} \left[\frac{q}{2\pi r} + \frac{q}{2\pi(d-r)} \right] dr$$

$$= \frac{q}{2\pi\varepsilon} \int_a^{d-a} \left(\frac{1}{r} + \frac{1}{d-r} \right) dr$$

$$= \frac{q}{2\pi\varepsilon} \left[2 \ln(d-a) - 2 \ln a \right]$$

$$= \frac{q}{\pi\varepsilon} \ln \left(\frac{d-a}{a} \right)$$

From this expression capacitance per unit length becomes

$$C = \frac{q}{V} = \frac{\pi\varepsilon}{\ln \left(\dfrac{d-a}{a} \right)} \quad \text{F/m} \tag{2.13}$$

Inductance per unit length is the next quantity to be found. The magnetic field at a point distance r from the center of a single long straight conductor can be expressed by equation (2.5). For a two wire system this becomes equal to the sum of the contribution from each wire

$$\mathbf{H} = \frac{I}{2\pi r} + \frac{I}{2\pi(d-r)} \tag{2.14}$$

From equations (2.2) and (2.14) magnetic flux becomes

$$\mathbf{B} = \frac{\mu I}{2\pi} \left(\frac{1}{r} + \frac{1}{d-r} \right)$$

Hence the magnetic flux outside the conductors

$$\phi = \frac{\mu I}{2\pi} \int_a^{d-a} \left(\frac{1}{r} + \frac{1}{d-r} \right) dr$$

$$= \frac{\mu I}{2\pi} \ln \left(\frac{d-a}{a} \right)$$

Compare this with the capacitance calculation given in this section. From the above expression for flux the inductance per unit length is

$$L = \frac{\phi}{I} = \frac{\mu}{\pi} \ln \left(\frac{d-a}{a} \right) \tag{2.15}$$

Equations (2.13) and (2.15) can now be combined to give the characteristic impedance of a parallel wire system

$$Z_o = \left(\frac{L}{C} \right)^{1/2} = \left(\frac{\mu}{\varepsilon} \right)^{1/2} \frac{1}{\pi} \ln \left(\frac{d-a}{a} \right)$$

this can be rewritten, after converting to $\log_{10}$ and substituting for ε_o and μ_o with $\mu_r = 1$ and $\varepsilon_r = 1$, as

$$Z_o = 276 \log \left(\frac{d-a}{a} \right)$$

for $d \gg a$ this reduces to

$$Z_o = 276 \log \left(\frac{d}{a} \right) \tag{2.16}$$

It should be noted that if the spacing to radius ratio d/a is made less than 10, equation (2.16) is no longer accurate since the charge distribution on the wires is non-uniform. Under these circumstances equation (2.16) is approximately 10 percent in error.

* * *

Example 2.3

Given a parallel wire transmission line with a spacing of 10 cm and a specified characteristic impedance of 600 Ω, calculate the required diameter of wire.

Solution

$$Z_o = 276 \log \left(\frac{d}{a} \right)$$

$d = 10$ cm, $Z_o = 600$ Ω

$$\therefore \quad \frac{d}{a} = 10^{600/276} = 149.25$$

$$\therefore \quad a = \frac{d}{149.25} = \frac{10}{149.25} = 0.06 \text{ cm}$$

Hence required diameter for conductors $= 2a = 0.12$ cm.

Example 2.4

Determine the spacing required for a parallel wire transmission line system if the wire diameter is 0.01 cm and the characteristic impedance of the line is (a) 600 Ω, (b) 150 Ω.

Solution

$$Z_o = 276 \log \left(\frac{d}{a}\right)$$

$$a = \frac{d}{2} = 0.005 \text{ cm}$$

(a) $600 = 276 \log \left(\frac{d}{0.005}\right)$

$$149.25 = \frac{d}{0.005}$$

$$\therefore \quad d = 0.7 \text{ cm}$$

(b) $150 = 276 \log \left(\frac{d}{0.005}\right)$

$$3.495 = \frac{d}{0.005}$$

$$\therefore \quad d = 0.017 \text{ cm}$$

* * *

It should be noted that for a characteristic impedance of 150 Ω this design example is not practical since the separation between the lines is only 0.007 cm.

Equations (2.13) and (2.15) can be used to find line capacitance per unit length and line inductance per unit length from which the characteristic impedance of flat twin transmission line can be found. Computer program TWIN, program 2.2, can also provide spacing to diameter ratios for a given characteristic impedance. As with program COAX a simple check is made on realizability of the structure. This check occupies lines 410–420. The rest of the program is straightforward with all the main variables defined at lines 190–230.

```
][ FORMATTED LISTING
FILE: PROGRAM 2.2 TWIN
PAGE-1


10   REM
20   REM    ---- PROGRAM TWIN ----
30   REM
40   REM   THIS PROGRAM FINDS
```

```
 50   REM    LINE INDUCTANCE,CAP
 60   REM    AND CHARAC. IMP'FOR
 70   REM    A GIVEN SYMMETRICAL
 80   REM    FLAT TWIN STRUCTURE.
 90   REM    LINE SPACING/DIA
100   REM    RATIO FOR A GIVEN
110   REM    CHARAC. IMP. CAN BE
120   REM    FOUND ALSO.
130   REM
140   REM    THIS PROGRAM IS
150   REM    VALID FOR SPACING
160   REM    TO DIA. RATIOS
170   REM    OF GREATER THAN 10
180   REM
190   REM    IDUCT=LINE INDUCTANCE
200   REM    CAP=LINE CAPACITANCE
210   REM    SPACE=LINE SPACING(CM)
220   REM    DIA=CONDUCTOR DIAMETER
230   REM    ZO=CHARAC. IMPEDANCE
240   REM
250   HOME
260   PRINT "TO FIND CHARAC. IMP ENTER 1"
270   PRINT "TO FIND RADIUS TO SPACING RATIO ENTER 2"
280   PRINT "TO ABORT ENTER 3"
290   INPUT K
300   IF K = 1 THEN
           340
310   IF K = 2 THEN
           750
320   IF K = 3 THEN
           920
330   GOTO 250
340   PRINT "ENTER DIA. OF CONDUCTOR"
350   PRINT "IN CMS."
360   INPUT DIA
370   LET DI = DIA
380   PRINT "ENTER LINE SPACING BETWEEN"
390   PRINT "CENTERS IN CMS."
400   INPUT SPACE
410   IF SPACE < = DI THEN
           430
420   GOTO 450
430   PRINT "*** REALISIBILITY ERROR ***"
440   GOTO 670
450   REM
460   REM   FIND LINE INDUCTANCE
470   LET IDUCT = 4E - 7 * ( LOG (2 * SPACE / DIA))
480   LET I =  INT (IDUCT * 1E9) / 1E9
490   PRINT
500   PRINT "********** RESULTS **********"
510   PRINT
520   PRINT "LINE INDUCTANCE (H/M) "I
530   REM
540   REM   FIND LINE CAPACITANCE
550   LET CAP = (3.14159 * 8.854E - 12) / ( LOG (2 * SPACE / DIA))
560   LET I =  INT (CAP * 1E13) / 1E13
570   PRINT
580   PRINT "LINE CAPACITANCE (F/M) "I
590   REM
600   REM   FIND CHARAC IMP
610   LET ZO =  SQR (IDUCT / CAP)
620   LET I =  INT (ZO * 100 + 0.5) / 100
630   PRINT
640   PRINT "CHARAC. IMP. (OHMS) "I
650   PRINT
660   PRINT "****************************"
670   PRINT
680   PRINT "DO YOU WANT ANOTHER GO ?"
690   PRINT "ENTER 1 IF YES ; 0 IF NO"
700   INPUT K
710   IF K = 1 GOTO 250
720   IF K = 0 GOTO 920
730   GOTO 670
```

```
740   REM
750   REM   FIND SPACING/DIA
760   PRINT
770   PRINT "ENTER CHARAC. IMPEDANCE"
780   INPUT ZO
790   LET SPDI = 0.5 * 10 ^ (ZO / 276)
800   LET I =  INT (SPDI * 100 + .5) / 100
810   PRINT
820   PRINT "********* RESULTS ********"
830   PRINT
840   PRINT "SPACING TO DIAMETER RATIO "I
850   IF I <  = 10 THEN
            870
860   GOTO 890
870   PRINT "INACCURACIES:--- SPACING TO"
880   PRINT "DIA. RATIO BELOW 10"
890   PRINT
900   PRINT "***************************"
910   GOTO 670
920   PRINT
930   PRINT "======= END OF PROGRAM ======"
940   END
```

END-OF-LISTING

```
]RUN
TO FIND CHARAC. IMP ENTER 1
TO FIND RADIUS TO SPACING RATIO ENTER 2
TO ABORT ENTER 3
?1
ENTER DIA. OF CONDUCTOR
IN CMS.
?0.32
ENTER LINE SPACING BETWEEN
CENTERS IN CMS.
?2

********* RESULTS *********

LINE INDUCTANCE (H/M) 1.01E-06

LINE CAPACITANCE (F/M) 1.1E-11

CHARAC. IMP. (OHMS) 302.88

***************************

DO YOU WANT ANOTHER GO ?
ENTER 1 IF YES ; 0 IF NO
?1
TO FIND CHARAC. IMP ENTER 1
TO FIND RADIUS TO SPACING RATIO ENTER 2
TO ABORT ENTER 3
?2

ENTER CHARAC. IMPEDANCE
?600

********* RESULTS ********

SPACING TO DIAMETER RATIO 74.62

**************************

DO YOU WANT ANOTHER GO ?
ENTER 1 IF YES ; 0 IF NO
?0

======= END OF PROGRAM ======
```

2.3   RECTANGULAR WAVEGUIDE

The normal method for investigating the propagation of waves in rectangular
waveguide is to find solutions of Maxwell's equations which satisfy the geometry of
the structure while also obeying fundamental electrical constraints. The solution of
Maxwell's equations for the rectangular waveguide problem is well documented.
Further information can be found from the references given in the further reading
section at the end of this chapter. In this section, however, we will deal with the final
equations pertinent to design applications.

In essence a rectangular waveguide is a metallic tube with the broad side
having a minimum internal dimension of one-half wavelength measured in free
space at the operating frequency of interest (figure 2.3). Ideally, the two vertical
metal supporting walls at the sides of the wave appear as a perfect conductor and
will act as short circuits providing a low impedance path to current. At the center
of the waveguide along axis $X-X'$ these short circuit walls will appear electrically
as open circuits (because of the distributed nature of the waveguide a physical
distance of one-quarter of a free-space wavelength is equivalent to an impedance
inversion, see chapter 5). This means that the centre of the waveguide along axis
$X-X'$ is a high impedance region. Therefore, signals can pass along the central axis
of the waveguide with minimum attenuation.

Assume that an electrical signal injected into a metallic waveguide will behave
like a ray of light. If the signal is injected as shown in figure 2.4 it will be reflected
back and forth across the waveguide at each metal surface. This increases the time
required for the signal to traverse the length of the waveguide when compared with
the time taken for a signal to travel the same horizontal distance but without
reflection. The measured wavelength along the length of the waveguide, the guide
wavelength λ_g, is therefore longer than the free-space wavelength λ_o. The lowest
frequency signal that a rectangular waveguide can accommodate is related to its

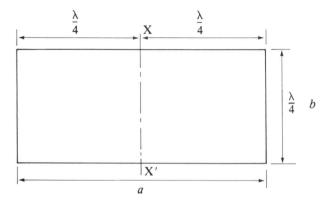

Figure 2.3   Cross-Sectional View of a Rectangular Waveguide

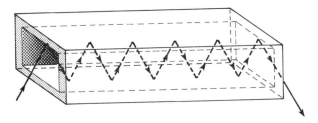

Figure 2.4   Internal Reflection in a Rectangular Waveguide

broad dimension a. The wavelength of this lowest frequency is referred to as the cut-off wavelength λ_c.

The governing equations for an air-filled rectangular waveguide are stated as

$$\lambda_c = \frac{2a}{m} \tag{2.17}$$

where λ_c is the cut-off wavelength, a is the rectangular waveguide broad dimension and m is the number of half wavelengths of voltage fields along the broad wall of the rectangular waveguide dimension a. A value of $m = 1$ indicates that the rectangular waveguide is operating in the dominant mode. This notation will be discussed later in this section.

$$\lambda_g = \frac{\lambda_o}{1 - (\lambda_o/\lambda_c)^2} \tag{2.18}$$

where λ_o is the free space wavelength and λ_g is the guide wavelength.

$$\left. Z_o \right|_{TE} = \frac{377}{[1 - (\lambda_o/\lambda_c)^2]^{\frac{1}{2}}} \ \Omega$$

or
$$\left. Z_o \right|_{TM} = 377 \left[1 - \left(\frac{\lambda_o}{\lambda_c} \right)^2 \right]^{\frac{1}{2}} \Omega \tag{2.19}$$

Where Z_o is the characteristic impedance for the waveguide and TE, TM denote transverse electric or transverse magnetic modes. Mode notation will be discussed later in this section.

Comparison of characteristic impedance in the TE and TM modes indicates that the TE mode characteristic impedance is greater than 377 Ω and approaches infinity as free space conditions are approached (λ_o/λ_c tending to one). For the TM mode of propagation the characteristic impedance is less than 377 Ω and tends to zero under free space conditions.

It should be noted that if the waveguide is filled with a dielectric with relative permittivity greater than unity, the cut-off wavelength can be increased by a factor equal to the square root of the relative permittivity. The new value computed for cut-off wavelength from equation (2.17) can be substituted into equations (2.18) and (2.19) in order to find guide wavelength and characteristic impedance.

* * *

Example 2.5

Consider a 10 GHz signal propagating in a rectangular waveguide with internal dimensions of 2.3×1 cm. Assuming the dominant mode for signal propagation find:

(a) cut-off wavelength;
(b) guide wavelength;
(c) characteristic wave impedance of the waveguide.

Solution

(a) $\lambda_o = \dfrac{c}{f} = \dfrac{3 \times 10^{10}}{10 \times 10^9} = 3$ cm = free space wavelength

$\lambda_c = \dfrac{2a}{m} = \dfrac{2 \times 2.3}{1} = 4.6$ cm = cut-off wavelength

where $m = 1$ dominant mode and $a = 2.3$ cm. Notice that since the cut-off wavelength is greater than the free space wavelength at this frequency the signal will propagate.

(b) $\lambda_g = \dfrac{\lambda_o}{[1-(\lambda_o/\lambda_c)^2]^{1/2}} = \dfrac{3}{[1-(3/4.6)^2]^{1/2}} = 3.96$ cm = guide wavelength

(c) $Z_0 = \dfrac{377}{[1-(\lambda_o/\lambda_c)^2]^{1/2}} = \dfrac{377}{0.758} = 497\ \Omega$ = characteristic impedance

Example 2.6

A rectangular waveguide has internal dimensions 2.8×1.2 cm. (a) Find the lowest frequency that will propagate in the waveguide for the dominant mode. (b) If the signal frequency is 8 GHz what is the highest mode that can be sustained by the waveguide. (c) If the signal frequency is 14 GHz and $m = 2$ calculate the cut-off wavelength, guide wavelength and characteristic impedance for the waveguide.

Solution

(a) $\lambda_c = \dfrac{2a}{m} = \dfrac{2 \times 2.8}{1} = 5.6$ cm

$f_c = \dfrac{c}{\lambda_c} = \dfrac{3 \times 10^{10}}{5.6} = 5.357$ GHz which is equal to the lowest frequency that can pass along a rectangular waveguide with the given dimensions operated in the dominant mode.

(b) For 8 GHz operation the free space wavelength

$$\lambda_o = \frac{c}{f_o} = \frac{3 \times 10^{10}}{8 \times 10^9} = 3.75 \text{ cm}$$

Since $3.75 < 5.6$ cm the dominant mode ($m = 1$) will propagate. For $m = 2$

$$\lambda_c = \frac{5.6}{2} = 2.8 \text{ cm}$$

Since $2.8 < 3.75$ then the $m = 2$ mode cannot be propagated.

(c) At 14 GHz

$$\lambda_o = \frac{3 \times 10^{10}}{14 \times 10^9} = 2.1 \text{ cm}$$

for $m = 2$, $\lambda_c = 2.8$ cm. Since $\lambda_o < \lambda_c$ this mode will propagate

$$\lambda_g = \frac{\lambda_o}{[1 - (\lambda_o/\lambda_c)^2]^{1/2}} = \frac{2.1}{[1 - (2.1/2.8)^2]^{1/2}} = 3.175 \text{ cm}$$

$$Z_o = \frac{377}{[1 - (\lambda_o/\lambda_c)^2]^{1/2}} = \frac{377}{0.438^{1/2}} = 569.9 \; \Omega$$

* * *

The term mode has been used freely throughout this section and will now be discussed in some detail. A TE (transverse electric) mode is defined as occurring when the signal in the waveguide has an electric field at right angles to the direction of propagation. Similarly when a signal has a magnetic field perpendicular to the direction of propagation a TM (transverse magnetic) mode is defined. For a rectangular waveguide to be of practical use waves must propagate along the length of the waveguide, this means that the electric field must be orientated in the vertical plane (plane X–X' in figure 2.3) for the transverse electric mode. Similarly for the transverse magnetic mode the magnetic field lies in the vertical plane. Each mode type can be more closely defined in terms of the number of half wavelengths of magnetic or electric field within the waveguide. This classification takes the form of subscripts m or n, i.e. TE$_{mn}$ or TM$_{mn}$ with the dominant mode occurring when $m = n = 1$. The dominant mode corresponds to the longest cut-off wavelength for a given waveguide geometry.

Equation (2.17), describing cut-off wavelength, can be rewritten in a more general form as

$$\lambda_c = \frac{2}{[(m/a)^2 + (n/b)^2]^{1/2}} \tag{2.20}$$

here m indicates the number of fields, electric or magnetic, displaying one-half wavelength of their period in width a; n represents the number of fields, electric or magnetic, displaying one half wavelength of their period in height b.

From equation (2.20) it can be seen that the cut-off wavelength depends on both the waveguide dimensions a, b and on the mode numbers m, n. Once cut-off wavelength has been calculated from equation (2.20), the value obtained can be substituted into equations (2.18) and (2.19) as before.

* * *

Example 2.7

A 24 GHz signal propagates in a rectangular waveguide with internal dimensions 1.58×0.79 cm operating in the TM_{11} mode. Find:

(a) cut-off frequency;
(b) guide wavelength;
(c) characteristic impedance.

Solution

(a) $\lambda_c = \dfrac{2}{[(m/a)^2 + (n/b)^2]^{\frac{1}{2}}}$ $m = 1,\ n = 1$

$\quad = \dfrac{2}{[(1/1.58)^2 + (1/0.79)^2]^{\frac{1}{2}}}$

$\quad = \dfrac{2}{[0.4 + 1.6]^{\frac{1}{2}}} = 1.41$ cm

$\lambda_o = \dfrac{c}{f} = \dfrac{3 \times 20^{10}}{12 \times 10^9} = 1.25$ cm

(b) $\lambda_g = \dfrac{\lambda_o}{[1 - (\lambda_c/\lambda_o)^2]^{\frac{1}{2}}} = \dfrac{1.25}{[1 - (1.25/1.41)^2]^{\frac{1}{2}}} = \dfrac{1.25}{0.46} = 2.7$ cm

(c) $Z_o = 377 \left[1 - \left(\dfrac{\lambda_c}{\lambda_o} \right)^2 \right]^{\frac{1}{2}} = 174.4 \ \Omega$

* * *

Under normal operating conditions only the lower order modes are exploited. Some lower order modes are illustrated in figure 2.5. In this diagram electric fields are represented by solid lines while magnetic fields are depicted with dotted lines. Notice how the electric and magnetic fields intersect at right angles, the two fields are said to be orthogonal.

Rectangular waveguides operating from about 1 GHz to 90 GHz are commercially available. The materials used for construction vary according to

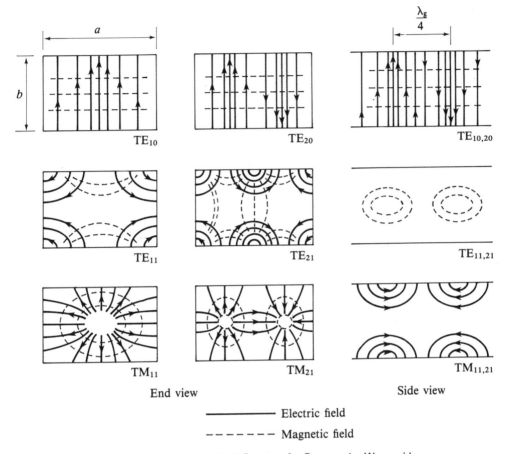

End view   Side view

———————— Electric field

— — — — — — Magnetic field

Figure 2.5 Low Order Mode Field Patterns for Rectangular Waveguide

application. For example where dimensional stability is required materials with low thermal expansion coefficients such as Invar or Covar are used. Where low attenuation is required the inside faces of the guide are coated with gold or silver. For mundane applications, brass, copper or aluminum are the most frequently used materials. Preliminary waveguide selection is normally in accordance with a letter denoting the frequency band of interest, e.g. G Band, 4–6 GHz; K Band 20–40 GHz. These frequency band allocations are subject to debate depending on the organization specifying the standard. Rectangular waveguides produced by manufacturers normally bear a WG or a JAN (Joint Army and Navy) prefix. As an example a rectangular waveguide constructed from brass and operating in the frequency range 8.2–12.4 GHz is designated as WG 16 or as JAN Type RG 52, both are equivalent.

A computer program RECTGUIDE, program 2.3, computes guide wavelength and characteristic impedance for a rectangular waveguide with known internal

```
][ FORMATTED LISTING
FILE: PROGRAM 2.3 RECTGUIDE
PAGE-1

    10   REM
    20   REM    ----- RECTGUIDE-----
    30   REM
    40   REM    THIS PROGRAM COMPUTES
    50   REM    GUIDE WAVELENGTH AND
    60   REM    CHARAC. IMP. FOR
    70   REM    RECTANGULAR WAVEGUIDE
    80   REM    OF KNOWN INTERNAL
    90   REM    DIMENSIONS, OPERATING
   100   REM    IN A KNOWN MODE.
   110   REM    A CHECK IS MADE TO SEE
   120   REM    IF THE GIVEN SIGNAL
   130   REM     WILL PROPAGATE IN THE
   140   REM    GUIDE SELECTED.
   150   REM
   160   REM    FRQ=SIGNAL FREQUENCY
   170   REM    CUT=CUTOFF WAVELTH.
   180   REM    M,N=MODE ORDERS
   190   REM    A,B=WAVEGUIDE SIZE
   200   REM    ZO=CHARAC. IMP.
   210   REM    GLAMB=GUIDE WAVELENGTH
   220   REM    LAMBDA=FREE-SPACE W-LTH
   230   REM
   240   REM    INPUT DATA
   250   HOME
   260   PRINT "FOR TM MODE ENTER 1"
   270   PRINT "FOR TE MODE ENTER 0"
   280   INPUT K
   290   IF K < 0 OR K > 1 THEN
            250
   300   PRINT
   310   PRINT "INPUT OPERATING FREQ. IN GHZ."
   320   INPUT FRQ
   330   PRINT
   340   PRINT "INPUT MODE M"
   350   INPUT M
   360   PRINT "INPUT MODE N"
   370   INPUT N
   380   PRINT
   390   PRINT "INPUT INTERNAL DIMENSIONS OF WAVEGUIDE"
   400   PRINT "IN CMS."
   410   PRINT
   420   PRINT "ENTER WIDTH"
   430   INPUT A
   440   PRINT "ENTER HEIGHT"
   450   INPUT B
   460   REM
   470   REM    FREE-SPACE W-LTH.
   480   PRINT
   490   PRINT "******** RESULTS ********"
   500   PRINT
   510   LET LAMBDA = 30 / FRQ
   520   LET L =   INT (LAMBDA * 100 + 0.5) / 100
   530   PRINT "FREE-SPACE WAVELTH. (CMS) "L
   540   REM   FIND CUTOFF FREQ.
   550   PRINT
   560   LET CUT = 2 / ( SQR ((M * M) / (A * A) + (N * N) / (B * B)))
   570   LET L =   INT (CUT * 100 + 0.5) / 100
   580   PRINT "CUTOFF WAVELTH. (CM) "L
   590   REM
   600   REM    CHECK FOR PROPAGATION
   610   IF LAMBDA > CUT THEN
            630
   620   GOTO 750
   630   PRINT
   640   PRINT "SIGNAL SELECTED WILL NOT"
   650   PRINT "PROPAGATE IN SELECTED WAVEGUIDE"
   660   PRINT
   670   PRINT "****************************"
   680   PRINT
   690   PRINT "DO YOU WANT ANOTHER GO ?"
   700   PRINT "ENTER 1 IF YES ; 0 IF NO"
```

```
710    INPUT I
720    IF I = 1 THEN
           250
730    IF I = 0 THEN
           930
740    GOTO 680
750    REM
760    REM   FIND GUIDE WAVELTH.
770    LET GLAM = LAMBDA /  SQR ((1 - ((LAMBA / CUT) ^ 2)))
780    LET L =  INT (GLAM * 100 + 0.5) / 100
790    PRINT
800    PRINT "GUIDE WAVELENGTH (CM) "L
810    REM
820    REM   FIND CHARAC IMP.
830    IF K = 0 THEN
           860
840    LET ZO = 377 *  SQR (1 - ((LAMBA / CUT) ^ 2))
850    GOTO 870
860    LET ZO = 377 /  SQR (1 - ((LAMBA / CUT) ^ 2))
870    LET L =  INT (ZO * 100 + 0.5) / 100
880    PRINT
890    PRINT "CHARAC. IMP. (OHMS) "L
900    PRINT
910    PRINT "**************************"
920    GOTO 680
930    PRINT
940    PRINT "******** END OF PROGRAM ********"
950    END
END-OF-LISTING


]RUN
FOR TM MODE ENTER 1
FOR TE MODE ENTER 0
?1

INPUT OPERATING FREQ. IN GHZ.
?8

INPUT MODE M
?1
INPUT MODE N
?1

INPUT INTERNAL DIMENSIONS OF WAVEGUIDE
IN CMS.

ENTER WIDTH
?4.2
ENTER HEIGHT
?2

******** RESULTS ********

FREE-SPACE WAVELTH. (CMS) 3.75

CUTOFF WAVELTH. (CM) 3.61

SIGNAL SELECTED WILL NOT
PROPAGATE IN SELECTED WAVEGUIDE

**************************

DO YOU WANT ANOTHER GO ?
ENTER 1 IF YES ; 0 IF NO
?1
FOR TM MODE ENTER 1
FOR TE MODE ENTER 0
?0

INPUT OPERATING FREQ. IN GHZ.
?8

INPUT MODE M
?1
```

```
INPUT MODE N
?1

INPUT INTERNAL DIMENSIONS OF WAVEGUIDE
IN CMS.

ENTER WIDTH
?7.4
ENTER HEIGHT
?3.6

******** RESULTS ********

FREE-SPACE WAVELTH. (CMS) 3.75

CUTOFF WAVELTH. (CM) 6.47

GUIDE WAVELENGTH (CM) 4.6

CHARAC. IMP. (OHMS) 462.47

**************************

DO YOU WANT ANOTHER GO ?
ENTER 1 IF YES ; 0 IF NO
?0

******** END OF PROGRAM ********
```

dimensions operating in a known mode. Equations (2.18), (2.19) and (2.20) provide the necessary design rules. A check is made to see if a signal of specified frequency will be transmitted in the waveguide for a particular order of mode. An example is given in the test run to illustrate a condition that results in a signal that cannot be propagated in the selected waveguide.

2.4   CIRCULAR WAVEGUIDE

Circular waveguide design is somewhat more complicated than rectangular waveguide design. This is due to the type of mathematical function needed to solve the particular form of Maxwell's equations applicable to the cylindrical geometry of this type of waveguide. The analysis of this problem is well represented in the literature, see the further reading section at the end of this chapter.

The TE_{mn} and TM_{mn} mode notations for the rectangular waveguide are similarly used for the cylindrical waveguide. There are changes in definition of the notation when compared with the rectangular waveguide case. In a circular section guide, m refers to the number of full period variations of the field around the perimeter of the waveguide. Subscript n refers to the number of half cycles of field along the radial direction of the guide.

Circular waveguide design involves the guide diameter and a parameter S dependent on the particular mode the guide is propagating. With the exception of the cut-off wavelength λ_c, equations (2.18) and (2.19) for rectangular section waveguide apply.

For circular waveguide the cut-off wavelength is given as

$$\lambda_c = \frac{\pi d}{S_{mn}} \tag{2.21}$$

where d is the mean waveguide diameter and S_{mn} is the nth root of a Bessel function

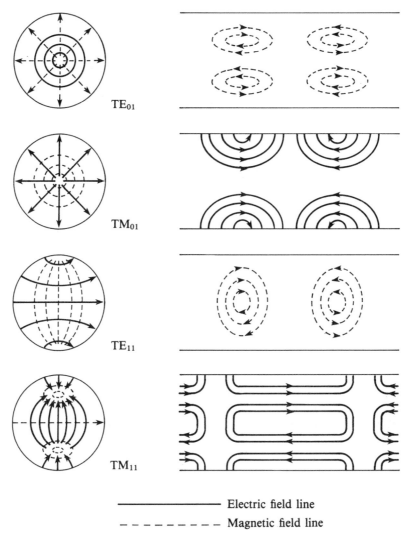

TE$_{01}$

TM$_{01}$

TE$_{11}$

TM$_{11}$

——————————— Electric field line

– – – – – – – – – Magnetic field line

Figure 2.6   Circular Waveguide Field Patterns for Low Order Modes

* * *

Example 2.8

Given a circular waveguide with internal diameter 6 cm carrying a signal at a frequency of 9 GHz and operating in the TE$_{11}$ mode, calculate:

   (a) cut-off wavelength;
   (b) guide wavelength;
   (c) characteristic impedance.

Table 2.2 S_{mn} for Circular Waveguide Fundamental Modes

Mode	S_{mn}	Mode	S_{mn}
TE_{01}	3.832	TM_{01}	2.405
TE_{11}	1.841	TM_{11}	3.832
TE_{21}	3.050	TM_{21}	5.136
TE_{02}	7.016	TM_{02}	5.520
TE_{12}	5.330	TM_{12}	7.016
TE_{22}	6.710	TM_{22}	8.420

of the first kind and nth order $J_n(x)$, i.e. S_{mn} is the nth root of $J_n(x) = 0$. This parameter is tabulated for common operating modes of circular section waveguide in table 2.2.

From table 2.2 and equation (2.21) it can be seen that the TE_{11} mode has the longest cut-off wavelength, lowest cut-off frequency of all the TE waves specified for a given diameter guide, which makes it the dominant mode. This cut-off frequency is lower than that found for the lowest-order TM wave propagating in a given diameter guide, i.e. the TE_{11} wave will propagate in a guide with diameter $1.841/2.405 = 0.76$ as big as that needed to support a TM_{01} wave of the same frequency.

Typical field patterns for some low order modes in a circular waveguide are as shown in figure 2.6. It can be seen that the TE_{01} and TM_{01} modes display rotational symmetry. Notice also how the TE_{01} mode is the dual of the TM_{01}. The rotational symmetry present in the TE_{01} and TM_{01} modes is preserved along the complete length of the guide. This symmetry can be exploited in a practical way. Suppose a signal is propagating in the TE_{01} mode along a circular section waveguide and it encounters some type of an obstruction or irregularity in the guide surface. The defect will almost certainly cause rotation of the waves within the guide. However, this will not matter to any great extent since the signal can still be received; it looks like the same signal to the receiving element. This means that the receiving efficiency is not reduced due to reflected energy. For other modes such as the TE_{11} mode this does not hold true since rotational symmetry does not occur. Therefore, for most applications a circular waveguide would be operated in the 01 mode. However, since this is not the fundamental mode for the guide there may be problems due to the generation of spurious signals. In practice these may have to be suppressed by modified design of the waveguide.

Solution

(a) $\lambda_o = \dfrac{c}{f} = \dfrac{3 \times 10^{10}}{9 \times 10^9} = \dfrac{30}{9} = 3.3$ cm

$m = 1$, $n = 1$

$\lambda_c = \dfrac{\pi \times 6}{S_{11}} = \dfrac{\pi \times 6}{1.841} = 10.24$ cm

$\lambda_o < \lambda_c$ thus signal will propagate

(b) $\lambda_g = \dfrac{\lambda}{[1 - (\lambda_0/\lambda_c)^2]^{1/2}} = \dfrac{3.3}{[1 - (0.32)^2]^{1/2}} = 3.48$ cm

(c) $Z_0|_{TE} = \dfrac{377}{[1 - (0.32)^2]^{1/2}} = 398 \ \Omega$

Example 2.9

For a circular waveguide with internal diameter 5 cm operating with a 12 GHz signal and propagating in TM_{01} mode calculate

(a) cut-off wavelength;
(b) guide wavelength;
(c) characteristic impedance.

Solution

(a) $\lambda_0 = \dfrac{c}{f} = \dfrac{3 \times 10^{10}}{12 \times 10^9} = \dfrac{30}{12} = 2.5$ cm

$m = 0, \ n = 1$

$\lambda_c = \dfrac{\pi \times 5}{2.405} = 7.88$ cm

$\lambda_0 < \lambda_c$ therefore signal will propagate

(b) $\lambda_g = \dfrac{\lambda_0}{[1 - (\lambda_0/\lambda_c)^2]^{1/2}} = \dfrac{2.5}{[1 - (0.31)^2]^{1/2}} = 2.64$ cm

(c) $Z_0 = 377[1 - (0.31)^2]^{1/2} = 358 \ \Omega$

* * *

Program 2.4 CIRCGUIDE allows the above examples to be computed automatically for a variety of circular waveguide modes selected from a given menu, lines 380–460 and 1110–1160.

```
][ FORMATTED LISTING
FILE: PROGRAM 2.4 CIRCGUIDE
PAGE-1


10   REM
20   REM      ----- CIRCGUIDE-----
30   REM
40   REM   THIS PROGRAM COMPUTES
50   REM   GUIDE WAVELENGTH AND
60   REM   CHARAC. IMP. FOR
70   REM   CIRCULAR WAVEGUIDE
80   REM   OF KNOWN INTERNAL
90   REM   DIAMETER, OPERATING
```

```
100   REM   IN A KNOWN MODE.
110   REM   A CHECK IS MADE TO SEE
120   REM   IF THE GIVEN SIGNAL
130   REM   WILL PROPAGATE IN THE
140   REM   GUIDE SELECTED.
150   REM
160   REM   FRQ=SIGNAL FREQUENCY
170   REM   CUT=CUTOFF WAVELTH.
180   REM   M,N=MODE ORDERS
190   REM   D=WAVEGUIDE DIAMETER
200   REM   ZO=CHARAC. IMP.
210   REM   GLAMB=GUIDE WAVELENGTH
220   REM   LAMBDA=FREE-SPACE W-LTH
230   REM
240   REM   INPUT DATA
250   DIM TE(2,2),TM(3,3)
260   FOR K = 1 TO 2
270       FOR J = 0 TO 2
280           READ TE(J,K)
290       NEXT J
300   NEXT K
310   FOR K = 1 TO 2
320       FOR J = 0 TO 2
330           READ TM(J,K)
340       NEXT J
350   NEXT K
360   READ TM(0,3)
370   HOME
380   PRINT "YOU CAN SELECT ONE OF THE"
390   PRINT "FOLLOWING MODES"
400   PRINT "TM 0,1:1,1:2,1:0,2:1,2:2,2:0,3"
410   PRINT "TE 0,1:1,1:0,2:1,2:2,2
420   PRINT
430   PRINT "FOR TM MODE ENTER 1"
440   PRINT "FOR TE MODE ENTER 0"
450   PRINT
460   INPUT K
470   IF K < 0 OR K > 1 THEN
          340
480   PRINT
490   PRINT "INPUT OPERATING FREQ. IN GHZ."
500   INPUT FRQ
510   PRINT
520   PRINT "INPUT MODE M"
530   INPUT M
540   PRINT "INPUT MODE N"
550   INPUT N
560   IF K = 1 THEN
          LET MEW = TM(M,N)
570   IF K = 0 THEN
          LET MEW = TE(M,N)
580   PRINT
590   PRINT "INPUT INTERNAL DIA. OF WAVEGUIDE (CMS) "
600   INPUT D
610   REM
620   REM   FREE-SPACE W-LTH.
630   PRINT
640   PRINT "******** RESULTS ********"
650   PRINT
660   LET LAMBDA = 30 / FRQ
670   LET L =   INT (LAMBDA * 100 + 0.5) / 100
680   PRINT "FREE-SPACE WAVELTH. (CMS) "L
690   REM   FIND CUTOFF FREQ.
700   PRINT
710   LET CUT = 3.14159 * D / MEW
720   LET L =   INT (CUT * 100 + 0.5) / 100
730   PRINT "CUTOFF WAVELTH. (CM) "L
740   REM
750   REM    CHECK FOR PROPAGATION
760   IF LAMBDA > CUT THEN
          780
770   GOTO 900
780   PRINT
790   PRINT "SIGNAL SELECTED WILL NOT"
800   PRINT "PROPAGATE IN SELECTED WAVEGUIDE"
```

```
810   PRINT
820   PRINT "***************************"
830   PRINT
840   PRINT "DO YOU WANT ANOTHER GO ?"
850   PRINT "ENTER 1 IF YES ; 0 IF NO"
860   INPUT I
870   IF I = 1 THEN
            370
880   IF I = 0 THEN
            1080
890   GOTO 830
900   REM
910   REM   FIND GUIDE WAVELTH.
920   LET GLAM = LAMBDA /  SQR ((1 - ((LAMBDA / CUT) ^ 2)))
930   LET L =  INT (GLAM * 100 + 0.5) / 100
940   PRINT
950   PRINT "GUIDE WAVELENGTH (CM) "L
960   REM
970   REM   FIND CHARAC IMP.
980   IF K = 0 THEN
            1010
990   LET ZO = 377 *  SQR (1 - ((LAMBA / CUT) ^ 2))
1000  GOTO 1020
1010  LET ZO = 377 /  SQR (1 - ((LAMBA / CUT) ^ 2))
1020  LET L =  INT (ZO * 100 + 0.5) / 100
1030  PRINT
1040  PRINT "CHARAC. IMP. (OHMS) "L
1050  PRINT
1060  PRINT "**************************"
1070  GOTO 830
1080  PRINT
1090  PRINT "******** END OF PROGRAM ********"
1100  END
1110  REM
1120  REM   TE MODE DATA
1130  DATA 3.832,1.841,3.05,7.016,5.33,6.71
1140  REM
1150  REM   TM MODE DATA
1160  DATA 2.405,3.832,5.136,5.52,7.016,8.42,8.654
```

END-OF-LISTING

```
]RUN
YOU CAN SELECT ONE OF THE
FOLLOWING MODES
TM 0,1:1,1:2,1:0,2:1,2:2,2:0,3
TE 0,1:1,1:0,2:1,2:2,2

FOR TM MODE ENTER 1
FOR TE MODE ENTER 0

?0

INPUT OPERATING FREQ. IN GHZ.
?11

INPUT MODE M
?0
INPUT MODE N
?1

INPUT INTERNAL DIA. OF WAVEGUIDE (CMS)
?4.5

******** RESULTS ********

FREE-SPACE WAVELTH. (CMS) 2.73

CUTOFF WAVELTH. (CM) 3.69

GUIDE WAVELENGTH (CM) 4.05

CHARAC. IMP. (OHMS) 559.82

**************************
```

```
DO YOU WANT ANOTHER GO ?
ENTER 1 IF YES ; 0 IF NO
?1
YOU CAN SELECT ONE OF THE
FOLLOWING MODES
TM 0,1:1,1:2,1:0,2:1,2:2,2:0,3
TE 0,1:1,1:0,2:1,2:2,2

FOR TM MODE ENTER 1
FOR TE MODE ENTER 0

?1

INPUT OPERATING FREQ. IN GHZ.
?8

INPUT MODE M
?2
INPUT MODE N
?2

INPUT INTERNAL DIA. OF WAVEGUIDE (CMS)
?12

******** RESULTS ********

FREE-SPACE WAVELTH. (CMS) 3.75

CUTOFF WAVELTH. (CM) 4.48

GUIDE WAVELENGTH (CM) 6.86

CHARAC. IMP. (OHMS) 205.98

*************************

DO YOU WANT ANOTHER GO ?
ENTER 1 IF YES ; 0 IF NO
?0

******** END OF PROGRAM ********
```

2.5   SINGLE STRIPLINE

At frequencies in excess of several hundred MHz, stripline or triplate transmission line is often used. Stripline has a number of advantages over coaxial or waveguide circuitry especially where active or passive devices such as Gunn diodes or mixer diodes are to be included as part of the circuit design or where large circuit bandwidth or miniaturization is required. It suffers from considerable disadvantages when compared with coaxial or waveguide in terms of isolation between circuits and power handling capability. The designer should be aware of the trade-offs that exist and should choose the medium most suited for a given application.

Stripline can be considered as a thin metal conductor of rectangular cross-section embedded in a uniform dielectric material which in turn is sandwiched between two ground planes (figure 2.7). The dominant mode of propagation for this structure is TEM (transverse electromagnetic). Stripline comes in a bewildering variety of dielectric materials with copper conductor on one or both sides of the center conductor. The thickness of the center conductor and dielectric material is also widely variable. The choice of a particular stripline material is often a difficult

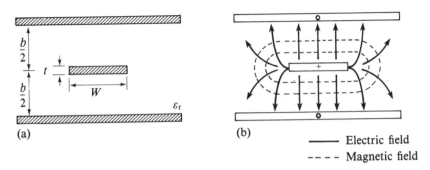

Figure 2.7   Single Stripline (a) Geometry (b) Field Profile

one and can normally only be resolved by experience based on designing the same circuit on a number of different stripline materials and assessing the effect of layout and manufacturing tolerances on overall circuit design. Computer program 2.5 SINGLE STRIP, will allow this to be done with ease.

Equations describing the characteristic impedance of stripline for a known geometry (this is the process known as analysis, i.e. given line width and dielectric constant find characteristic impedance) have been deduced by various people. Cohn [4] pioneered the analysis of stripline circuits with very thin center conductors using the technique of conformal mapping. Here the rectangular cross-section conductor of stripline is mathematically mapped to a new circular shape that can be studied using the technique developed at the beginning of this chapter for coaxial and flat twin transmission line. When the center conductor has a non-zero thickness, the effect is to reduce the line width to ground plane spacing ratio W/b calculated using Cohn's technique [4] for a particular characteristic impedance. For finite thickness conductor, Wheeler's formulae [5] are reported to be accurate to within 0.5 percent for $W/(b-t)$ less than 10. Wheeler's equations are restated here for convenience (equation (2.22)).

$$Z_0 = \frac{30}{(\varepsilon_r)^{1/2}} \ln \left(1 + \frac{4}{\pi} \cdot \frac{1}{m} \left\{ \frac{8}{\pi} \cdot \frac{1}{m} + \left[\left(\frac{8}{\pi} \cdot \frac{1}{m}\right)^2 + 6.27 \right]^{1/2} \right\} \right)$$

where

$$m = \frac{W}{b-t} + \frac{\Delta W}{b-t}$$

and

$$\frac{\Delta W}{b-t} = \frac{x}{\pi(1-x)} \left\{ 1 - 0.5 \ln \left[\left(\frac{x}{2-x}\right)^2 + \left(\frac{0.0796x}{W/b+1.1x}\right)^m \right] \right\}$$

with

$$m = 2 \bigg/ \left(1 + \frac{2}{3} \frac{x}{1-x}\right) \quad \text{and} \quad x = t/b$$

(2.22)

```
]( FORMATTED LISTING
FILE: PROGRAM 2.5 SINGLE STRIP
PAGE-1


    10    REM
    20    REM   ----- SINGLE STRIP -----
    30    REM
    40    REM
    50    REM   THIS PROGRAM CAN BE
    60    REM   USED FOR THE SYNTHESIS/
    70    REM ANALYSIS OF STRIPLINE
    80    REM TRANSMISSION LINE PLACED
    90    REM SYMMETRICALLY BETWEEN
   100    REM GROUND PLANES
   110    REM
   120    REM T=LINE THICKNESS (CM)
   130    REM B=GROUND PLANE SEP.
   140    REM IN CMS.
   150    REM ZO=CHARAC. IMP.
   160    REM GLAM=GUIDEWAVELTH.
   170    REM ER=REL.DIELECTRIC
   180    REM FRQ=FREQUENCY (GHZ)
   190    REM TB=CONDUCTOR THICKNESS
   200    REM TO GROUND PLANE RATIO
   210    REM WB=LINE WIDTH TO GROUND
   220    REM PLANE RATIO
   230    REM W=LINE WIDTH (CM)
   240    REM
   250    REM INPUT ROUTINE
   260    HOME
   270    PRINT
   280    PRINT "FOR SYNTHESIS ENTER 0"
   290    PRINT "FOR ANALYSIS ENTER 1"
   300    INPUT K
   310    IF K < 0 OR K > 1 THEN
             260
   320    PRINT
   330    PRINT "DESIRED OPERATING FREQ. (GHZ)"
   340    INPUT FRQ
   350    PRINT
   360    PRINT "INPUT REL. DIELECTRIC"
   370    INPUT ER
   380    PRINT
   390    PRINT "**** INPUT LINE GEOMETRY ****"
   400    PRINT
   410    PRINT "FIRST ENTER LINE THICKNESS"
   420    PRINT "TO GROUND PLANE SPACING RATIO (T/B)"
   430    INPUT TB
   440    PRINT "NEXT ENTER GROUND PLANE SPACING (B)"
   450    INPUT B
   460    LET PI = 3.14159
   470    IF K = 0 THEN
             750
   480    REM
   490    REM ANALYSIS SECTION
   500    PRINT "ENTER LINE WIDTH TO GROUND"
   510    PRINT "PLANE SPACING RATIO (W/B)"
   520    INPUT WB
   530    GOSUB 1070
   540    PRINT
   550    PRINT "****** ANALYSIS RESULTS ******"
   560    PRINT
   570    LET Z2 =  INT (Z1 * 100 + 0.5) / 100
   580    LET GLAM = 30 / FRQ /  SQR (ER)
   590    LET G =  INT (GLAM * 100 + 0.5) / 100
   600    PRINT "FOR W/B= "WB"  T/B= "TB
   610    PRINT "AND B= "B" CMS"
   620    PRINT "CHARAC. IMP.="Z2"OHMS"
   630    PRINT "GUIDE WAVELTH.="G"CMS"
   640    PRINT "FOR AN OPERATING FREQUENCY OF"FRQ"GHZ"
   650    PRINT
   660    PRINT "*****************************"
```

```
670   REM
680   PRINT
690   PRINT "DO YOU WANT ANOTHER GO"
700   PRINT "ENTER 1 IF YES ; 0 IF NO"
710   INPUT L
720   IF L = 1 THEN
          260
730   IF L = 0 THEN
          1040
740   GOTO 690
750   REM
760   REM   SYNTHESIS SECTION
770   LET WB = 0
780   PRINT "INPUT DESIRED CHARAC. IMPEDANCE"
790   INPUT ZO
800   PRINT
810   PRINT
820   PRINT " --------- WORKING ----------"
830   LET WB = WB + 0.01
840   GOSUB 1070
850   IF Z1 < = (ZO + 2 /  SQR (ER)) AND Z1 > = (ZO - 2 /  SQR (ER)) THEN
          870
860   GOTO 830
870   PRINT
880   PRINT "******** SYNTHESIS RESULTS *******"
890   PRINT
900   LET Z1 =  INT (Z1 * 100 + 0.5) / 100
910   PRINT "FOR A CHARAC. IMP. "Z1" OHMS"
920   PRINT "WITH T/B ="TB" AND B="B" CMS"
930   LET W =  INT (WB * 10000) / 10000
940   PRINT "THE REQUIRED LINE TO WIDTH SPACING RATIO (W/B)= "W
950   LET LW =  INT (WB * B * 100 + 0.5) / 100
960   PRINT "LINE WIDTH= "LW" CMS"
970   LET GLAM = 30 / FRQ /  SQR (ER)
980   LET G =  INT (GLAM * 100 + 0.5) / 100
990   PRINT "GUIDE WAVELTH.="G" CMS."
1000  PRINT "FOR AN OPERATING FREQ. OF "FRQ" GHZ"
1010  PRINT
1020  PRINT "*******************************"
1030  GOTO 670
1040  PRINT
1050  PRINT "****** END OF PROGRAM ******"
1060  END
1070  LET W = B * WB
1080  LET T = B * TB
1090  LET M = 2 / (1 + 0.6666 * TB / (1 - TB))
1100  LET DW = (0.0796 * TB) / (WB + 1.1 * TB) ^ M
1110  LET DW = DW + (TB / (2 - TB)) ^ 2
1120  LET DW = 0.5 *  LOG (DW)
1130  LET DW = TB / PI / (1 - TB) * (1 - DW)
1140  LET W1 = W / (B - T) + DW
1150  LET Z =  SQR ((8 / PI / W1) ^ 2 + 6.27) + 8 / PI / W1
1160  LET Z = Z * 4 / PI / W1 + 1
1170  LET Z = 30 *  LOG (Z)
1180  LET Z1 = Z /  SQR (ER)
1190  RETURN
```

END-OF-LISTING

]RUN

FOR SYNTHESIS ENTER 0
FOR ANALYSIS ENTER 1
?0

DESIRED OPERATING FREQ. (GHZ)
?1

INPUT REL. DIELECTRIC
?4

**** INPUT LINE GEOMETRY ****

```
FIRST ENTER LINE THICKNESS
TO GROUND PLANE SPACING RATIO (T/B)
?0.175
NEXT ENTER GROUND PLANE SPACING (B)
?0.01
INPUT DESIRED CHARAC. IMPEDANCE
?25


  --------- WORKING ----------

******** SYNTHESIS RESULTS *******

FOR A CHARAC. IMP. 25.83 OHMS
WITH T/B =.175 AND B=.01 CMS
THE REQUIRED LINE TO WIDTH SPACING RATIO (W/B)= .97
LINE WIDTH= .01 CMS
GUIDE WAVELTH.=15 CMS.
FOR AN OPERATING FREQ. OF 1 GHZ

******************************

DO YOU WANT ANOTHER GO
ENTER 1 IF YES ; 0 IF NO
?1

FOR SYNTHESIS ENTER 0
FOR ANALYSIS ENTER 1
?1

DESIRED OPERATING FREQ. (GHZ)
?1

INPUT REL. DIELECTRIC
?4

**** INPUT LINE GEOMETRY ****

FIRST ENTER LINE THICKNESS
TO GROUND PLANE SPACING RATIO (T/B)
?0.175
NEXT ENTER GROUND PLANE SPACING (B)
?0.01
ENTER LINE WIDTH TO GROUND
PLANE SPACING RATIO (W/B)
?0.8

****** ANALYSIS RESULTS ******

FOR W/B= .8  T/B= .175
AND B= .01 CMS
CHARAC. IMP.=29.25OHMS
GUIDE WAVELTH.=15CMS
FOR AN OPERATING FREQUENCY OF1GHZ

****************************

DO YOU WANT ANOTHER GO
ENTER 1 IF YES ; 0 IF NO
?0

****** END OF PROGRAM ******
```

where t = conductor thickness. Normally most designers will find a requirement for a computer-aided scheme that allows for the synthesis (given characteristic impedance and dielectric constant find line width to ground plane spacing ratio) of stripline circuitry. Equation (2.22) can be used as part of an iterative scheme so that synthesis can be carried out. This is easily done by allowing the width to spacing ratio W/b to increase by a small amount $\Delta W/b$ after each evaluation of (2.22). In

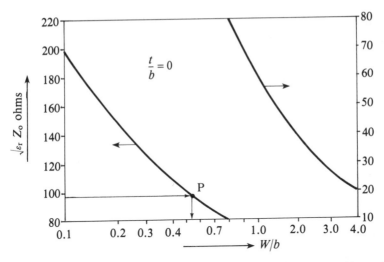

Figure 2.8 Characteristic Impedance Curve for Stripline (Source: Howe, H., 'Stripline Circuit Design,' 1974. Reproduced by Permission of Artech House, 888 Washington Street, Dedham, Massachusetts, USA 02026)

this way for each value guessed for W/b the characteristic impedance of the stripline can be calculated. This is then compared to the demanded characteristic impedance and if the computed value lies within $\pm 2/(\varepsilon_r)^{\frac{1}{2}}$ the program is terminated. The actual value of characteristic impedance achieved for the selected W/b ratio is printed out.

Equation (2.22) is somewhat complicated for hand calculation, but is fairly simple to implement on a microcomputer. Computer program 2.5 SINGLE STRIP allows both analysis and synthesis of the symmetrically disposed stripline structure shown in figure 2.7. For preliminary design a graphical approach may be preferred. The graphical approach recommended would be to prepare a graph of characteristic impedance versus line width to ground plane spacing ratio for the stripline material under consideration. This graph can be generated a priori using program SINGLE STRIP. One such graph has been prepared for theoretical stripline $(t/b = 0)$ (figure 2.8) and is used in worked example 2.10.

* * *

Example 2.10

A section of transmission line one-quarter wavelength long at an operating frequency of 4 GHz is to be constructed from stripline. The characteristic impedance of the line is to be 50 ohms. The relative permittivity of the stripline material to be used is 4. The center conductor of the stripline may be assumed to be ideally thin.

Solution

This is a synthesis problem, i.e. given $Z_o = 50$, $\varepsilon_r = 4$, find W/b for $t/b = 0$. Consider figure 2.8. Here $Z_o\sqrt{\varepsilon_r} = 50 \times \sqrt{4} = 100$. This means that point P on the curve given in figure 2.8 gives the required ratio W/b. This is read off the graph as approximately 0.51. Hence

$$\frac{W}{b} = 0.51$$

To find the guide wavelength, first compute freespace wavelength at the desired operating frequency

$$\lambda_o = \frac{c}{f} = \frac{3 \times 10^{10}}{4 \times 10^9} = 7.5 \text{ cm}$$

Next find the guide wavelength at this frequency.

$$\lambda_g = \frac{\lambda_o}{\sqrt{\varepsilon_r}} = \frac{7.5 \text{ cm}}{\sqrt{4}} = 3.75 \text{ cm}$$

and from this physical length for a one-quarter guide wavelength section

$$\frac{\lambda_g}{4} = \frac{3.75}{4} = 0.94 \text{ cm}$$

This completes the design.

*   *   *

It is worth emphasizing at this point that more than one type of transmission line may be required in order to implement a particular circuit to desired specifications, for example coaxial and stripline circuitry may have to be employed. For this reason the designer should keep his ideas as open ended and as flexible as possible.

2.6   SINGLE MICROSTRIP LINE

Like stripline, microstrip has become an attractive transmission line medium, especially where integration with chip devices and lumped elements is desired. Most of the preliminary comments made about stripline in comparison to other transmission line media made in the last section apply equally to microstrip with one notable exception. This exception is a very important one, namely, the isolation between neighboring circuits is very poor. This is due to the open geometry of microstrip. In figure 2.9, the physical construction of a microstrip line together with the field patterns around it are described.

    From Fig. 2.9 it can be seen that although the geometry of microstrip is extremely simple (a conductor of width W and thickness t suspended a distance h

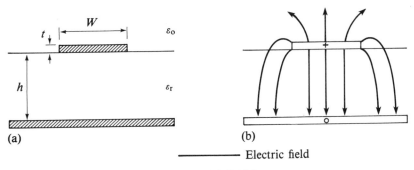

Figure 2.9   Microstrip Line (a) Geometry (b) Field Pattern

above a conducting ground plane by a uniform dielectric of relative permittivity ε_r which provides strength and rigidity to the structure) the field profiles around the lines are complex. The problem of theoretically modeling the fields in microstrip arises due to the fact that some of the fields are contained between the dielectric material and the ground plane where they are more concentrated than those that lie in the air above the top conductor. Therefore, the propagating mode along the strip is not TEM but is quasi-TEM, a term coined to indicate that the field patterns are similar to those for stripline but are slightly distorted because of the mixed dielectric (air $\varepsilon_r = 1$ and dielectric with relative permittivity $\varepsilon_r \neq 1$). At low frequencies calculations based on quasi-TEM mode propagation are reasonably accurate, but at higher frequencies a component of field along the length of the line becomes more prominent and has to be taken into account. These high frequency effects lead to dispersion, where line impedance and effective dielectric constant become slowly varying functions of operating frequency.

Calculations for microstrip usually involve a quantity known as effective dielectric constant (ε_{eff}). The effective dielectric constant for microstrip is useful since it is related to the proportion of field in the air relative to that in the dielectric. From chapter 1 any TEM transmission line will have associated with it a phase velocity V_p. Remember that V_p is the phase velocity of the signal traveling along the line and is equal to

$$V_p = \frac{1}{(LC)^{1/2}}$$

With the dielectric filling for the line removed, the phase velocity becomes equal to the velocity of an electromagnetic wave propagating in air c

$$c = \frac{1}{(LC_{air})^{1/2}}$$

where $c \approx 3 \times 10^{10}$ m/s, L is the inductance per unit length of dielectrically-loaded line (nH/m) which is equal to the inductance per unit length of air-spaced line (nH/m), C_{air} is the capacitance of air-spaced line (pF/m) and C is the capacitance of dielectrically loaded line (pF/m).

From the above

$$\frac{c}{V_p} = \left(\frac{C}{C_{air}}\right)^{1/2}$$

hence

$$\frac{C}{C_{air}} = \left(\frac{c}{V_p}\right)^2 = \varepsilon_{eff} \tag{2.23}$$

where ε_{eff} is the effective microstrip permittivity.

For microstrip lines that are very wide ($W \to \infty$) the structure will behave like a parallel plate capacitor such that nearly all of the electric fields will be trapped under the center conductor, therefore ε_{eff} will become almost equal to ε_r. For very narrow lines ($W \to 0$) the electric field is shared approximately equally between the air and dielectric. Therefore ε_{eff} is the average of ε_r for the dielectric and the air, i.e. $\varepsilon_{eff} \approx \frac{1}{2}(\varepsilon_r + 1)$. The range of ε_{eff} can be stated as

$$\frac{1}{2}(\varepsilon_r + 1) < \varepsilon_{eff} < \varepsilon_r$$

For any wave propagating in a transmission line the velocity of propagation is

$$c = f\lambda_o \quad \text{for free space}$$
$$V_p = f\lambda_g \quad \text{for dielectrically loaded line}$$

Substituting these expressions into equation (2.23) gives

$$\varepsilon_{eff} = \left(\frac{\lambda_o}{\lambda_g}\right)^2$$

hence

$$\lambda_g = \frac{\lambda_o}{(\varepsilon_{eff})^{1/2}} = \frac{c}{f(\varepsilon_{eff})^{1/2}} \tag{2.24}$$

where λ_g is the guide wavelength microstrip line.

A number of closed form expressions for the analysis of microstrip lines have appeared in the literature. Most of these are based on formulae that have been curve fitted to data gathered from experiment or more usually from elaborate numerical computation. Closed form formula are absolutely necessary for computer-aided design especially if a large number of calculations are to be carried out as would be the case in an optimization exercise. General purpose expressions for the analysis of a practical range of microstrip lines ($0.05 < W/h < 20$ and $\varepsilon_r < 16$) have been reported by Hammerstadt [7]. These are restated below for lines with zero thickness.

$W/h < 1$

$$Z_o = \frac{60}{(\varepsilon_{eff})^{1/2}} \ln\left(\frac{8h}{W} + 0.25\frac{W}{h}\right)$$

where

$$\varepsilon_{eff} = \frac{\varepsilon_r + 1}{2} + \frac{\varepsilon_r - 1}{2}\left[\left(1 + \frac{12h}{W}\right)^{-1/2} + 0.041\left(1 - \frac{W}{h}\right)^2\right]$$

also

$$W/h \geqslant 1$$

$$Z_0 = \frac{120\pi}{(\varepsilon_{\text{eff}})^{\frac{1}{2}} \left[\dfrac{W}{h} + 1.393 + 0.667 \ln \left(\dfrac{W}{h} + 1.4444 \right) \right]}$$

where

$$\varepsilon_{\text{eff}} = \frac{\varepsilon_r + 1}{2} + \frac{\varepsilon_r - 1}{2} \left(1 + 12 \frac{h}{W} \right)^{-\frac{1}{2}} \tag{2.25}$$

Both the value obtained for Z_0 and ε_{eff} using equation (2.25) are found to a maximum relative error of less than 1 percent over the ranges for W/h and ε_r specified above.

The equation dual to equation (2.25) is given below as equation (2.26), this can be used for single microstrip line synthesis with about the same accuracy as equation (2.25). For

$$A < 1.52$$

$$\frac{W}{h} = \frac{8 \exp (A)}{\exp (2A) - 2}$$

For

$$A \geqslant 1.52$$

$$\frac{W}{h} = \frac{2}{\pi} \left\{ b - 1 - \ln (2B - 1) + \frac{\varepsilon_r - 1}{2\varepsilon_r} \left[\ln (B - 1) + 0.39 - \frac{0.61}{\varepsilon_r} \right] \right\}$$

where

$$A = \frac{Z_0}{60} \left(\frac{\varepsilon_r + 1}{2} \right)^{\frac{1}{2}} + \frac{\varepsilon_r - 1}{\varepsilon_r + 1} \left(0.23 + \frac{0.11}{\varepsilon_r} \right)$$

$$B = \frac{377\pi}{2Z_0 (\varepsilon_r)^{\frac{1}{2}}} \tag{2.26}$$

For lines with non-zero thickness $t/h \neq 0$, a correction factor can be applied to equations (2.25) and (2.26). The correction factors given below in equation (2.27) are expressed in terms of effective width W^1

$$\frac{W}{h} > \frac{1}{2\pi}$$

$$\frac{W^1}{h} = \frac{W}{h} + \frac{t}{\pi h} + \frac{t}{\pi h} \left(1 + \ln \frac{2h}{t} \right)$$

$$\frac{W}{h} < \frac{1}{2\pi}$$

$$\frac{W^1}{h} = \frac{W}{h} + \frac{t}{\pi h} \left(1 + \ln \frac{4\pi W}{t} \right) \tag{2.27}$$

With the constraint that $t < h$ and $t < W/2$ equation (2.27) can be used in conjunction with equation (2.25) and (2.26) by replacing W/h with W^1/h.

The effects of dispersion on the effective permittivity of a microstrip material is usually quite small for frequencies below 10 GHz and can normally be neglected. For cases where dispersion is important the work of Getsinger [8] should be studied. Getsinger has produced a number of simple equations that when used in conjunction with equations (2.26) and (2.27) allow the effects of frequency on characteristic impedance and line width to be predicted.

Typical line widths that can be fabricated accurately limit the characteristic impedance of microstrip to about 125 ohms while wide line leads to microstrip transmission sections with low characteristic impedance of about 20 ohms. The lower limit is determined by radiation losses and also by the possibility of transverse modes being set up across the line. A number of substrate materials are available for microstrip circuit design. Of these only two types find major application at frequencies up to 18 GHz.

(1) Alumina based material usually with a dielectric constant of 8–10, this type of substrate is normally used for production runs with 99.5% pure material leading to reproducible results.
(2) Plastic substrate materials, polyolefin or woven glass fiber normally used for experimental or preproduction circuits and having relative permittivity in the range 2–3.

It should be noted that alumina circuits can be designed to a first approximation from existing plastic circuits and vice versa by appropriate scaling of circuit dimensions. Further, it shoud be noted that alumina circuits are subject to a

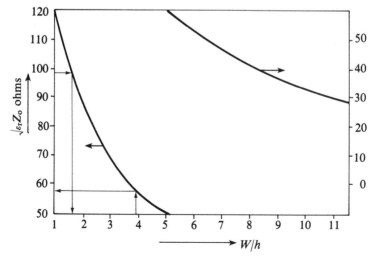

Figure 2.10 Characteristic Impedance as a Function of Width to Height Ratio for Single Microstrip Transmission Line, Negligible Conductor Thickness

reduction in size when compared with plastic based substrates due to the increased dielectric constant hence increased effective permittivity (see equation (2.24)).

Since the substrate height for microstrip laminates is usually quite small some form of metallic enclosure is used. This provides mechanical strength, heatsinking for active devices and protection against the atmosphere. Unfortunately the metal enclosure affects the design rules given by equations (2.25)–(2.27). A metallic enclosure provides electromagnetic shielding of the circuit on the outside of the enclosure by deflecting electric fields. On the inside of the enclosure the fringe field lines will be terminated before they extend to infinity. This tends to increase the density of the field lines in the air space above the line and the enclosure. When the top and sides of the metal enclosure are removed to a distance about five or six times the substrate height and line width respectively, the effect of the enclosure is negligible on the microstrip parameters computed in equations (2.25)–(2.27). The proximity effect caused by metal conductors near to microstrip lines can be studied using program 3.1 RELGRID given in the next chapter.

Design graphs can be generated from either equations (2.25) or (2.26). These design graphs can be used for approximate analysis or synthesis of a microstrip transmission line (figure 2.10).

*   *   *

Example 2.11

Using the design curves given in figure 2.10 calculate the shape (width to height) ratio for a 70 Ω microstrip transmission line constructed on a substrate material with a relative dielectric constant of 2. For a line fabricated on a substrate with relative dielectric constant of 10 find the characteristic impedance for a shape ratio of 4. Assume that the conductors have zero thickness.

Solution

The required shape ratios and characteristic impedances may now be found from the design curves given in Fig. 2.10. These values are then inserted into table 2.3.

Table 2.3

Characteristic Impedance	70 Ω	18 Ω
ε_r	2	10
$\sqrt{\varepsilon_r}.\, Z_o$	99	57
Shape Ratio (W/h)	1.6	4

It should be noted for each case that a manipulation involving the term $\sqrt{\varepsilon_r}.\, Z_o$ is required before characteristic impedance or shape ratio can be obtained.

*   *   *

For the analysis or synthesis of a microstrip transmission line with relative dielectric constant less than 16 and finite line thickness, computer program 2.6 SINGLE MIC will perform all the necessary computations. It should be noted that for this program the usable range of shape ratios over which relative accuracy of better than 1 percent can be maintained is 0.65–20. This range of values is normally sufficient for most practical line designs.

```
][ FORMATTED LISTING
FILE: PROGRAM 2.6 SINGLE MIC
PAGE-1


     10    REM
     20    REM    --SINGLE MICROSTRIP--
     30    REM
     40    REM    THIS PROGRAM CAN
     50    REM    BE USED FOR THE
     60    REM    ANALYSIS OR SYNTHESIS
     70    REM    .OF SINGLE MICROSTRIP
     80    REM    LINE. PROVIDED ER IS
     90    REM    LESS THAN 16, AND
    100    REM    0.65<=W/H<=20.
    110    REM    FINITE CONDUCTOR
    120    REM    THICKNESS IS INCLUDED
    130    REM
    140    REM W/H=WIDTH/HEIGHT
    150    REM H=DIE. THICKNESS(MM)
    160    REM W=LINE WIDTH(MM)
    170    REM ER=REL. DIE. CONST.
    180    REM EEF=EFFECTIVE DIE.
    190    REM CONSTANT
    200    REM T=COND. THICKNESS(MM)
    210    REM
    220    REM    INPUT ROUTINE
    230    HOME
    240    PRINT "ENTER 1 FOR SYNTHESIS"
    250    PRINT "ENTER 0 FOR ANALYSIS"
    260    INPUT K
    270    IF K < 0 OR K > 1 THEN
              230
    280    PRINT
    290    PRINT "===== INPUT LINE GEOMETRY ====="
    300    PRINT
    310    PRINT "INPUT CONDUCTOR THICKNESS (MM)"
    320    INPUT T
    330    PRINT "INPUT DIELECTRIC HEIGHT (MM)"
    340    INPUT H
    350    PRINT "ENTER REL. DIE CONSTANT"
    360    INPUT ER
    370    LET TH = T / H
    380    LET PI = 3.14159
    390    IF K = 1 THEN
              570
    400    REM
    410    REM    ANALYSIS ROUTINE
    420    PRINT
    430    PRINT "INPUT LINE WIDTH (MM)"
    440    INPUT W
    450    LET WH = W / H
    460    PRINT
    470    PRINT "***** ANALYSIS RESULTS *****"
    480    GOSUB 1060
    490    GOSUB 740
    500    PRINT "LINE WIDTH (MM) "W
    510    PRINT
    520    PRINT "EFFECTIVE DIELECTRIC CONSTANT "EEF
    530    PRINT "CHARAC. IMP. "ZO" OHMS"
    540    PRINT
    550    PRINT "******************************"
```

```
560    GOTO 1130
570    REM
580    REM   SYNTHESIS ROUTINE
590    PRINT
600    PRINT "ENTER CHARAC. IMP. (OHMS)"
610    INPUT ZO
620    GOSUB 910
630    PRINT
640    PRINT "***** SYNTHESIS ROUTINE *****"
650    GOSUB 1060
660    PRINT "CHARAC. IMP. "ZO" OHMS"
670    PRINT "LINE WIDTH "W" MMS."
680    PRINT
690    PRINT "*****************************"
700    GOTO 1130
710    PRINT
720    PRINT "******* END OF PROGRAM *******"
730    END
740    REM
750    REM   ANALYSIS ROUTINE
760    IF WH >  = 1 / 2 / PI THEN
           790
770    LET WHE = WH + TH / PI * (1 +  LOG (4 * PI * W / T))
780    GOTO 800
790    LET WHE = WH + TH / PI * (1 +  LOG (2 / TH))
800    IF WH >  = 1 THEN
           850
810    LET EEF = (1 /  SQR (1 + 12 / WH) + 0.04 * (1 - WH) ^ 2)
820    LET EEF = (ER + 1) / 2 + ((ER - 1) / 2) * EEF
830    LET ZO = 60 /  SQR (EEF) *  LOG (8 / WH + 0.25 * WH)
840    GOTO 880
850    LET EEF = (ER + 1) / 2 + (((ER - 1) / 2) /  SQR (1 + 12 / WH))
860    LET ZO = 120 * PI /  SQR (EEF)
870    LET ZO = ZO / (WH + 1.393 + 0.667 *  LOG (WH + 1.444))
880    LET ZO =  INT (ZO * 100 + 0.5) / 100
890    LET EEF =  INT (EEF * 100 + 0.5) / 100
900    RETURN
910    REM
920    REM   SYNTHESIS ROUTINE
930    LET B = 377 * PI / 2 / ZO /  SQR (ER)
940    LET A = ((ER - 1) / (ER + 1)) * (0.23 + 0.11 / ER)
950    LET A = A + ZO / 60 *  SQR ((ER + 1) / 2)
960    IF A <  = 1.52 THEN
           1020
970    LET WH =  LOG (B - 1) + 0.39 - 0.61 / ER
980    LET WH = (ER - 1) / 2 / ER * WH
990    LET WH = WH + (B - 1 -  LOG (2 * B - 1))
1000   LET WH = 2 / PI * WH
1010   GOTO 1030
1020   LET WH = 8 *  EXP (A) / ( EXP (2 * A) - 2)
1030   LET W = WH * H
1040   LET W =  INT (W * 100 + 0.5) / 100
1050   RETURN
1060   REM
1070   PRINT
1080   PRINT "CONDUCTOR THICKNESS (MMS) "T
1090   PRINT "DIELECTRIC HEIGHT (MMS) "H
1100   PRINT "REL. DIE. CONSTANT "ER
1110   PRINT
1120   RETURN
1130   REM
1140   REM   SERVICE ROUTINE
1150   PRINT
1160   PRINT "DO YOU WANT ANOTHER GO? "
1170   PRINT "ENTER 1 FOR YES ; 0 FOR NO"
1180   INPUT L
1190   IF L = 1 THEN
           230
1200   IF L = 0 THEN
           710
1210   GOTO 1130
```

END-OF-LISTING

```
]RUN
ENTER 1 FOR SYNTHESIS
ENTER 0 FOR ANALYSIS
?1

===== INPUT LINE GEOMETRY =====

INPUT CONDUCTOR THICKNESS (MM)
?0.001
INPUT DIELECTRIC HEIGHT (MM)
?0.1
ENTER REL. DIE CONSTANT
?2.3

ENTER CHARAC. IMP. (OHMS)
?50

***** SYNTHESIS ROUTINE *****

CONDUCTOR THICKNESS (MMS) 1E-03
DIELECTRIC HEIGHT (MMS) .1
REL. DIE. CONSTANT 2.3

CHARAC. IMP. 50 OHMS
LINE WIDTH .3 MMS.

*****************************

DO YOU WANT ANOTHER GO?
ENTER 1 FOR YES ; 0 FOR NO
?1
ENTER 1 FOR SYNTHESIS
ENTER 0 FOR ANALYSIS
?0

===== INPUT LINE GEOMETRY =====

INPUT CONDUCTOR THICKNESS (MM)
?0.001
INPUT DIELECTRIC HEIGHT (MM)
?0.1
ENTER REL. DIE CONSTANT
?2.3

INPUT LINE WIDTH (MM)
?0.3

***** ANALYSIS RESULTS *****

CONDUCTOR THICKNESS (MMS) 1E-03
DIELECTRIC HEIGHT (MMS) .1
REL. DIE. CONSTANT 2.3

LINE WIDTH (MM) .3

EFFECTIVE DIELECTRIC CONSTANT 1.94
CHARAC. IMP. 50 OHMS

*****************************

DO YOU WANT ANOTHER GO?
ENTER 1 FOR YES ; 0 FOR NO
?0

******* END OF PROGRAM *******
```

2.7   EVEN AND ODD MODE EXCITATION

Consider stripline or microstrip line conductors of equal width placed parallel to each other (figure 2.11). Since the electric fields associated with these conductors are

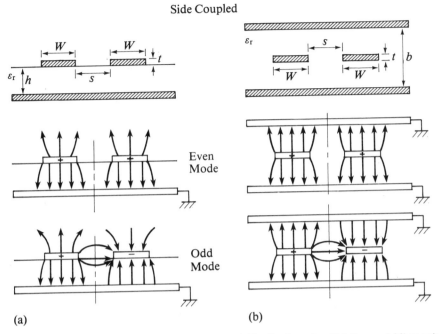

Figure 2.11   Even and Odd Mode Electric Field Distributions for Stripline and Microstrip
(a) Microstrip (b) Stripline

not closely bound to the conductors, fringe fields will exist, the exact distribution of which will depend on the potentials and geometry of the conductors and on the dielectric substrate used.

The fringing fields can be exploited to form circuits such as directional couplers. Directional couplers form the basis for many different types of transmission line circuits such as balanced mixers or modulators (see also chapter 5). Many types of filter circuits can be constructed by exploiting the coupling between lines. Normally this is achieved by arranging line sections that are resonant at a selected frequency and whose spacing gives the desired response.

Consider intuitively how the electric field lines will arrange themselves around conductors in the stripline and microstrip configurations given in figure 2.11. For the pairs of parallel edge coupled lines shown, two possibilities exist, either both center conductors can be set at positive potentials say $+1$ volt (even mode excitation) or one conductor can be set to $+1$ volt while the other is held at -1 volt (odd mode excitation). The axis for symmetry for the even mode excitation case is called the magnetic wall and for the odd mode case the electric wall. All parallel edge coupled lines exhibit this even and odd mode excitation property. This is useful since the degree of coupling between lines can be specified in terms of even and odd mode impedance [9]. Equation (2.28) below gives useful relationships for side coupled lines.

Coupling factor $= C$ dB

$$= 20 \log_{10} \text{ (voltage coupling factor)}$$

$$= -20 \log_{10} C_{\text{o}}$$

Normalized even mode impedance $Z_{\text{oe}} = \left(\dfrac{1 + C_{\text{o}}}{1 - C_{\text{o}}} \right)^{\frac{1}{2}}$ \hfill (2.28a)

Normalized odd mode impedance $Z_{\text{oo}} = \left(\dfrac{1 - C_{\text{o}}}{1 + C_{\text{o}}} \right)^{\frac{1}{2}}$ \hfill (2.28b)

and finally for a matched condition within quarter wave line sections with loose coupling, say greater than 6 dB

$$Z_{\text{o}} = (Z_{\text{oe}} Z_{\text{oo}})^{\frac{1}{2}} \hfill (2.28c)$$

The above expressions are strictly true only for TEM propagation e.g. in coaxial or stripline where even and odd mode propagation velocities are equal. In the microstrip case two different phase velocities exist, one for the even mode v_{pe} and one for the odd mode v_{po}. These expressions can still be used, however, provided calculations of coupled line guide wavelengths are empirically adjusted. This will be further discussed in section 2.7.2 on coupled microstrip line design. From figure 2.11 it can be seen that for the odd mode case more electric field lines are concentrated in the gap between the upper conductors than in the even mode case. This means that the tightest coupling will occur for the odd mode case. It should also be noted that from the definition given for odd mode impedance the currents flowing in each conductor will be in opposite directions. In the odd mode propagation of energy down one line is coupled into the parallel line and travels in the reverse direction. Odd mode coupling is therefore sometimes called reverse coupling while for the even mode the term forward coupling can be applied.

2.7.1 Coupled Striplines

The simplest case of coupling for stripline circuits is the symmetrical side coupled case shown in Fig. 2.11. Here two transmission lines placed halfway between ground planes are positioned close enough to each other so that mutual coupling between the lines can occur. The analysis and synthesis equations for this configuration are given below and are attributed to the work of Cohn [10].

Closed form analysis equation ($t < 0.1$ and $W/b > 0.35$):

$$(\varepsilon_{\text{r}})^{\frac{1}{2}} Z_{\text{oe}} = 30\pi (b - t) \left/ \left(W + \frac{bC}{2\pi} A_{\text{e}} \right) \right.$$

$$(\varepsilon_{\text{r}})^{\frac{1}{2}} Z_{\text{oo}} = 30\pi (b - t) \left/ \left(W + \frac{bC}{2\pi} A_{\text{o}} \right) \right.$$

where

$$A_e = 1 + \frac{\ln (1 + \tanh \theta)}{0.6932}; \quad A_o = 1 + \frac{\ln (1 + \coth \theta)}{0.6932} \tag{2.29}$$

$$\theta = \frac{\pi s}{2b}$$

and

$$C\left(\frac{t}{b}\right) = 2 \ln \left(\frac{2b - t}{b - t}\right) - \frac{t}{b} \ln \left[\frac{t(2b - t)}{(b - t)^2}\right]$$

Closed form synthesis equation:

$$\frac{W}{b} = \frac{2}{\pi} \tanh^{-1}[k_e k_o]$$

$$\frac{s}{b} = \frac{2}{\pi} \tanh^{-1}\left[\frac{1 - k_o}{1 - k_e} \left(\frac{k_e}{k_o}\right)^{\frac{1}{2}}\right]$$

with

$$k_{e,o} = \left[1 - \left(\frac{\exp (\pi x) - 2}{\exp (\pi x) + 2}\right)^4\right]^{\frac{1}{2}} \quad \text{from the range } 1 \leqslant x \leqslant \infty \tag{2.30}$$

and

$$k_{e,o} = \left[\frac{\exp (\pi/x) - 2}{\exp (\pi/x) + 2}\right]^2 \quad \text{for the range } 0 \leqslant x \leqslant 1$$

where

$$x = \frac{Z_{oo,e}(\varepsilon_r)^{\frac{1}{2}}}{30\pi}$$

In equations (2.29) and (2.30) the subscripts o,e represent odd and even mode propagation respectively.

Equations (2.29) and (2.30) are available in graphical form and are presented in the form of nomograms giving W/b ratios and s/b ratios for known Z_{oe}, Z_{oo} [11]. The above equations have been implemented in computer program 2.7 CSTRIP so that analysis and synthesis of side (parallel edge) coupled stripline circuits can be facilitated.

For stripline circuits using parallel edge coupling it is difficult to obtain tight coupling (coupling that is greater than -10 dB or at the very most -6 dB). This is due mainly to problems in etching narrow line spacings in the stripline material. Loose coupling implies that only narrow band circuits can be constructed. For broad band circuitry tight coupling, say less than -6 dB, is normally required and broadside coupled parallel sections or variable overlap sections are needed. Broadside coupled parallel sections can provide coupling of -3 dB or greater while

]| FORMATTED LISTING
FILE: PROGRAM 2.7 CSTRIP
PAGE-1

```
   10   REM
   20   REM --COUPLED STRIPLINE--
   30   REM
   40   REM THIS PROGRAM COMPUTES
   50   REM THE PARAMETERS REQUIRED
   60   REM FOR THE ANALYSIS OR
   70   REM OR SYNTHESIS OF
   80   REM COUPLED STRIPLINES.
   90   REM BEST ACCURACY OCCURS
  100   REM FOR T/B<0.1 AND W>0.35
  110   REM
  120   REM W=LINE WIDTH (MMS)
  130   REM S=LINE SPACING (MMS)
  140   REM B=GROUND PLANE SPACING
  150   REM GIVEN IN MMS
  160   REM ER=REL. DIE. CONST.
  170   REM ZE=EVEN MODE IMP.
  180   REM ZO=ODD MODE IMP.
  190   REM
  200   HOME
  210   PRINT "FOR ANALYSIS ENTER 1"
  220   PRINT "FOR SYNTHESIS ENTER 0"
  230   INPUT P
  240   IF P < 0 OR P > 1 THEN
           200
  250   LET PI = 3.141592
  260   PRINT
  270   PRINT "INPUT GROUND PLANE SPACING (MMS)"
  280   INPUT B
  290   PRINT "INPUT CONDUCTOR THICKNESS (MMS)"
  300   INPUT T
  310   PRINT "INPUT RELATIVE DIELECTRIC CONST."
  320   INPUT ER
  330   IF P = 0 THEN
              440
  340   PRINT "INPUT LINE WIDTH (MMS)"
  350   INPUT W
  360   PRINT "INPUT LINE SPACING (MMS)"
  370   INPUT S
  380   PRINT
  390   PRINT "**** ANALYSIS RESULTS ****"
  400   PRINT
  410   PRINT "LINE WIDTH "W" MMS"
  420   PRINT "LINE SPACING "S" MMS"
  430   GOTO 600
  440   PRINT "INPUT REQUIRED COUPLING (DB)"
  450   INPUT DB
  460   PRINT "INPUT COUPLER IMPEDANCE (OHMS)"
  470   INPUT Z
  480   PRINT
  490   PRINT "**** SYNTHESIS RESULTS ****"
  500   PRINT
  510   LET X = DB / 20
  520   LET COUPLE = 10 ^ X
  530   LET F1 = (1 + COUPLE) / (1 - COUPLE)
  540   LET F2 = 1 / F1
  550   LET ZE = Z *  SQR (F1)
  560   LET ZO = Z *  SQR (F2)
  570   PRINT "REQUIRED COUPLING " INT (COUPLE * 10000 + 0.5) / 10000" OR"DB" DB"
  580   PRINT "EVEN MODE IMPEDANCE " INT (ZE * 100 + 0.5) / 100" OHMS"
  590   PRINT "ODD MODE IMPEDANCE " INT (ZO * 100 + 0.5) / 100" OHMS"
  600   PRINT "GROUND PLANE SPACING "B" MMS"
  610   PRINT "CONDUCTOR THICKNESS "T" MMS"
  620   PRINT "REL. DIELECTRIC CONSTANT "ER
  630   IF P = 0 THEN
           GOSUB 740
  640   IF P = 1 THEN
           GOSUB 1010
  650   PRINT
```

```
660   PRINT "DO YOU WANT ANOTHER GO?"
670   PRINT "ENTER 1 IF YES ; 0 IF NO"
680   INPUT P
690   IF P < 0 OR P > 1 THEN
         650
700   IF P = 1 THEN
         200
710   PRINT
720   PRINT "**** END OF PROGRAM ****"
730   END
740   REM
750   REM   SYNTHESIS ROUTINE
760   FOR I = 1 TO 2
770       IF I = 1 THEN
             LET Z = ZE
780       IF I = 2 THEN
             LET Z = ZO
790       LET X = Z * SQR (ER) / 30 / PI
800       LET L = EXP (PI * X)
810       LET M = EXP (PI / X)
820       IF X < = 1 AND X > = 0 THEN
             850
830       LET K = SQR (1 - ((L - 2) / (L + 2)) ^ 4)
840       GOTO 860
850       LET K = ((M - 2) / (M + 2)) ^ 2
860       IF I = 1 THEN
             LET KE = K
870       IF I = 2 THEN
             LET KO = K
880   NEXT I
890   LET WB = LOG ((1 + ( SQR (KE * KO))) / (1 - ( SQR (KE * KO)))) / PI
900   LET A = (1 - KO) / (1 - KE)
910   LET A = A * SQR (KE / KO)
920   LET SB = LOG ((1 + A) / (1 - A)) / PI
930   LET W = WB * B
940   LET S = SB * B
950   PRINT
960   PRINT "LINE WIDTH " INT (W * 1000 + 0.5) / 1000" MMS"
970   PRINT "LINE SPACING " INT (S * 1000 + 0.5) / 1000" MMS"
980   PRINT
990   PRINT "*********************"
1000  RETURN
1010  REM
1020  REM ANALYSIS ROUTINE
1030  LET CF = 2 * LOG ((2 * B - T) / (B - T))
1040  LET CF = CF - T / B * LOG ((T * (2 * B - T)) / (B - T) / (B - T))
1050  LET TH = PI * S / 2 / B
1060  LET AO = 1 + LOG (1 + ( EXP ( - TH) / ( EXP (TH) - EXP ( - TH)) * 2 + 1
         )) / LOG (2)
1070  LET AE = 1 + LOG (1 + ( - EXP ( - TH) / ( EXP (TH) + EXP ( - TH)) * 2
         + 1)) / LOG (2)
1080  LET Z = 30 * PI * (B - T) / SQR (ER)
1090  LET ZE = Z / (W + B * CF / 2 / PI * AE)
1100  LET ZO = Z / (W + B * CF / 2 / PI * AO)
1110  PRINT
1120  PRINT "EVEN MODE IMPEDANCE " INT (ZE * 100 + 0.5) / 100" OHMS"
1130  PRINT "ODD MODE IMPEDANCE " INT (ZO * 100 + 0.5) / 100" OHMS"
1140  LET C = 20 / 2.303 * LOG ( ABS ((ZE - ZO) / (ZE + ZO)))
1150  PRINT "COUPLING " INT (C * 100 + 0.5) / 100" DB"
1160  PRINT
1170  PRINT "*********************"
1180  RETURN


END-OF-LISTING


]RUN
FOR ANALYSIS ENTER 1
FOR SYNTHESIS ENTER 0
?1

INPUT GROUND PLANE SPACING (MMS)
?10
```

```
INPUT CONDUCTOR THICKNESS (MMS)
?0.005
INPUT RELATIVE DIELECTRIC CONST.
?2
INPUT LINE WIDTH (MMS)
?9
INPUT LINE SPACING (MMS)
?0.02

**** ANALYSIS RESULTS ****

LINE WIDTH 9 MMS
LINE SPACING .02 MMS
GROUND PLANE SPACING 10 MMS
CONDUCTOR THICKNESS 5E-03 MMS
REL. DIELECTRIC CONSTANT 2

EVEN MODE IMPEDANCE 59.35 OHMS
ODD MODE IMPEDANCE 22.49 OHMS
COUPLING -6.93 DB

**********************

DO YOU WANT ANOTHER GO?
ENTER 1 IF YES ; 0 IF NO
?1
FOR ANALYSIS ENTER 1
FOR SYNTHESIS ENTER 0
?0

INPUT GROUND PLANE SPACING (MMS)
?10
INPUT CONDUCTOR THICKNESS (MMS)
?0.005
INPUT RELATIVE DIELECTRIC CONST.
?2
INPUT REQUIRED COUPLING (DB)
?-10
INPUT COUPLER IMPEDANCE (OHMS)
?50

**** SYNTHESIS RESULTS ****

REQUIRED COUPLING .3162 OR-10 DB
EVEN MODE IMPEDANCE 69.37 OHMS
ODD MODE IMPEDANCE 36.04 OHMS
GROUND PLANE SPACING 10 MMS
CONDUCTOR THICKNESS 5E-03 MMS
REL. DIELECTRIC CONSTANT 2

LINE WIDTH 7.213 MMS
LINE SPACING .387 MMS

**********************

DO YOU WANT ANOTHER GO?
ENTER 1 IF YES ; 0 IF NO
?0

**** END OF PROGRAM ****
```

variable overlap sections are normally used for the intermediate region between broadside and side coupled lines, -6 dB to -3 dB. A discussion of these line configurations and relevant design data is given by Howe [11].

2.7.2 Coupled Microstrip

For coupled lines on microstrip, only side coupling is acceptable due to the planar nature of the material. The problems associated with the design of single striplines exist for coupled mictrostrip lines and are complicated by the existence of odd and even mode dependent velocity parameters. Most coupled line design techniques involve the use of analysis formulae in an iterative scheme designed to allow synthesis. Two methods are normally used to generate coupler data. The first and most accurate for loosely coupled lines was formulated by Weiss [12] and uses a complex numerical scheme based on the simultaneous solution of a number of equations related to the charge distribution on the coupled lines. This technique provides families of useful design curves that can be used for preliminary coupler synthesis. The second approach formulated by Schwarzmann [13] provides semi-empirical analysis equations. The latter paper should be consulted for the relevant design equations. Schwarzmann's equations are claimed to be about 97 percent accurate and form the basis of program 2.8 CMIC. This enables relatively easy computation but at the expense of accuracy.

Consider figure 2.11(a). The electric field distribution for coupled microstrip displays a large proportion of fringe fields. From section 2.7 odd mode excitation was shown intuitively to give tightest coupling, this, in combination with Schwarzmann's equations gives the basis of a design procedure for parallel edge coupled lines. The approach used by Schwarzmann is to investigate odd and even mode propagation separately by dividing the total capacitance in each case into a number of components that could be more easily analyzed. For each mode of operation the total line capacitance is divided into three sections.

(1) A parallel plate capacitor between the lower surface of the upper conductor and the ground plane.
(2) The fringing capacitance at the edges of the conductor.
(3) The capacitance between the upper surface of the upper conductor and the ground plane. Each capacitance is a function of line width W or conductor spacing s or both. The design procedure for coupled line synthesis is outlined below:

(a) decide on a coupling factor and coupler impedance;
(b) compute even and odd mode impedance Z_{oe}, Z_{oo} using Schwarzmann's equations;
(c) solve $Z_{oe} = f(C_{oe})$
$\qquad Z_{oo} = f(C_{oo})$ for W and s
$\qquad C_{oo}$, $C_{oe} =$ total odd and even mode capacitance;
(d) finally, evaluate length of coupling region. This part requires an empirical assumption (see later in this section).

Program 2.8 CMIC uses the above technique to implement Schwarzmann's analysis formulae in a synthesis role. First the program computes Z_{oe}, Z_{oo} from given coupling factor and coupler impedance. It then solves Z_{oe} in terms of line spacing by assuming line width is known. This value of line spacing is replaced in the equation describing Z_{oo} giving Z_{oo} in terms of line width only.

Now line width is allowed to vary in a known manner and from the expression for Z_{oo} a series of test results for Z_{oo} become available. These test values are compared with the demand value of Z_{oo} obtained from the coupling equation (2.28).

```
]{ FORMATTED LISTING
FILE: PROGRAM 2.8 CMIC
PAGE-1


    10  REM
    20  REM    **COUPLED MICROSTRIP**
    30  REM
    40  REM    THIS PROGRAM CAN BE
    50  REM    USED FOR SYNTHESIS
    60  REM    OR FOR ANALYSIS
    70  REM    OF PARALLEL EDGE
    80  REM    COUPLED MICROSTRIP
    90  REM    LINES CONSTRUCTED
   100  REM    ON A GIVEN MATERIAL
   110  REM    WITH A REQUIRED
   120  REM    ELECTRICAL COUPLING
   130  REM    GIVEN IN DB.
   140  REM
   150  REM ER=REL. DIE. CONST.
   160  REM H=DIE. THICKNESS(MMS)
   170  REM T=COND. THICKNESS(MMS)
   180  REM DB=COUPLING(DB)
   190  REM FRQ=FREQUENCY(GHZ)
   200  REM ZO=COUPLER IMP.
   210  REM ZE=EVEN MODE IMP.
   220  REM Z1=ODD MODE IMP.
   230  REM GO=GUIDE WAVELTH.
   240  REM ODD MODE(CMS)
   250  REM GE=GUIDE WAVELTH.
   260  REM EVEN MODE(CMS)
   270  REM
   280  REM  INPUT DATA
   290  HOME
   300  PRINT
   310  PRINT "DO YOU REQUIRE ANALYSIS OR SYNTHESIS"
   320  PRINT "FOR SYNTHESIS ENTER 1"
   330  PRINT "FOR ANALYSIS ENTER 0"
   340  INPUT P
   350  IF P < 0 OR P > 1 THEN
            290
   360  PRINT "ENTER REL. DIE. CONSTANT"
   370  INPUT ER
   380  PRINT "ENTER DIE. THICKNESS(MMS)"
   390  INPUT H
   400  PRINT "ENTER CONDUCTOR THICKNESS(MMS)"
   410  INPUT T
   420  PRINT "ENTER FREQUENCY (GHZ)"
   430  INPUT FRQ
   440  IF P = 1 THEN
            500
   450  PRINT "INPUT LINE SPACING (MMS)"
   460  INPUT S
   470  PRINT "INPUT LINE WIDTH (MMS)"
   480  INPUT W
   490  IF P = 0 THEN
            540
   500  PRINT "ENTER LINE COUPLING(DB)"
   510  INPUT DB
   520  PRINT "ENTER COUPLER IMPEDANCE(OHMS)"
   530  INPUT ZO
   540  PRINT
   550  IF P = 0 THEN
            PRINT "**** ANALYSIS RESULTS ****"
   560  IF P = 0 THEN
            580
   570  PRINT "**** SYNTHESIS RESULTS ****"
   580  PRINT
   590  PRINT "REL. DIE. CONST. "ER
   600  PRINT "DIELECTRIC THICKNESS "H" MMS"
   610  PRINT "CONDUCTOR THICKNESS "T" MMS"
   620  PRINT "OPERATING FREQUENCY "FRQ" GHZ"
   630  IF P = 0 THEN
            730
```

```
640    PRINT "COUPLER IMPEDANCE "ZO" OHMS"
650    LET X = DB / 20
660    LET CO = 10 ^ X
670    LET C =  INT (CO * 10000 + 0.5) / 10000
680    PRINT "COUPLING COEFF "C" OR "DB" DB"
690    LET F1 = (1 + CO) / (1 - CO)
700    LET F2 = 1 / F1
710    LET ZE = ZO *  SQR (F1)
720    LET Z0 = ZO *  SQR (F2)
730    LET FRQ = FRQ * 1E9
740    LET B = 376.6 /  SQR (ER)
750    LET D = 1 / 3 /  SQR (ER)
760    LET F = 1.35 /  LOG (4 * H / T)
770    LET K = 1
780    IF P = 0 THEN
             880
790    LET W = 0
800    LET M = 0
810    LET STE = 0.01
820    IF K > 1000 THEN
             850
830    GOTO 870
840    PRINT
850    PRINT "ITERATION LIMIT EXCEEDED"
860    GOTO 1610
870    LET W = W + STE
880    LET A = W / H
890    LET G = (A + 1) ^ 2
900    LET AA = D * (A - (1 / G))
910    LET BB = D / G
920    IF P = 0 THEN
             1200
930    LET CC = A + AA + 0.5 * D * A - BB + 1.5 * F
940    LET DD = 0.5 * D * A + 0.5 * F
950    LET X = ((B / ZE) - CC) / DD
960    LET Y = 1 / X
970    LET XX = 4 * (Y - 1)
980    LET S = W / XX
990    IF S > 0 THEN
             1020
1000   LET TE = 1E10
1010   GOTO 1080
1020   LET Z1 = 2 * AA
1030   LET M = M + 1
1040   LET Z2 = 4 * D / ((S / W) + 1)
1050   LET Z3 = 1.35 /  LOG (4 * S / (3.141592 * T))
1060   LET TE = B / (A + Z1 + F + Z2 + Z3)
1070   IF M = 1 THEN
               LET STE = STE / 10
1080   LET CH = (TE - ZO) / ZO
1090   IF ( ABS (CHECK)) < = 0.01 THEN
             1230
1100   IF CH = 0 THEN
             1230
1110   IF CH > 0 THEN
             1180
1120   IF CH < 0 THEN
             1140
1130   IF M = 1 THEN
               LET STE = STE * 10
1140   LET W = W - STE
1150   LET STE = STE / 10
1160   LET K = K + 1
1170   GOTO 820
1180   LET K = K + 1
1190   GOTO 820
1200   REM
1210   REM FIND ODD MODE GUIDE
1220   REM WAVELTH.
1230   LET K1 = ER / (2.998E8 * 376.6)
1240   LET C1 = 2 * K1 * AA
1250   LET C2 = 8 * D * K1 / ((S / W + 1))
1260   LET CP = K1 * W / H
1270   LET CF = K1 * 2.7 /  LOG (4 * H / T)
```

```
1280   LET CD = K1 * 2.7 /  LOG (4 * S / (3.141592 * T))
1290   LET CO = CP + 0.5 * C1 + 0.5 * CF + 0.5 * C2 + 0.5 * CD
1300   LET AL = 1 / (1 + ((C1 + C2) * (1 - (1 /  SQR (ER))) / (2 * CO)))
1310   LET AK = 1 /  SQR (1 + AL * AL * (ER - 1))
1320   LET GO =  INT (3E10 * AK / FRQ * 100 + 0.5) / 100
1330   REM
1340   REM   FIND EVEN MODE
1350   REM   WAVELTH.
1360   LET C4 = K1 * (F + F / (1 + W / 4 / S))
1370   LET C5 = K1 * ((D * W / H) + (D * W / H / (1 + W / 4 / S)) - BB * 2)
1380   LET C6 = CP + 0.5 * C1 + 0.5 * CF + 0.5 * C4 + 0.5 * C5
1390   LET A1 = 1 / (1 + ((C1 + C5) * (1 - (1 /  SQR (ER))) / (2 * C6)))
1400   LET A1 = 1 /  SQR (1 + A1 ^ 2 * (ER - 1))
1410   LET GE =  INT (3E10 * A1 / FRQ * 100 + 0.5) / 100
1420   IF P = 0 THEN
            1440
1430   GOTO 1470
1440   LET ZE = 376.6 * K1 /  SQR (ER) / C6
1450   LET Z0 = 376.6 * K1 /  SQR (ER) / CO
1460   LET C = 20 / 2.303 *  LOG ( ABS ((ZE / Z0 - 1) / (ZE / Z0 + 1)))
1470   LET Z0 =  INT (Z0 * 10000 + 0.5) / 10000
1480   LET ZE =  INT (ZE * 10000 + 0.5) / 10000
1490   LET W =  INT (W * 1000 + 0.5) / 1000
1500   LET S =  INT (S * 1000 + 0.5) / 1000
1510   PRINT
1520   PRINT "LINE WIDTH "W" MMS"
1530   PRINT "LINE SPACING "S" MMS"
1540   PRINT "ODD MODE IMP. "Z0"OHMS"
1550   PRINT "EVEN MODE IMP. "ZE" OHMS"
1560   PRINT "EVEN MODE GUIDE WAVELTH "GE" CMS."
1570   PRINT "ODD MODE GUIDE WAVELTH "GO" CMS."
1580   IF P = 0 THEN
            PRINT "COUPLING ACHEIVED " INT (C * 100 + 0.5) / 100" DB"
1590   PRINT
1600   PRINT "***************************"
1610   PRINT
1620   PRINT "DO YOU WANT ANOTHER GO?"
1630   PRINT "ENTER 1 IF YES ; 0 IF NO"
1640   INPUT L
1650   IF L = 1 THEN
            290
1660   IF L = 0 THEN
            1680
1670   IF L < 0 OR L > 1 THEN
            1610
1680   PRINT
1690   PRINT "**** END OF PROGRAM ****"
1700   END
```

END-OF-LISTING

] RUN

```
DO YOU REQUIRE ANALYSIS OR SYNTHESIS
FOR SYNTHESIS ENTER 1
FOR ANALYSIS ENTER 0
?1
ENTER REL. DIE. CONSTANT
?2.3
ENTER DIE. THICKNESS(MMS)
?0.16
ENTER CONDUCTOR THICKNESS(MMS)
?0.08
ENTER FREQUENCY (GHZ)
?0.5
ENTER LINE COUPLING(DB)
?-20
ENTER COUPLER IMPEDANCE(OHMS)
?50
```

**** SYNTHESIS RESULTS ****

```
REL. DIE. CONST. 2.3
DIELECTRIC THICKNESS .16 MMS
CONDUCTOR THICKNESS .08 MMS
OPERATING FREQUENCY .5 GHZ
COUPLER IMPEDANCE 50 OHMS
COUPLING COEFF .1 OR -20 DB

LINE WIDTH .377 MMS
LINE SPACING .251 MMS
ODD MODE IMP. 45.2267OHMS
EVEN MODE IMP. 55.2771 OHMS
EVEN MODE GUIDE WAVELTH 41.06 CMS.
ODD MODE GUIDE WAVELTH 41.05 CMS.

***************************

DO YOU WANT ANOTHER GO?
ENTER 1 IF YES ; 0 IF NO
?1

DO YOU REQUIRE ANALYSIS OR SYNTHESIS
FOR SYNTHESIS ENTER 1
FOR ANALYSIS ENTER 0
?0
ENTER REL. DIE. CONSTANT
?2.3
ENTER DIE. THICKNESS(MMS)
?0.17
ENTER CONDUCTOR THICKNESS(MMS)
?0.05
ENTER FREQUENCY (GHZ)
?1
INPUT LINE SPACING (MMS)
?0.5
INPUT LINE WIDTH (MMS)
?3

**** ANALYSIS RESULTS ****

REL. DIE. CONST. 2.3
DIELECTRIC THICKNESS .17 MMS
CONDUCTOR THICKNESS .05 MMS
OPERATING FREQUENCY 1 GHZ

LINE WIDTH 3 MMS
LINE SPACING .5 MMS
ODD MODE IMP. 10.6455OHMS
EVEN MODE IMP. 9.8859 OHMS
EVEN MODE GUIDE WAVELTH 20.72 CMS.
ODD MODE GUIDE WAVELTH 20.51 CMS.
COUPLING ACHEIVED -28.63 DB

***************************

DO YOU WANT ANOTHER GO?
ENTER 1 IF YES ; 0 IF NO
?0

**** END OF PROGRAM ****
```

Depending on the error value obtained, the line width is changed accordingly to give a smaller error between the demanded and calculated Z_{oo} values. In this way a value for line width is found that satisfies the value of Z_{oo} to within one percent of the required value. This value can then be substituted into the same equation and line spacing found. Finally, knowing line width and line spacing the odd mode guide wavelength can be obtained.

Selecting odd mode guide wavelength can cause problems due to the unequal phase velocities that exist in the microstrip medium under different excitation conditions, ε_{eff} is different in both cases since the electric field distributions are different for even and odd mode cases. The difference in even and odd mode propagation velocities leads to reduced coupler directivity (ability to reject signals entering at an output port reaching an input port). For tight coupling this difference increases. To overcome this an arithmetic mean of the odd and even mode guide wavelength can be used to approximately give the required guide wavelength.

Hand calculation for approximate synthesis is carried out using graphical presentation of test data obtained from the program CMIC or better still from the graphs produced by Weiss [12]. Graphs of even and odd mode impedance as a function of shape ratio for varying s/h can be plotted and, from these, approximate synthesis obtained. Repeating the exercise for air dielectric gives a reference set of curves that, when used in conjunction with the microstrip material curves, allows guide wavelength to be calculated. This follows from the considerations given below:

$$Z_{o,e} = V_{po,e} L$$

$$Z_{air_{o,e}} = cL$$

hence

$$V_{po,e} = c \frac{Z_{o,e}}{Z_{air_{o,e}}}$$

therefore

$$\lambda_{go,e} = \frac{c}{f} \frac{Z_{o,e}}{Z_{air_{o,e}}} \qquad\qquad (2.31)$$

where f is the operating frequency.

A very approximate value for guide wavelength would use the value corresponding to a single microstrip line of width W. This can, however, result in up to 10 percent error in length.

With side coupled microstrip lines the tightest coupling that can be obtained is about -10 dB due to difficulties in fabricating narrow line spacings and the open geometry of the microstrip. Tightest coupling will occur for line lengths of one-quarter guide wavelength.

*   *   *

Example 2.12

Given the graph shown in figure 2.12 for a coupled microstrip line constructed on a material with known relative dielectric constant, calculate the line spacing and line width required for -10 dB coupling between lines. Assume dielectric thickness to be 0.25 mm and 1 GHz operating frequency.

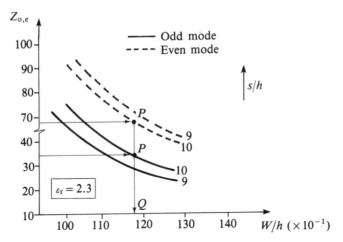

Figure 2.12   Microstrip Coupler Design Curves

Solution

From equation (2.28) calculate odd and even mode characteristic impedance.

$$-10 = 20 \log_{10} C_0 \rightarrow C_0 = 0.32$$

Normalized

$$Z_{oe} = \left(\frac{1 + C_0}{1 - C_0}\right)^{\frac{1}{2}} = \left(\frac{1 + 0.32}{1 - 0.32}\right)^{\frac{1}{2}} = 1.4$$

$$\therefore \quad Z_{oe} = Z_o \times 1.4 = 50 \times 1.4 = 70 \ \Omega$$

Normalized

$$Z_{oo} = \left(\frac{1 - C_0}{1 + C_0}\right)^{\frac{1}{2}} = \left(\frac{0.68}{1.32}\right)^{\frac{1}{2}} = 0.72$$

$$\therefore \quad Z_{oo} = 50 \times 0.72 = 36 \ \Omega$$

Remember to check whether this is applicable to a loosely coupled line

$$Z_o = (Z_{oo} Z_{oe})^{\frac{1}{2}} = (70 \times 36)^{\frac{1}{2}} = 50.2 \approx 50 \ \Omega$$

Next calculate from figure 2.12 where the required shape ratios W/h and s/h lie. To do this mark Z_{oe}, Z_{oo} on the vertical axis. Then read these values across until two points are found that lie vertically above each other and at identical line spacings, points P in figure 2.12. A line dropped through these points gives the required line spacing to height ratio Q.

$$P = 10.0 = s/h \rightarrow s = 10 \times 0.25 = 2.5 \text{ mm}$$

$$Q = 11.7 = W/h \rightarrow W = 11.7 \times 0.25 = 2.9 \text{ mm}$$

from equation 2.25

$$\varepsilon_{\text{eff}} \approx \frac{2.3 + 1}{2} + \frac{2.3 - 1}{2}\left(1 + \frac{3}{2.9}\right)^{-\frac{1}{2}}$$

$$\approx 2.11$$

$$\therefore \quad \lambda_g \approx \frac{c}{f}(2.11)^{-\frac{1}{2}} = \frac{3 \times 10^{10}}{1 \times 10^9} \times 0.69 = 21 \text{ cm}$$

* * *

After testing, some adjustment of this circuit will be necessary. Altering line spacing or line width alters both even and odd mode characteristic impedance. This in turn will vary the coupling coefficient. To keep coupling constant, both line spacing and line width have to be varied simultaneously. As a guide, if line spacing is increased then line width should be increased also; in practice, experience of tuning a particular circuit will indicate the amount by which to trim line widths or spacings.

2.8   FURTHER READING

1   Ramo, S., Whinnery, J. R. and Van Duzer, T., *Fields and Waves in Communication Electronics*, Wiley, 1967.
This text provides an excellent introduction to the theory of electromagnetic propagation. Chapter 4 deals with the solution of Maxwell's equations.

2   Lamont, H. R. L., *Wave Guides*, Methuens Monographs on Physical Subjects, 1953.
This book may be difficult to obtain but it is certainly worthy of note. It explains waveguide theory in a clear and concise fashion, containing a chapter on transmission line theory for general cross-section waveguide.

3   Glazier, E. V. D and Lamont, H. R. L., *The Services Textbook of Radio, Volume 5, Transmission and Propagation*, H.M.S.O., 1958.
Gives a number of fold out sheets with color charts depicting the electric and magnetic field distributions within circular and rectangular cross-section waveguides for various operating modes. Contains a large section on antenna design.

4   Cohn, S., 'Problems in Strip Transmission Lines', *Microwave Theory and Techniques*, 3(2), March. 1955, 119–26.
This is the classic Stripline paper in which Cohn developed his widely used design equations. Data obtained from these equations is presented graphically.

5   Wheeler, H. A., 'Transmission Line Properties of Parallel Strips Separated by Dielectric Sheets', *Microwave Theory and Techniques*, **13**(2), March 1965, 172–85. (Copyright © 1965 IEEE.)
Gives an insight into the use of conformal transformations for the reduction of transmission line problems.

6   Bahl, I. J. and Garg, R., 'Simple and Accurate Formulas for a Microstrip with Finite Thickness', *Proceedings of the I.E.E.E.*, **5**(11), Nov. 1977, 1611–12.
Contains a number of useful references together with closed form expressions for the design of single microstrip lines.

7   Hammerstadt, E. O., 'Equations for Microstrip Circuit Design', *Proceedings European Microwave Conference*, 1975, 262–72.
The equations presented in this paper represent some of the most widely used in industry. They are accurate enough for most engineering applications.

8   Getsinger, W. J., 'Microstrip Dispersion Model', *I.E.E.E. Transactions on Microwave Theory and Techniques*, **21**(1), Jan. 1973, 34–9.
In this paper Getsinger develops an alternative model for microstrip. This model eases analysis difficulties and allows a simple expression for microstrip dispersion to be developed.

9   Edwards, T. C., *Foundations for Microstrip Circuit Design*, Wiley, 1981.
This book has an appendix that deals with the analysis of parallel-coupled lines. The book itself contains a wealth of microstrip design information and is highly recommended to any serious microstrip designer.

10   Cohn, S. B., 'Shielded Coupled-Strip Transmission Line', *I.R.E. Transactions Microwave Theory and Techniques*, **3**, Oct. 1955, 29–38.
Cohn develops a number of relationships for coupled striplines. This paper is an extension of reference [5].

11   Howe, H., *Stripline Circuit Design*, Artech House, 1974.
Howe's book is an essential prerequisite for those involved in stripline design. Contains analysis and synthesis formulae and component design information.

12   Weiss, J. A., 'Microwave Propagation for Coupled Pairs of Microstrip Transmission Lines', *Advances in Microwaves*, **8**, 1974, 295–320.
This is the classic microstrip analysis paper, contains 29 excellent references and gives a computer listing in FORTRAN language of MSTRIP, a microstrip analysis program based on Weiss's analysis of the material by the use of Green's functions.

13   Schwarzmann, A., 'Approximate Solutions for a Coupled Pair of Microstrip Lines in Microwave Integrated Circuits', *Microwave Journal*, **12**(5), May 1969, 79–82.
Schwarzmann in this paper presents a review of his own work, gives his design formulae and presents some experimental results.

14   Bahl, I. J. and Trivedi, D. K., 'A Designers Guide to Microstrip Line', *Microwaves*; **16** (5), May 1977, 174–82.
A review paper covering the field of single microstrip line design. Copious formulae including those for loss and dispersion are quoted. The paper includes 48 references.

3

Simple Numerical Solutions of Laplace's Equation

3.1   GENERAL REMARKS

This chapter deals with the static solution of Laplace's equation using a discrete approximation. The finite difference approach to Laplace's equation is useful in situations where numerical analytical expertise is limited. This technique is readily understood and is useful when non-standard line geometries have to be assessed. Computerized approaches are developed that are useful for the generation of data for preliminary designs for special line configurations or for investigating effects like undercutting in the production of planar transmission lines. Sometimes detailed investigation of charge distribution about a particular part of a line is required or perhaps only the potential difference between two points is needed. These criteria limit the usefulness of the finite difference approach by imposing sometimes intolerable constraints on execution time and computer storage requirement. Detailed investigations normally require high resolution grids with associated increased computer memory requirements and number of calculations required. An alternative method is suggested in this chapter, this uses a random walk technique to solve Laplace's equation for individual points within a given transmission line geometry. Also discussed is a simple application of a powerful numerical technique known as the method of moments. This is particularly valuable for the investigation of line segments in three dimensions.

3.1.1 Introduction to Laplace's Equation

For TEM or quasi-TEM transmission lines it has been suggested (section 1.1) that the capacitance per unit length for this type of transmission line will be independent of operating frequency. This means that for any transmission line assumed to be operating in a mode that approximates to TEM it is possible to define characteristic impedance for that line in terms of capacitance values established under static conditions only.

94

Select an arbitrary cross-section for the transmission line which is assumed to consist of a center conductor held at 1 volt and surrounded by a conducting outer conductor that is connected to ground. With any dielectric structural support removed between the center conductor and the outer conductor the characteristic impedance of the air-spaced line will be $Z_o{}^{air}$ such that

$$Z_o{}^{air} = \left(\frac{L}{C_{air}}\right)^{1/2} = cL = \frac{1}{cC_{air}}$$

where C_{air} is the capacitance per unit length of air-spaced line, L is the inductance per unit length of air-spaced line which is equal to the inductance per unit length of dielectric filled line and c is the velocity of light in air $\approx 3 \times 10^{10}$ cm/s. Hence

$$L = \frac{Z_o{}^{air}}{c} = \frac{1}{c^2 C_{air}}$$

For the same line but with spacing dielectric of relative permittivity ε_r added

$$Z_o = \left(\frac{1}{c^2 C_{air}} \frac{1}{C}\right) = \frac{1}{c}(C_{air}C)^{-1/2} \qquad (3.1)$$

where Z_o is the characteristic impedance of dielectrically loaded line and C is the capacitance per unit length of dielectrically loaded line.

Equation (3.1) shows that characteristic impedance for any arbitrary shaped transmission line with dielectric filling can be found from an evaluation of line capacitances, with and without dielectric filling.

The problem of finding characteristic impedance now becomes an exercise in finding line capacitance. To do this Laplace's equation is used. This equation is partially derived below, the potential difference between two points $\Delta\phi$ a distance **d** apart is

$$\Delta\phi = -\mathbf{E} \cdot \mathbf{d} \qquad (3.2)$$

where **d** and **E** are the vector quantities representing distance and electric field in three-dimensional space. These can be rewritten as

$$\mathbf{d} = \Delta x \mathbf{i} + \Delta y \mathbf{j} + \Delta z \mathbf{k}$$

$$\mathbf{E} = E_x \mathbf{i} + E_y \mathbf{j} + E_z \mathbf{k}$$

where $\mathbf{i}, \mathbf{j}, \mathbf{k}$ are unit vectors.

Expanding the scalar dot product in equation (3.2) using the above expressions gives

$$\Delta\phi = -(E_x \Delta x + E_y \Delta y + E_z \Delta z) \qquad (3.3)$$

Since $\Delta\phi$ is a function of $\Delta x, \Delta y$ and Δz then

$$\Delta\phi = \frac{\partial\phi}{\partial x} \Delta x + \frac{\partial\phi}{\partial y} \Delta y + \frac{\partial\phi}{\partial z} \Delta z \qquad (3.4)$$

Comparing equations (3.3) and (3.4)

$$E_x = -\frac{\partial \phi}{\partial x}, \text{etc.}$$

hence

$$\mathbf{E} = -\left(\frac{\partial \phi}{\partial x}\mathbf{i} + \frac{\partial \phi}{\partial y}\mathbf{j} + \frac{\partial \phi}{\partial z}\mathbf{k}\right) \tag{3.5}$$

If permittivity is constant throughout the line then $\mathbf{D} = \varepsilon\mathbf{E}$. If the amount of flux in a given volume emitting from a point charge ρ can be written as

$$\frac{\partial D_x}{\partial x} + \frac{\partial D_y}{\partial y} + \frac{\partial D_z}{\partial z} = \rho$$

then equation (3.5) becomes

$$\mathbf{D} = -\varepsilon\left(\frac{\partial \phi}{\partial x}\mathbf{i} + \frac{\partial \phi}{\partial y}\mathbf{j} + \frac{\partial \phi}{\partial z}\mathbf{k}\right) \tag{3.6}$$

Let

$$\frac{\partial}{\partial x} + \frac{\partial}{\partial y} + \frac{\partial}{\partial z}$$

be written as ∇ (pronounced grad or del), then equations (3.5) and (3.6) can be written as

$$\mathbf{E} = -\nabla\phi \quad \text{and} \quad \mathbf{D} = -\varepsilon\nabla\phi$$

Also, since $\nabla \cdot \mathbf{D} = \rho$ from Gauss' law, then $\nabla \cdot (-\varepsilon\nabla\phi) = \rho$. ε is constant so that the final expression

$$\nabla^2\phi = -\frac{\rho}{\varepsilon}$$

results in rectangular coordinates

$$\frac{\partial^2 \phi}{\partial x^2} + \frac{\partial^2 \phi}{\partial y^2} + \frac{\partial^2 \phi}{\partial z^2} = -\frac{\rho}{\varepsilon} \tag{3.7}$$

Equation (3.7), which relates the potential change at any given location to the charge density at that location, is known as Poisson's equation.

In the special case of a charge-free region, such as the transmission line structures of interest in this section, equation (3.7) reduces to

$$\frac{\partial^2 \phi}{\partial x^2} + \frac{\partial^2 \phi}{\partial y^2} + \frac{\partial^2 \phi}{\partial z^2} = 0 \tag{3.8}$$

or

$$\nabla^2\phi = 0$$

This is known as Laplace's equation and is the tool by which the capacitance parameters necessary for equation (3.1) can be be established.

3.2   FINITE DIFFERENCE FORM OF LAPLACE'S EQUATION

The finite difference form of a differential equation is an approximation that allows solution of the equation under investigation by digital computer via an iterative procedure. When the finite difference method is applied to the solution of Laplace's equation for transmission line problems the problem becomes that of solving the governing partial differential equation for the voltage distribution over a two- or three-dimensional space governed by line geometry subject to known boundary conditions. Once the voltage distribution over the entire space is known, the electric field can be found and, from this, line capacitance estimated. Once line capacitance has been found, it is a simple matter to establish effective dielectric constant and characteristic line impedance.

3.2.1 Single Dielectric, Uniform Grid

The first task to be completed when framing a problem in terms of finite difference equations is to find the governing finite difference equations representing the partial differential equation of interest over the region to be examined. For the specific case of Laplace's equation applied to transmission line problems this involves finding a spacial voltage distribution over a finite area. In order to apply the finite difference solution to this problem, it is necessary for the region under investigation to be dissected into a fine mesh. The intersection between lines forming the mesh are called node points. It is the values of potential at these node points that have to be calculated.

The simplest grid structure that can be examined is the uniform two-dimensional grid consisting of a number of square cells as shown in figure 3.1 for two dimensions. Cartesian coordinates are the most appropriate for describing this region. Examination of any four adjacent cells (figure 3.1) shows that in two dimensions a five-pointed star is formed by the intersection of the mesh lines forming the cells; similarly in three dimensions a seven-pointed star would be formed.

Using the notation given in figure 3.1 and rewriting Laplace's equation for convenience in two dimensions in cartesian coordinates, a modified form of equation (3.8) results

$$\frac{\partial^2 \phi}{\partial x^2} + \frac{\partial^2 \phi}{\partial y^2} = 0$$

In order to solve this equation, it has first to be reduced to a suitable numerical form, i.e. a finite difference form. For the uniform grid specified in figure 3.1 let

$$
\begin{array}{ll}
x_{i+1} - x_i & \quad y_{i+1} - y_i \\
= x_i - x_{i-1} = \Delta_x & \quad = y_j - y_{j-1} = \Delta_y
\end{array}
$$

$$\text{and } \Delta x = \Delta y = \Delta$$

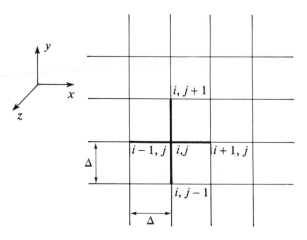

Figure 3.1   Uniform Two-Dimensional Grid

at a node point i, j the voltage can be written as $\phi_{i,j}$ along a straight line connecting i, j and $i+1$, j

$$\frac{\partial \phi}{\partial x} \approx \frac{\phi_{i+1,j} - \phi_{i,j}}{\Delta}$$

along the line connecting $i-1$, j and i, j the equation

$$\frac{\partial \phi}{\partial x} \approx \frac{\phi_{i,j} - \phi_{i-1,j}}{\Delta}$$

holds true.

The last two expressions derived above form the second derivative $\partial^2 \phi / \partial x^2$, this being approximated as

$$\frac{\partial^2 \phi}{\partial x^2} = \frac{(\phi_{i+1,j} - \phi_{i,j})/\Delta - (\phi_{i,j} - \phi_{i-1,j})/\Delta}{\Delta}$$

$$= \frac{\phi_{i+1,j} + \phi_{i-1,j} - 2\phi_{i,j}}{\Delta^2}$$

Similarly, in the y direction

$$\frac{\partial^2 \phi}{\partial y^2} = \frac{\phi_{i,j+1} + \phi_{i,j-1} - 2\phi_{i,j}}{\Delta^2}$$

From these two expressions the discrete form of Laplace's equation in two dimensions after direct substitution of the above expressions for the second derivatives can be written as

$$\phi_{i+1,j} + \phi_{i-1,j} + \phi_{i,j+1} + \phi_{i,j-1} - 4\phi_{i,j} = 0 \tag{3.9}$$

Equation (3.9) represents the finite difference form of Laplace's equation for a uniform dielectric. From equation (3.9) it can be seen that the voltage at any particular node on the grid denoted by i, j can be written as

$$\phi_{i,j} = \frac{\phi_{i+1,j} + \phi_{i-1,j} + \phi_{i,j+1} + \phi_{i,j-1}}{4} \qquad (3.10)$$

Thus, provided the voltages on the right-hand side of equation (3.10) are known or they can be found, the voltage at node i, j is forthcoming. One method of solution is to solve equation (3.10) by an iterative procedure. The iteration method scans each node on the grid in turn and replaces the voltage at each node with the average of the voltages on the four points closest to it. This process is repeated a sufficient number of times until the difference obtained in node voltages between successive iterations is reduced to some preset tolerance limit. It should be noted that nodes that lie at metal surfaces will be at fixed voltages, and, therefore, give the required boundary conditions. For three-dimensional problems equation (3.10) can be extended and becomes

$$\phi_{i,j,k} = \frac{\phi_{i-1,j,k} + \phi_{i+1,j,k} + \phi_{i,j-1,k} + \phi_{i,j+1,k} + \phi_{i,j,k-1} + \phi_{i,j,k+1}}{6}$$

The derivation used above to find equation (3.10) is rather simplistic. A more rigorous solution procedure capable of handling non-uniform grid spacings and curved boundaries, to be discussed in section 3.2.2, will now be described. Consider once again Laplace's equation but this time in three-dimensional cartesian form:

$$\frac{\partial^2 \phi}{\partial x^2} + \frac{\partial^2 \phi}{\partial y^2} + \frac{\partial^2 \phi}{\partial y^2} = 0$$

A Taylor series expansion provides the tool necessary for generating the required finite difference equations.

The Taylor series expansion of a function $g(x)$ in the region around $x = a$ can be found by setting $x = a + \Delta$ where Δ is small relative to a. In this way it is possible to write an infinite series

$$g(x) = g(a + \Delta)$$

$$= g(a) + \Delta \left. \frac{\partial g(a)}{\partial x} \right|_a + \frac{\Delta^2}{2!} \left. \frac{\partial^2 g(a)}{\partial x^2} \right|_a + \frac{\Delta^3}{3!} \left. \frac{\partial^3 g(a)}{\partial x^3} \right|_a + \cdots$$

For the problem under consideration here, i.e. the solution of Laplace's equation, the potential as a function in the x direction will first be found. Here the Taylor series expansion can be expressed as

$$\phi(a + \Delta) = \phi(a) + \Delta \left. \frac{\partial \phi}{\partial x} \right|_a + \frac{\Delta^2}{2} \frac{\partial^2 \phi}{\partial x^2} + \cdots$$

Let 'a' be the centre of the seven-pointed star representing the mesh line intersecting the uniformly spaced grid lines in three dimensions as shown in figure 3.2. Here i, j, k can be set to zero for convenience of notation. Therefore, each point of the star can be represented by the integers 0–6 inclusive (i, j, $k = 0$, $i + 1$, j, $k = 2$, etc.).

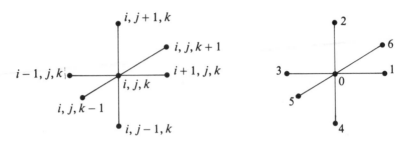

Figure 3.2   Seven-Pointed Star Notation

$\phi(a + \Delta)$ becomes equal to the potential at point 1, ϕ_1 and $\phi(a)$ become equal to the potential at point zero ϕ_0 thus

$$\phi_1 = \phi_0 + \Delta \left.\frac{\partial \phi}{\partial x}\right|_0 + \frac{\Delta^2}{2} \left.\frac{\partial^2 \phi}{\partial x^2}\right|_0 + \ldots$$

similarly for $\phi(a - \Delta)$

$$\phi_3 = \phi_0 - \Delta \left.\frac{\partial \phi}{\partial x}\right|_0 + \frac{\Delta^2}{2} \left.\frac{\partial^2 \phi}{\partial x^2}\right|_0 - \ldots$$

adding these two expressions gives

$$\phi_1 + \phi_3 = 2\phi_0 + \Delta^2 \left.\frac{\partial^2 \phi}{\partial x^2}\right|_0 + \text{higher order terms in even powers of } \Delta$$

If Δ has been chosen to be sufficiently small then Δ^4 and higher terms will rapidly become small so that they can be neglected. Neglecting these terms will, of course, lead to error in the value calculated for potential distribution throughout the field. Thus it should always be remembered that the solution obtained by the finite difference approach is an approximate one; the degree of approximation depending on the coarseness of the grid selected to generate the nodes used to represent the potentials that exist within the geometry of the structure under investigation.

Returning now to the problem in hand, it is easy to see from the last expression derived above that

$$\left.\frac{\partial^2 \phi}{\partial x^2}\right|_0 \approx \frac{\phi_1 + \phi_3 - 2\phi_0}{\Delta^2}$$

Similarly it is easy to show that

$$\left.\frac{\partial^2 \phi}{\partial y^2}\right|_0 \approx \frac{\phi_2 + \phi_4 - 2\phi_0}{\Delta^2}$$

and

$$\left.\frac{\partial^2 \phi}{\partial z^2}\right|_0 \approx \frac{\phi_5 + \phi_6 - 2\phi_0}{\Delta^2}$$

The summation of these three equations gives the finite difference approximation for Laplace's equation in three dimensions specified in cartesian coordinates

$$\nabla^2 \phi = \frac{1}{\Delta^2}(\phi_1 + \phi_2 + \phi_3 + \phi_4 + \phi_5 - 6\phi_0) = 0$$

for two-dimensional fields

$$\phi_5 = \phi_6 = \left.\frac{\partial \phi}{\partial z}\right|_0 = \left.\frac{\partial^2 \phi}{\partial z^2}\right|_0 = 0$$

hence

$$\phi_0 = \frac{\phi_1 + \phi_2 + \phi_3 + \phi_4}{4}$$

which is the same as equation (3.10) derived by the less rigorous method first used.

As an example, consider the use of the finite difference equation (3.10) for the solution of the potential distribution in the simple problem stated graphically in figure 3.3. Here a number of straight perfectly conducting metal plates separated from each other by very thin ideal insulators are held at different potentials. The potentials on these metal plates represent fixed boundary conditions and do not change with respect to time. The space enclosed by the conductors is divided into twelve square subsections. This forms six internal nodes denoted (i)–(vi) inclusive, these nodes are not physically connected to the metal plates or insulators, and fourteen external nodes denoted a–n inclusive, representing fixed boundary conditions.

The solution proceeds as follows: First set all the internal nodes to zero potential. Then apply equation (3.10), which is directly applicable to the geometry of this problem to each internal node. The following calculations result for the first few steps:

node (i)   first iteration $\frac{1}{4}(0 + 0 + 0 + 15) = 3.75$
node (ii)   first iteration $\frac{1}{4}(0 + 5 + 0 + 3.75) = 2.19$
node (iii)   first iteration $\frac{1}{4}(3.75 + 0 + 0 + 15) = 4.69$

Notice how each time a node value is computed the most recent values of node voltages at the four closest nodes excluding boundary potentials are used. The above procedure for evaluating nodes is repeated over and over again until the potentials at each node start to converge to some predetermined level, say, for example, when the difference between node potentials calculated at successive iterations is equal to a few percent or when the RMS value of the difference of all the node potentials evaluated at successive passes is reduced to some tolerable level predetermined by the designer. The process may be terminated when the total capacitance of the structure under investigation has converged to some acceptable level. In figure 3.3 potentials calculated at successive iterations for each internal node are shown along the diagonal placed in each grid square. Examination of the difference between successive values for node potentials shows convergence. It is worthy of note that

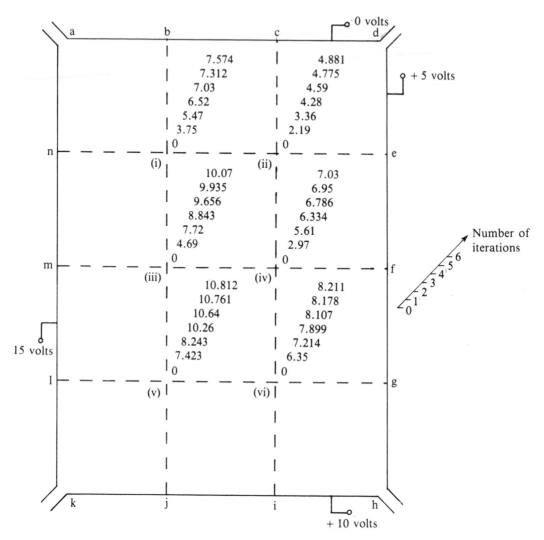

Figure 3.3   Example of Finite Difference Problem Solved by Iteration

since the calculations for node potentials were obtained using a fixed number of decimal places, then rounding off errors occur. These errors add to those already introduced by expressing partial differential equations in finite difference form.

3.2.2 Single Dielectric, Non-Uniform Grids

The Taylor series expansion method of the last section is particularly amenable for finding finite difference formulae when non-uniform (asymmetric) meshes are used.

The asymmetric or irregular star, as it is often known, is useful when the field in a particular subregion of a larger region is to be studied in greater detail than the rest of the region. Rather than use a fine grid over the entire problem area it is computationally more efficient to reduce the mesh size only in the region in which detailed information is required, while at the same time having a coarser grid elsewhere. This normally requires a transition of the grid size in the coarse to the fine regions. The transition region will consist of a number of irregular stars. The situation is illustrated in figure 3.4.

Applying the Taylor series to the irregular five-point star denoted in figure 3.4 and expanding for ϕ_1 in the x direction gives

$$\phi_1 = \phi_0 - \Delta_1 \left.\frac{\partial \phi}{\partial x}\right|_0 + \frac{\Delta_1{}^2}{2} \frac{\partial^2 \phi}{\partial x^2} + \cdots \qquad (3.11)$$

For ϕ_3 the expansion becomes

$$\phi_3 = \phi_0 + \Delta_2 \left.\frac{d\phi}{\partial x}\right|_0 + \frac{\Delta_2{}^2}{2} \frac{\partial^2 \phi}{\partial x^2} + \cdots \qquad (3.12)$$

From these expressions

$$\frac{\partial \phi}{\partial x}$$

can be found to be

$$\left.\frac{\partial \phi}{\partial x}\right|_0 = \frac{\phi_0 - \phi_1}{\Delta_1} + \text{higher order terms}$$

and

$$\left.\frac{\partial \phi}{\partial x}\right|_0 = \frac{\phi_3 - \phi_0}{\Delta_2} + \text{higher order terms}$$

Unfortunately this time the higher order terms cannot be ignored since, if they were, a fairly large error would result. This means that, in order to find the appropriate

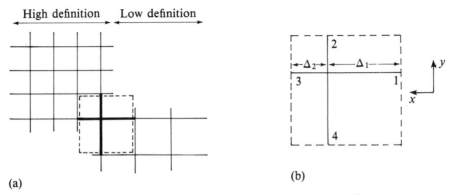

High definition   Low definition

(a)   (b)

Figure 3.4   Irregular Grid Structure (a) Mixed Grid (b) General Irregular Star

finite difference equation, a more convoluted approach than that used before must be applied. Here equations (3.11) and (3.12) are multiplied by as yet unknown coefficients A and B. Adding the resulting expressions results in equation (3.13). Here

$$A(\phi_1 - \phi_0) + B(\phi_3 - \phi_0) = \frac{\partial\phi}{\partial x}\bigg|_0 (B\Delta_2 - A\Delta_1) + \frac{1}{2}\frac{\partial^2\phi}{\partial x^2}\bigg|_0 (B\Delta_2^2 + A\Delta_1^2) + \cdots$$

(3.13)

In order to find

$$\frac{\partial\phi}{\partial x}\bigg|_0$$

it is necessary to choose coefficients A and B such that

$$A = -\frac{B\Delta_2^2}{\Delta_1^2}$$

substituting this into the above expression gives

$$\frac{\partial\phi}{\partial x}\bigg|_0 = \frac{\Delta_1^2(\phi_3 - \phi_0) - \Delta_2^2(\phi_1 - \phi_0)}{\Delta_1\Delta_2(\Delta_1 + \Delta_2)} + \text{higher order terms}$$

Neglecting the higher order terms gives the desired expression for

$$\frac{\partial\phi}{\partial x}\bigg|_0$$

Applying an identical technique in order to find

$$\frac{\partial^2\phi}{\partial x^2}\bigg|_0$$

with the substitution for coefficient A of

$$A = B\frac{\Delta_2}{\Delta_1}$$

gives

$$\frac{\partial^2\phi}{\partial x^2}\bigg|_0 = 2\left[\frac{\Delta_2(\phi_1 - \phi_0) + \Delta_1(\phi_3 - \phi_0)}{\Delta_1\Delta_2(\Delta_1 + \Delta_2)}\right] + \text{other terms}$$

(3.14)

Again, it is usual practice to drop these other terms. At this point it is of interest to note what happens when $\Delta_2 = \Delta_1 = \Delta$. That is to say, when the irregular star is transformed to a regular one. In this case equation (3.14) reduces to

$$\frac{\partial^2\phi}{\partial x^2}\bigg|_0 = \frac{\phi_1 + \phi_3 - 2\phi_0}{\Delta^2}$$

which is an identical expression to that derived for the regular five-pointed star in the last section.

Expressions similar to equation (3.14) exist for the y direction so that the finite difference approximation to Laplace's equation for irregular grids can be found. It should be noted that, in the derivation of equation (3.14), the third derivate of ϕ_0 was neglected. This leads to a reduction of accuracy of these formulae when compared with those obtained for uniform grids.

Remember that the uniform grid equations can be used for both fine and coarse grid sizes in the high and low definition regions of the field and that the irregular star formulae is applied at the interface between these regions.

3.2.3   Single Dielectric, Curved Boundaries

A very useful feature to build into any finite difference solution program is the ability to handle curved boundaries over which potential values are to be specified. This means that a specially prepared grid with very fine line spacings need not be selected for use over an entire region in order that it will adequately define the specified boundary (figure 3.5(a)).

Consider the curved boundary shown in figure 3.5(b). The Taylor series expansion gives

for node 1

$$\phi_1 = \phi_0 + a\,\Delta\,\left.\frac{\partial\phi}{\partial x}\right|_0 + \frac{a^2\,\Delta^2}{2}\left.\frac{\partial^2\phi}{\partial x^2}\right|_0 + \frac{a^3\,\Delta^3}{6}\frac{\partial^3\phi}{\partial x^3} + \cdots$$

for node 3

$$\phi_3 = \phi_0 - \Delta\,\left.\frac{\partial\phi}{\partial x}\right|_0 + \frac{\Delta^2}{2}\left.\frac{\partial^2\phi}{\partial x^2}\right|_0 + \frac{\Delta^3}{6}\frac{\partial^3\phi}{\partial x^3} + \cdots$$

From these

$$\phi_1 + a\phi_3 = (1+a)\phi_0 + \tfrac{1}{2}a\,\Delta^2(1+a)\left.\frac{\partial^2\phi}{\partial x^2}\right|_0 + \text{higher order terms}$$

Once again neglecting higher order terms, and thereby introducing residual error into the computation, gives

$$\left.\frac{\partial^2\phi}{\partial x^2}\right|_0 = \frac{1}{\Delta^2}\left[\frac{2\phi_1}{a(1+a)} + \frac{2\phi_3}{(1+a)} - \frac{2\phi_0}{a}\right]$$

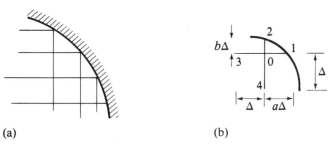

(a)   (b)

Figure 3.5   Curved Boundary Grid (a) Curved Boundary Fitted with High Resolution Uniform Grid (b) Curved Boundary Grid

Similarly, in the y direction

$$\frac{\partial^2 \phi}{\partial y^2}\bigg|_0 = \frac{1}{\Delta^2}\left[\frac{2\phi_2}{b(1+b)} + \frac{2\phi_4}{(1+b)} - \frac{2\phi_0}{b}\right]$$

from which the relevant finite difference equations immediately follow. Once again it should be noted that if $a = b = 1$ then the expressions given above for the curved boundary specified in cartesian coordinates reduce to the simple symmetric five-point star shown before.

All of the expressions discussed so far in this chapter have been derived exclusively for problems whose geometry is best cited in the form of rectangular, i.e. cartesian, coordinates. This specification is particularly useful when stripline or microstrip structures are being investigated, since both have rectangular geometry. However, when coaxial structures are to be investigated a rectangular coordinate system is not really the most appropriate coordinate system to use. In fact, because of the nature of coaxial systems cylindrical coordinates would be the most useful for these types of problems since they reflect the geometry of the cable.

3.2.4 Single Dielectric, Cylindrical Coordinates

In order to develop finite difference equations for Laplace's equation in terms of cylindrical coordinates it will first be necessary to define the cylindrical coordinate system and then to state Laplace's equation in terms of these coordinates. Figure 3.6 shows a typical cylinder section specified in cylindrical coordinates. In this system any point in space is defined as lying at a position r, $\angle\theta$, z. This means that

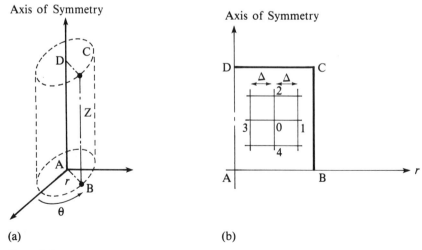

(a)   (b)

Figure 3.6   Finite Difference Grid for Cylindrical Coordinates (a) Cylindrical Coordinate System (b) Cell Assignment

Laplace's equation in three dimensions can be written in cylindrical coordinates as

$$\nabla^2\phi = \frac{\partial^2\phi}{\partial r^2} + \frac{1}{r}\frac{\partial\phi}{\partial r} + \frac{\partial^2\phi}{\partial z^2} = 0$$

Here axial symmetry has been exploited so that there is no variation in potential in the θ direction (see figure 3.6(b)). Comparison with the equations previously derived in section 3.2.1 for the regular five-pointed star yields

$$\frac{\partial^2\phi}{\partial r^2} = \frac{\phi_1 + \phi_3 - 2\phi_0}{\Delta}$$

$$\frac{1}{r}\frac{\partial\phi}{\partial r} = \frac{1}{r}\left(\frac{\phi_1 - \phi_3}{2\Delta}\right)$$

and

$$\frac{\partial^2\phi}{\partial z^2} = \frac{\phi_4 + \phi_2 - 2\phi_0}{\Delta^2}$$

Examination of the expression for the second term on the right-hand side of Laplace's equation expressed in cylindrical coordinates indicates a possible problem area. For node points close to the axis of symmetry, i.e. as $r \to 0$, $(1/r)(\partial\phi/\partial r)$ will become extremely large. This is highly undesirable in a computational sense. One way around the problem is to invoke a rather elegant piece of mathematics known as l'Hôpitals rule. This rule states that

$$\underset{x \to k}{\text{limit}}\left|\frac{G(x)}{F(x)}\right| = \underset{x \to k}{\text{limit}}\left|\frac{G(x)}{F'(x)}\right|$$

which is very useful since in this case if $G(x)$ is made equal to $\partial\phi/\partial r$, $F(x)$ is set equal to r, and if x, equivalent now to r, is allowed to tend to zero, then

$$\underset{r \to 0}{\text{limit}}\left.\frac{\partial\phi}{\partial r}\right/r = \frac{\partial^2\phi}{\partial r^2}$$

Therefore, along the axis of symmetry an alternative form of the finite difference equation in cylindrical coordinates can now be found.

$$\left.\nabla^2\phi\right|_{r=0} = 2\frac{\partial^2\phi}{\partial r^2} + \frac{\partial^2\phi}{\partial z^2}$$

Another simplification is worthy of note for points along the axis of symmetry. In figure 3.6 if the basic cell shown is moved so that mesh line 2–4 lies along the axis of symmetry then $\phi_1 = \phi_3$. Gathering together the pieces of information gleaned about cells close to the axis of symmetry leads to a new expression

$$\left.\nabla^2\phi\right|_{r=0} = 4\phi_1 + \phi_4 + \phi_2 + 6\phi_0$$

and

$$\left.\nabla^2\phi\right|_{r\neq0} = \phi_1\left(1 + \frac{\Delta}{2r}\right) + \phi_3\left(1 - \frac{1}{2r}\right) + \phi_4 + \phi_2 - 4\phi_0$$

These last two equations give the relevant finite difference approximations to Laplace's equation written in cylindrical coordinates. Note that as r becomes very large the cylindrical coordinate system reduces to the simple five-point symmetrical star described earlier in cartesian coordinates.

So far, all the work done has been involved with setting up equations for systems that contain only one dielectric. What happens when dielectrics of more than one type, mixed dielectrics, are present in the structure? This will be discussed in the next section.

3.2.5 Mixed Dielectric

When a mixed dielectric, that is more than one dielectric, is present in the transmission line then equation (3.10) governing the potential at any point in the field must be modified to cope with this new situation. For ease of explanation, a two-dimensional problem having a uniform grid size is described in figure 3.7.

Here a mixed dielectric consisting of two differing dielectric materials has been selected. The interface between these materials is chosen to be a straight line as would be the case for a microstrip transmission line. Also, it has been arranged that the dielectric line boundary lies along a mesh line. The grid size is assumed to be Δ and the star points have been numbered relative to the center of the star, assumed to lie at point i,j. Consider Laplace's equation written for a homogeneous dielectric

$$\nabla^2 \phi = 0$$

For the general case of an inhomogeneous dielectric it is necessary to use a more general form of Laplace's equation.

Returning to section 3.1.1, Gauss's law was stated to be given in point form as

$$\nabla \cdot \mathbf{D} = \rho \qquad \text{where } \mathbf{D} = \varepsilon \mathbf{E}$$

However, since for a charge free region $\rho = 0$ then

$$\nabla \cdot \mathbf{D} = 0$$

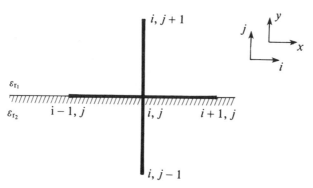

Figure 3.7   Mixed Dielectric Problem

or expressed in rectangular coordinates in two dimensions

$$\frac{\partial D_x}{\partial x} + \frac{\partial D_y}{\partial y} = 0$$

This expression allows a valid finite difference equation that can be applied along a straight dielectric boundary to be found.

Above the dielectric boundary in the ε_{r_1} region the y component of the flux density is given as

$$D_y = \frac{-\varepsilon_0 \varepsilon_{r_1}(\phi_{i,j+1} - \phi_{i,j})}{\Delta}$$

below the dielectric interface in the ε_{r_2} region

$$D_y = \frac{-\varepsilon_0 \varepsilon_{r_2}(\phi_{i,j} - \phi_{i,j-1})}{\Delta}$$

This means that the rate of change of electric flux density normal to the dielectric interface can be approximated by

$$\frac{\partial D_y}{\partial y} = \frac{D_y \text{ above} - D_y \text{ below}}{\Delta}$$

$$= -\frac{\varepsilon_0}{\Delta}[\varepsilon_{r_1}(\phi_{i,j+1} - \phi_{i,j}) - \varepsilon_{r_2}(\phi_{i,j} - \phi_{i,j-1})] \qquad (3.15)$$

A similar expression exists along the x direction but what value of ε_r should be used at points $i-1,j$ or $i-1,j$? Since this cannot be determined at this stage, assume a hypothetical value ε_{r_3} so that the rate of change of flux density along the dielectric interface can be written as

$$\frac{\partial D_x}{\partial y} = \frac{\varepsilon_0 \varepsilon_{r_3}}{\Delta}(\phi_{i+1,j} + \phi_{i-1,j} - 2\phi_{i,j}) \qquad (3.16)$$

Adding equations (3.15) and (3.16) and equating the result to zero gives

$$-\phi_{i,j}(\varepsilon_{r_1} + \varepsilon_{r_2} + 2\varepsilon_{r_3}) + \varepsilon_{r_1}\phi_{i,j+1} + \varepsilon_{r_2}\phi_{i,j-1} + \varepsilon_{r_3}\phi_{i+1,j} = 0$$

so that the equation governing the potential at points along the dielectric boundary becomes

$$\phi_{i,j} = \frac{\varepsilon_{r_1}\phi_{i,j+1} + \varepsilon_{r_2}\phi_{i,j-1} + \varepsilon_{r_3}(\phi_{i+1,j} + \phi_{i-1,j})}{\varepsilon_{r_1} + \varepsilon_{r_2} + 2\varepsilon_{r_3}} \qquad (3.17)$$

This equation is used only along the dielectric boundary. Above the boundary or below it equation (3.10) is used with its right-hand side multiplied by the appropriate value of relative dielectric constant ε_{r_1} or ε_{r_2} depending on which region the computation is being done. All that remains now is to select a value for ε_{r_3} so that the previous equation can be usefully applied. One choice would be to allow the node potentials along the x-axis to contribute an equal amount to $\phi_{i,j}$ as those along the y-axis, in this case

$$\varepsilon_{r_1}\phi_{i,j+1} + \varepsilon_{r_2}\phi_{i,j-1} = \varepsilon_{r_3}(\phi_{i-1,j} + \phi_{i+1,j})$$

so that

$$\varepsilon_{r_3} = \frac{\varepsilon_{r_1}\phi_{i,j+1} + \varepsilon_{r_2}\phi_{i,j-1}}{\phi_{i-1,j} + \phi_{i+1,j}} \tag{3.18}$$

Before equation (3.17) could be used, equation (3.18) would have to be evaluated in each iteration loop and the resulting value of ε_{r_3} substituted into equation (3.17) which could then be used in the normal way. One excellent way for evaluating the modified forms of the finite difference representations of Laplaces's equation for regions containing mixed dielectrics was developed by Wensley and Parker [6]. They took a symmetrical five-point star and embedded it in a sandwich of different dielectric materials (figure 3.8). Here, eight wedges each with different dielectric constant are catered for.

By showing that for Laplace's equation written in various forms as $\varepsilon \nabla^2 \phi = \nabla \cdot \varepsilon \nabla \phi = \operatorname{div}(\varepsilon \operatorname{grad} \phi) = 0$, Wensley and Parker were able to show by summing the eight components of $\varepsilon \operatorname{grad} \phi = 0$ that

$$(\varepsilon_{r_1} + \varepsilon_{r_8})(\phi_{i+1,j} - \phi_{i,j}) + (\varepsilon_{r_2} + \varepsilon_{r_3})(\phi_{i,j+1} - \phi_{i,j})$$

$$+ (\varepsilon_{r_4} + \varepsilon_{r_5})(\phi_{i-1,j} - \phi_{i,j}) + (\varepsilon_{r_6} + \varepsilon_{r_7})(\phi_{i,j-1} - \phi_{i,j}) = 0 \tag{3.19}$$

This equation is invaluable since it allows mixed dielectric problems containing up to eight different dielectrics to be handled with ease.

If, as in the case of a microstrip transmission line, there exists a straight interface between two dissimilar dielectric materials then equation (3.19) can be used to find a finite difference equation that is valid for the calculation of potential values on the dielectric boundary.

Let the boundary between dissimilar dielectrics lie along the line $i-1,j$ and $i+1,j$ (see figure 3.7), so that ε_{r_1} to ε_{r_4} in figure 3.8 become equal and are assigned

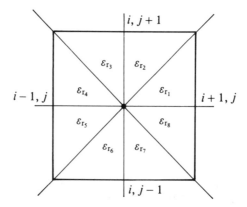

Figure 3.8   Multiple Mixed Dielectrics

the value ε_{r_1}; similarly, let ε_{r_5} to ε_{r_8} be set equal to some value, say ε_{r_2}. Then from equation (3.19) it is evident that

$$(\varepsilon_{r_1} + \varepsilon_{r_2})(\phi_{i+1,j} - \phi_{i,j}) + (2\varepsilon_{r_1})(\phi_{i,j+1} - \phi_{i,j})$$
$$+ (\varepsilon_{r_1} + \varepsilon_{r_2})(\phi_{i-1,j} - \phi_{i,j}) + (2\varepsilon_{r_2})(\phi_{i,j-1} - \phi_{i,j}) = 0$$

hence

$$4\phi_{i,j}(\varepsilon_{r_1} + \varepsilon_{r_2}) = \phi_{i+1,j}(\varepsilon_{r_1} + \varepsilon_{r_2}) + 2\varepsilon_{r_1}(\phi_{i,j+1}) + \phi_{i-1,j}(\varepsilon_{r_1} + \varepsilon_{r_2}) + 2\varepsilon_{r_2}(\phi_{i,j-1})$$

Let $2K = \varepsilon_{r_1} + \varepsilon_{r_2}$ so that

$$\phi_{i,j} = \frac{1}{4K}[K(\phi_{i+1,j} + \phi_{i-1,j}) + \varepsilon_{r_1}\phi_{i,j+1} + \varepsilon_{r_2}\phi_{i,j-1}]$$

As for equation (3.17), this expression is valid for the evaluation of the potential of node points of a uniform five-point star that lie along a straight interface between two dissimilar dielectrics.

If equation (3.18) is re-examined and it is noted that in practice the potentials at adjacent node points are very similar, then equation (3.18) reduces to

$$2\varepsilon_{r_3} = \varepsilon_{r_1} + \varepsilon_{r_2}$$

so that equation (3.17) becomes equivalent to that derived above for the same problem by application of the Wensley–Parker relationship. This then justifies the assumption made earlier when deriving equation (3.17) that the node potentials in the x direction would contribute an equal amount to $\phi_{i,j}$ as those along the y-axis.

3.2.6 Faster Solution Methods for Finite Difference Equations

In section 3.2.1 hand calculation enabled the iterative solution of the finite difference equation governing the potential at the internal nodes of a fixed boundary potential problem in two dimensions. In the example no attempt was made to use the value of potential calculated at the $(k-1)$th iteration in the computation of the potential, the new potential at the kth iteration. The classical method for solving finite difference equations by the iteration method described in section 3.2.1 can be speeded up considerably by using a modified form of the finite difference equation given as equation (3.10). The modified form of equation (3.10) is given below for two dimensions and a uniform dielectric as

$$\phi_{i,j}\Big|_{\text{new}} = R\left(\frac{\phi_{i+1,j} + \phi_{i-1,j} + \phi_{i,j+1} + \phi_{i,j-1}}{4}\right) + (1-R)\phi_{i,j}\Big|_{\text{old}}$$

This equation can, of course, be modified for mixed dielectric, curved boundaries, etc., with ease. In this equation, the coefficient R is called the residue and has the effect of speeding up the rate of convergence of the node potentials. It is of note that for $R = 1$ the modified finite difference equation becomes equal to equation (3.10). This is illustrated by the graph shown in figure 3.9, where it can be clearly

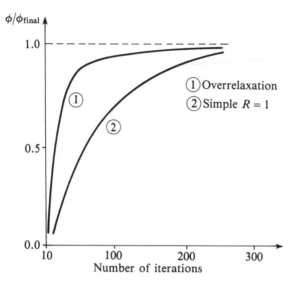

Figure 3.9   Sample Finite Difference Convergence Rates

seen that the feedback introduced by the coefficient R has somewhat enhanced the rate of convergence for the iterative solution of the governing finite difference equation. The residue controls the amount of feedback occurring.

For residue values of less than or equal to one, the iteration is said to be underrelaxed, for R greater than one overrelaxed; because of this, this technique is known as point relaxation or just relaxation and is used in preference to the simple iteration method for the solution of finite difference equations (see figure 3.10).

Although no optimum value for R exists for all classes of finite difference problems, the number 1.5 offers the prospect of good convergence rates leading to efficient computation under most circumstances.

The relaxation method for solving finite difference equations is very popular because it is fairly easy to program on a digital computer. However, another solution method does exist and solving systems of finite difference equations by this alternative technique should always be considered especially if large numbers of node points are under consideration. When large numbers of node points are calculated, finite difference solutions require large numbers of calculations to satisfactorily complete a computation to some required degree of accuracy. An alternative to the iteration or relaxation methods for the solution of finite difference equations involves the use of matrix arithmetic as described below.

Consider once more the example structure shown earlier in figure 3.3. For each node it is possible to write a finite difference equation. There are six in total for this problem

$$4V_{(i)} = V_b + V_{(ii)} + V_{(iii)} + V_n$$
$$4V_{(ii)} = V_c + V_e + V_{(iv)} + V_{(i)}$$

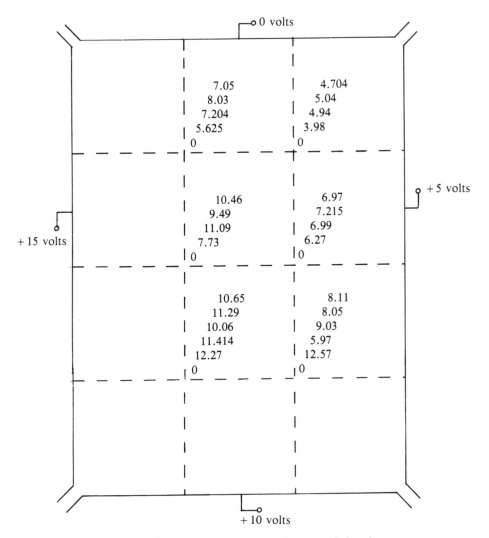

Figure 3.10   Example Finite Difference Problem Solved by Relaxation

$$4V_{(iii)} = V_{(i)} + V_{(iv)} + V_{(v)} + V_m$$

$$4V_{(iv)} = V_{(ii)} + V_f + V_{(vi)} + V_{(iii)}$$

$$4V_{(v)} = V_{(iii)} + V_{(vi)} + V_j + V_l$$

$$4V_{(vi)} = V_{(iv)} + V_g + V_i + V_{(v)}$$

Rearranging these six governing equations so that nodes which represent boundary potentials, i.e. nodes a–n, appear on the right-hand side and maintaining the order in which they appear gives

$$4V_{(i)} - V_{(ii)} - V_{(iii)} + 0V_{(iv)} + 0V_{(v)} + 0V_{(vi)} = V_b + V_n$$

$$- V_{(i)} + 4V_{(ii)} + 0V_{(iii)} - V_{(iv)} + 0V_{(v)} + 0V_{(vi)} = V_c + V_e$$

$$- V_{(i)} + 0V_{(ii)} + 4V_{(iii)} - V_{(iv)} - V_{(v)} + 0V_{(vi)} = V_m$$

$$0V_{(i)} - V_{(ii)} - V_{(iii)} + 4V_{(iv)} + 0V_{(v)} - V_{(vi)} = V_f$$

$$0V_{(i)} + 0V_{(ii)} - V_{(iii)} + 0V_{(iv)} + 4V_{(v)} - V_{(vi)} = V_j + V_l$$

$$0V_{(i)} + 0V_{(ii)} + 0V_{(iii)} - V_{(iv)} - V_{(v)} + 4V_{(vi)} = V_g + V_i$$

This looks rather clumsy but when gathered together forms a matrix equation as shown below.

$$
\begin{bmatrix}
-4 & 1 & 1 & 0 & 0 & 0 \\
1 & -4 & 0 & 1 & 0 & 0 \\
1 & 0 & -4 & 1 & 1 & 0 \\
0 & 1 & 1 & -4 & 0 & 1 \\
0 & 0 & 1 & 0 & -4 & 1 \\
0 & 0 & 0 & 1 & 1 & -4
\end{bmatrix}
\begin{bmatrix}
V_{(i)} \\
V_{(ii)} \\
V_{(iii)} \\
V_{(iv)} \\
V_{(v)} \\
V_{(vi)}
\end{bmatrix}
= -1
\begin{bmatrix}
V_b + V_n \\
V_c + V_e \\
V_m \\
V_f \\
V_j + V_e \\
V_g + V_i
\end{bmatrix}
$$

This matrix equation can be expressed in the form

$$\mathbf{AX} = \mathbf{B}$$

where the column vector $\mathbf{X}$ represents the unknown potential distribution that requires to be solved and $\mathbf{B}$ represents the fixed boundary potentials. Examination of the $\mathbf{A}$ matrix shows it to be a square matrix of order $n \times n$ where n is the number of internal nodes in the structure, six in this case. It is also noted that the matrix is symmetric about the leading diagonal. This is due to the fact that, for the example, selected mixed dielectrics are not present so that the same form of finite difference equation is valid throughout the structure. Also note the large number of zero elements in the matrix. In this case nearly 40 percent of all the matrix elements are zero. This will, in general, be true for any large system of finite difference equations so that when the system of matrix equations are to be solved, special techniques can be used.

So how does the solution proceed? Examination of the matrix equation generated for the problem in hand shows that since all of the coefficients of $\mathbf{A}$ and $\mathbf{B}$ matrices are known then the unknown column vector $\mathbf{X}$ containing the desired potential values associated with the electric field distribution within the structure can be found by inversion of matrix $\mathbf{A}$ followed by premultiplication of matrix $\mathbf{B}$.

In this way

$$\mathbf{AX} = \mathbf{B}$$

becomes

$$\mathbf{A}^{-1}\mathbf{A}\mathbf{X} = \mathbf{A}^{-1}\mathbf{B} \qquad \text{hence } \mathbf{I}\mathbf{X} = \mathbf{A}^{-1}\mathbf{B}$$

since $\mathbf{I}$ is the identity matrix then the resulting equation represents the desired solution

$$\mathbf{X} = \mathbf{A}^{-1}\mathbf{B}$$

Due to the sparse nature of matrix $\mathbf{A}$ and the fact that most of its elements lie in a well-defined band, very elaborate and computationally efficient algorithms will allow the inverse $\mathbf{A}$ matrix $\mathbf{A}^{-1}$ to be swiftly evaluated. Ultimately this means that provided enough computer memory is available for storage of the inverse $\mathbf{A}$ matrix together with any intermediate storage needed to convert the $\mathbf{A}$ matrix into its inverse form, then the calculation of node potentials can be exceedingly fast compared to the speed of the solutions offered by the relaxation or iteration methods. The main sources of error introduced by this technique are round-off error due to the finite word length used for computation together with truncation error associated with the finite difference approximation to the governing partial differential equations. Errors due to selecting the number of iteration cycles that occur for the iteration or relaxation methods are not a problem when this method of solution is employed.

There is, however, an unfortunate aspect about this technique which arises due to the computation efficiency of the matrix inversion algorithms used. These algorithms tend to be rather complex in that they pack the data contained in the $\mathbf{A}$ matrix to preserve computer memory space and enhance execution time with the result that the computer subroutines capable of performing them are written in a sophisticated way in order to speed up their performance. This means that the development time required for an individual programmer to obtain the same efficiency as afforded by the professionally produced algorithms represents a huge investment in time. This limits the matrix inversion solution technique mainly to mainframe computers where these types of programs are available as user packages in subroutine libraries [13].

3.3 EVALUATION OF LINE CAPACITANCE

In the preceding sections it was shown how systems of finite difference equations could be set up for various types of geometry and dielectric materials. It was also shown how the resulting system of equations could be solved to find the potential distribution within the space created by a given system of conductors, each conductor held at known potentials thereby forming the necessary boundary conditions needed to complete the specification of the system of equations governing the structure under investigation. Once the potential distribution pertinent to this system of equations is solved, it is usually necessary in transmission line problems to be able to calculate the electric field intensity. Once this is known it is possible to find the line capacitance and hence characteristic impedance and guide wavelength. In order to show how this can be achieved, consider the following argument.

In general, capacitance can be expressed in terms of the total charge Q that exists in a given structure

$$C = \frac{Q}{\phi_0}$$

Here ϕ_0 is the voltage on the center conductor and is normally set to $+1$ volt. The total energy U stored in the system can be written as

$$U = \tfrac{1}{2} C \phi_0^2$$

Equally, this can be expressed in terms of the total energy contained in the electric field in an enclosed volume V as

$$U = \tfrac{1}{2} \int_V \varepsilon |E|^2 \, dV \qquad \text{for } \phi_0 = 1 \text{ volt}$$

These last two expressions can be equated to yield

$$C = \int_V \varepsilon |E|^2 \, dV$$

For a two-dimensional system this can be reduced to

$$C = \int_A \varepsilon |E|^2 \, dA$$

From this expression it is clear that provided the electric field distribution across the structure is known, then the desired capacitance values needed to estimate characteristic impedance and guide wavelength, for a particular line from equations (3.1), (2.11) and (2.24), can be found.

What now has to be done is to evaluate approximate electric field distribution from the known potential distribution. Approximate values for E_x can be found by averaging the rate of change of potential taken at node points j and $j+1$

$$E_x \approx -\tfrac{1}{2} \left(\frac{\phi_{i+1,j} - \phi_{i,j}}{\Delta} + \frac{\phi_{i+1,j+1} - \phi_{i,j+1}}{\Delta} \right)$$

$$= -\tfrac{1}{2}\Delta (\phi_{i,j+1} - \phi_{i,j} + \phi_{i+1,j+1} - \phi_{i+1,j})$$

and for E_y

$$E_y \approx -\frac{1}{2\Delta} (\phi_{i,j+1} - \phi_{i,j} + \phi_{i+1,j+1} - \phi_{i+1,j})$$

The energy stored in the electric field in this area can be written using the energy equation for a small increment as

$$\Delta U = \tfrac{1}{2}\varepsilon (E_x^2 + E_y^2)\, \Delta^2$$

$$= \frac{\varepsilon}{4} [(\phi_{i,j} - \phi_{i+1,j+1})^2 + (\phi_{i+1,j} - \phi_{i,j+1})^2] \qquad\qquad (3.20)$$

after substitution for E_x, E_y and Δ

To obtain total energy the contributions of each small area above are summed across all the squares in the grid.

$$U = \sum_{i=1}^{i_{max}-1} \sum_{j=1}^{j_{max}-1} \Delta U$$

with ϕ_0 set equal to $+1$ volt, total capacitance C can now be found since

$$C = \frac{2U}{\phi_0{}^2}$$

$$\therefore \quad C = 2U$$

3.4   MICROSTRIP EXAMPLE

3.4.1 Single Line Evaluation

Figure 3.11 shows a microstrip line housed in a metal enclosure. The dielectric height is selected to be 12 grid units; the width of the conductor to be 30 grid units, and the conductor thickness to be one unit. A uniform square grid is appropriate for the solution of this problem.

By evaluating the capacitance of the structure shown in figure 3.11 with and then without the dielectric substrate material, line impedance and guide wavelength under static conditions can be evaluated, using the methods discussed previously. Program 3.1 RELGRID enables these calculations to be carried out automatically from a line specification such as that shown in figure 3.11. Program RELGRID requires two complete computer runs, one with and one without a dielectric substrate. From program 2.6 the characteristic impedance of the line specified in figure 3.11 is 50 ohms. With program RELGRID a value of 46 ohms is produced, these values are within 8 percent. To reduce the difference between these values a finer grid structure could be superimposed on the structure. This would, of course, have to be paid for in terms of increased computation time. Program RELGRID can also be used for the investigation of coupled microstrip line pairs.

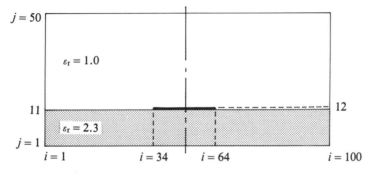

Figure 3.11   Encased Microstrip Line

```
][ FORMATTED LISTING
FILE: PROGRAM 3.1 RELGRID
PAGE-1


  10   REM
  20   REM  --- RELAXATION GRID ---
  30   REM
  40   REM THIS PROGRAM USES
  50   REM THE FINITE DIFFERENCE
  60   REM APPROACH FOR A STATIC
  70   REM SOLUTION OF LAPLACE'S
  80   REM EQUATION.
  90   REM FOR THE PURPOSE OF
 100   REM DEMONSTRATION THIS
 110   REM PROGRAM HAS BEEN
 120   REM SET UP TO COMPUTE
 130   REM CHARAC. IMP. AND
 140   REM EFFECTIVE PERM.
 150   REM FOR A SINGLE
 160   REM MICROSTRIP LINE
 170   REM HELD IN A METAL
 180   REM ENCLOSURE.
 190   REM MODS. GIVEN FOR
 200   REM COUPLED LINES
 210   REM
 220   REM ITRY = MAX NO OF ITS
 230   REM X1,Y1 = LOWER COND.
 240   REM CORNER
 250   REM X2,Y2 = UPPER COND.
 260   REM CORNER
 270   REM RES=DAMPING COEFF.
 280   REM DAMPING COEFF.
 290   REM ER=REL. DIE. CONST.
 300   REM ACC=CONVERGENCE ACC.
 310   REM
 320   REM FOR THIS ARRAY
 330   REM MAX BOX SIZE
 340   REM IS 100*50
 350   REM FOR ZOE SET
 360   REM DIMV(101,50)
 370   REM
 380   REM PROGRAM IS TERMINATED
 390   REM WHEN LINE CAP.
 400   REM IS SEEN TO CONVERGE
 410   REM
 420   DIM V(100,50)
 430   HOME
 440   LET MEW = 12.57E - 7
 450   LET EO = 8.854E - 12
 460   HOME
 470   PRINT "INPUT RELATIVE DIELECTRIC CONSTANT"
 480   INPUT ER
 490   LET ACC = 0.01
 500   LET D = 0
 510   LET RES = 1.5
 520   PRINT "INPUT MAX NO. OF ITERATIONS ALLOWED"
 530   INPUT ITRY
 540   REM
 550   REM SET UP METAL CASE
 560   PRINT "I/P DIMENSIONS OF METAL CASE"
 570   PRINT "FIRST X-COORD THEN Y-COORD"
 580   INPUT X,Y
 590   REM
 600   REM SET UP CENTER COND.
 610   PRINT "I/P DIMENSIONS OF CENTER CONDUCTOR"
 620   PRINT "RELATIVE TO ENCLOSURE ORIGIN"
 630   PRINT "FIRST X-COORDS"
 640   INPUT X1,X2
 650   PRINT "NOW Y-COORDS"
 660   INPUT Y1,Y2
 670   HOME
 680   PRINT
 690   PRINT "**********************"
 700   PRINT
```

```
710    PRINT "RELATIVE DIELECTRIC CONSTANT "ER
720    PRINT "MAX. NO. OF ITERATIONS "ITRY
730    PRINT
740    PRINT "CASE DIMENSIONS "X" X "Y:
       PRINT
750    PRINT "CENTER CONDUCTOR:---"
760    PRINT "X-COORDS "X1","X2
770    PRINT "Y-COORDS "Y1","Y2
780    PRINT
790    REM SET COND. TO 1 VOLT
800    REM FOR ZOO SET
810    REM FORI=(X-1-(X2-X1))
820    REM TO(X-1)
830    FOR I = X1 TO X2
840        FOR J = Y1 TO Y2
850            LET V(I,J) = 1
860        NEXT J
870    NEXT I
880    REM
890    LET E = (1 + ER) / 2
900    REM
910    REM
920    REM SET ITERATION LOOP
930    FOR L7 = 1 TO 2
940        IF L7 = 2 THEN
               LET ER = 1
950        FOR L4 = 1 TO ITRY
960            LET E1 = 0
970            LET E2 = 0
980            REM   FOR ZOE SET
990            REM FOR I=2 TO X
1000           REM AND INSERT NEW LINE
1010           REM V(101,J)=V(99,J)
1020           FOR I = 2 TO (X - 1)
1030               FOR J = 2 TO (Y - 1)
1040                   IF V(I,J) = 1 THEN
                           1140
1050                   IF (J - Y1) = 0 THEN
                           1070
1060                   GOTO 1090
1070                   LET VCAL = (V(I,J + 1) + ER * V(I,J - 1) + E * (V(I + 1,J
                       ) + V(I - 1,J))) / (1 + ER + 2 * E) * RES + (1 - RES) * V
                       (I,J)
1080                   GOTO 1100
1090                   LET VCAL = (V(I + 1,J) + V(I - 1,J) + V(I,J + 1) + V(I,J
                       - 1)) / 4 * RES + (1 - RES) * V(I,J)
1100                   LET D1 = (VCAL - V(I,J)) ^ 2
1110                   IF D1 > E1 THEN
                           LET E1 = D1
1120                   LET E2 = E2 + D1
1130                   LET V(I,J) = VCAL
1140               NEXT J
1150           NEXT I
1160           IF  INT ( INT (L4 / 10) * 10 - L4) = 0 THEN
                   1180
1170           GOTO 1340
1180           LET CAP = 0
1190           REM FIND CAPACITANCE
1200           FOR I = 1 TO (X - 1)
1210               FOR J = 1 TO (Y - 1)
1220                   LET L5 = (V(I,J) - V(I + 1,J + 1)) ^ 2
1230                   LET L6 = (V(I + 1,J) - V(I,J + 1)) ^ 2
1240                   LET LOT = L5 + L6
1250                   IF J < Y1 THEN
                           LET LOT = LOT * ER
1260                   LET CAP = CAP + LOT
1270               NEXT J
1280           NEXT I
1290           LET CAP = CAP * EO / 2
1300           REM CONVERGENCE CHECK
1310           IF L4 < 20 THEN
                   1340
1320           IF  ABS (CAP - D) < ACC *  ABS (CAP) THEN
                   1350
1330           LET D = CAP
```

```
1340       NEXT L4
1360       LET D = CAP
1370       LET CFINAL = D
1380       IF (L7 - 1) <  = 0 THEN
                    1460
1390       LET ZO =  SQR (EO * MEW / CFINAL / CKEEP)
1400       LET EEF = CKEEP / CFINAL
1410       PRINT
1420       PRINT "CHARACTERISTIC IMPEDANCE= " INT (ZO * 100 + 0.5) / 100" OHMS"
1430       PRINT
1440       PRINT "EFFECTIVE DIELECTRIC CONST.= " INT (EEF * 100 + 0.5) / 100
1450       GOTO 1480
1460          LET CKEEP = D
1470   NEXT L7
1480   PRINT
1490   PRINT "**********************"
1500   PRINT
1510   END
```

END-OF-LISTING

```
]RUN
INPUT RELATIVE DIELECTRIC CONSTANT
?2.3
INPUT MAX NO.ITERATIONS ALLOWED
?200
I/P DIMENSIONS OF METAL CASE
FIRST X-COORD THEN Y-COORD
?100
??50
I/P DIMENSIONS OF CENTER CONDUCTOR
RELATIVE TO ENCLOSURE ORIGIN
FIRST X-COORDS
?32
??42
NOW Y-COORDS
?12
??13


**********************

RELATIVE DIELECTRIC CONSTANT 2.3
MAX. NO. OF ITERATIONS 200


CASE DIMENSIONS 100 X 50


CENTER CONDUCTOR:---
X-COORDS 32,42
Y-COORDS 12,13


CHARACTERISTIC IMPEDANCE = 86.32 OHMS

EFFECTIVE DIELECTRIC CONST. = 1.74

**********************
```

3.4.2 Coupled Line Evaluation

With coupled lines, even and odd mode impedances have to be calculated separately. For odd mode excitation one line has $+1V$ applied and the other has -1 volt applied. The odd mode characteristic impedance can then be found from the calculated capacitance of either line to ground, figure 3.12 shows the situation.

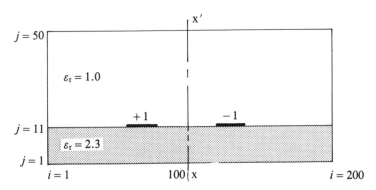

Figure 3.12   Coupled Microstrip Line

Since the structure is symmetric in terms of electrical energy distribution about X–X′ only half the energy in the structure need be calculated since each line will contribute equally to the total stored energy in the complete structure (see figure 2.11).

By exploiting this symmetry aspect, coupled line calculations can be carried out that require approximately the same amount of computer storage space as required for single line calculations. The same program that was used for single line computation can be used here. All that needs to be done is to relocate the left-hand conductor. The appropriate modifications are annotated in program RELGRID.

For calculation of even mode excitation the solution is slightly more complex than in the case of odd mode excitation. Consider the properties of the axis of symmetry in figure 3.12 for even mode excitation. Here, with both conductors raised to a potential of $+1$ volt, the electric field profiles that exist in the left-hand side of the enclosure have a mirror image in the right-hand side. This means that if the potential distribution is known for all grid points $i = 0$ up to and including the axis of symmetry, then it is known for all i. Since it is now required to know the potential distribution from $i = 0$ to the axis of symmetry, it is necessary to extend the mesh by one more grid unit so that the finite difference equations can be applied. Again suitable modifications to program RELGRID are given in the program listing. The execution time for program RELGRID may be several hours due to the large number of calculations that must be performed.

3.5 MONTE CARLO ALTERNATIVE

The relaxation grid method for the solution of Laplace's equation becomes cumbersome when three-dimensional analysis is required, or when fine detail around a particular section of line is needed. The main problem is normally one of computer memory storage. An alternative numerical solution scheme exists that can be used to solve the finite difference equations given by equations (3.10) and (3.17). Unlike the previous method any potential at a specified location in the transmission line geometry can be solved without having to evaluate potential across a complete mesh of node points. The method is therefore useful for detailed potential investigations about a specific area. Since it is independent of computer storage requirements it is also useful when only the potential difference between two or more points in a system is required.

The technique involves the use of a stochastic method called probabilistic potential theory. This is derived from a branch of mathematics called Monte Carlo modeling [7]. Consider what would happen to a 'particle' if it was placed at one of the node points on the grid in figure 3.1.

If the particle experiences Brownian motion, i.e. it is allowed to jump around in an unbiased fashion (the probability of a jump in one direction being the same as the probability of a jump in another direction) the probability of reaching a particular node is the average of the probabilities for the particle at any of the four equally spaced points around that node. Let $P_{i,j}$ be the probability of reaching the node placed at i,j then for two dimensions

$$P_{i,j} = \frac{P_{i+1,j} + P_{i-1,j} + P_{i,j+1} + P_{i,j-1}}{4} \tag{3.21}$$

For three dimensions the factor 4 on the bottom line of equation (3.21) is replaced by 6. Comparing equation (3.21) with equations (3.10) and (3.17), shows these to be very similar. Hammersley [7] suggested that the analogy between the finite difference equations and the simple probability equation could be used to solve Laplace's equation, the scheme is as follows (see figure 3.11). If a particle that is released from a specific node point and allowed to experience a random walk encounters a grounded conductor, it is assigned a numerical value of zero. A new particle is then released from the same node point; if this particle, after following a random walk, encounters a conductor with $+1$ volt on it, it is given a value of one. If it encounters a grounded conductor it is assigned a value of zero as before. If the process is repeated a large number of times to establish a measure of confidence in the results obtained, it is possible, on calculating the ratio of the number of particles that reached the $+1$ conductor to the total number released from the node of interest, to determine the potential at the node of interest. In essence the Monte Carlo method is being used to solve the system of simultaneous equations governing the line on a node by node basis rather than over the complete grid as was the case in the finite difference solution. Herein lies the simplicity and advantage of the Monte Carlo method: (i) the potential at specific points within the structure can be examined independently, and (ii) the numerical computation involved is trivial.

As an aside, in order to gain some familiarity with this technique consider figure 3.13. Here the Monte Carlo method is employed to find the area occupied by the plane shape. In figure 3.13 random points are distributed uniformly in two dimensions; of these, 20 fall within the boundary of the plane shape under investigation while the total number of random points used is 60. This enables the area to be calculated as $20/60 = 0.33$ (the true area is 0.3). Obviously the more points

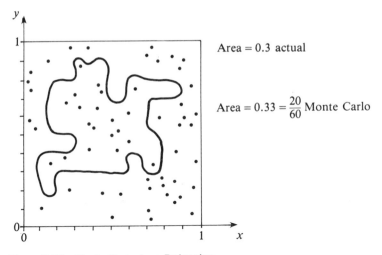

Area = 0.3 actual

Area = $0.33 = \frac{20}{60}$ Monte Carlo

Figure 3.13   Monte Carlo Area Estimation

used in this experiment the greater the accuracy of the result. Be warned, however, in all Monte Carlo type calculations a law of diminished returns applies so that the effort needed to get an increase in accuracy of, say, 5 the number of particles that would have to be released would increase by a factor of 25.

Returning to the problem in hand. In order to compute characteristic imped-ance or guide wavelength, line capacitance is required. The approach detailed in section 3.3 for the evaluation of line capacitance is somewhat inconvenient if the potential at all node points of the relaxation within the line geometry are not known. An alternative approach is to use Gauss's law for the evaluation of line capacitance. This method has several distinct advantages when used in conjunction with the Monte Carlo modeling technique.

3.6   GAUSS'S EQUATION AND APPLICATIONS

The use of Gauss's equation is based on knowing the electric flux density over an entire closed region. This lends itself to the Monte Carlo technique since by evaluating the potential at a number of points around a conductor the electric field strength perpendicular to the conductor can be evaluated.

For perfect conductors the electric field lines always intersect the surface of the conductor at right angles. Gauss's law says that by summing the electric flux density over a closed surface surrounding a conductor the total charge Q on that conductor is then known. Stated mathematically

$$Q = \int_V \varepsilon \mathbf{E} \cdot d\mathbf{v} = \oint_c \mathbf{E} \, dc \quad \text{in two dimensions}$$

where c is a closed contour surrounding the perimeter of the conductor. Capacitance can be found since

$$C = \frac{Q}{\phi_0} = \int_c \varepsilon E \, dc \qquad \phi_0 = 1 \tag{3.22}$$

The electric field component normal to the conductor of interest $\mathbf{E}$ is found from a knowledge of the potential distribution since

$$\mathbf{E} = \left[\frac{\partial \phi}{\partial x} \mathbf{i} + \frac{\partial \phi}{\partial y} \mathbf{j} + \frac{\partial \phi}{\partial z} \mathbf{k} \right] = \nabla \phi$$

Equation (3.22) involves an integration around an enclosed region. This is advantageous especially where the potential distribution has been computed from a probabilistic approach. The probabilistic method tends to give data that can be considered 'noisy' due to the stochastic nature of the method. Integration of this data will filter out some of this noise. What now remains is to develop a numerical approach for solving equation (3.22). Consider figure 3.14. If a set of values are known for the voltages around the conductor, then the electric field normal to the conductor's surface can be calculated.

The method for calculating the electric field strengths parallel and normal to the conductor is as follows for a unity grid spacing

$$E_x(I, J) = \phi_{i+1, j} - \phi_{i, j}$$
$$E_y(I, J) = \phi_{i, j+1} - \phi_{i, j}$$

It is now possible to compute using equation (3.22) the total charge around the center conductor (table 3.1). Notice how in table 3.1 the correct value of dielectric in the region of interest is selected, with an average value of ε_{r_1} and ε_{r_2} used for the dielectric interface as suggested in section 3.2.5.

Also note how, for the first and last ordinate values in each summation in the charge calculation, only half of the numerical value is used. This is done since the simplest form of numerical integration is called the trapizoidal rule and requires the first and last ordinate values to be halved.

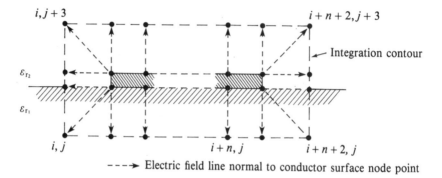

Figure 3.14   Conductor Profile for Gauss Evaluation of Charge

Table 3.1 Total Charge Calculation

1	lower $= (\varepsilon_o)(2.3)(0.05/2 + 0.3 + 0.2 + 0.2 + 0.3 + 0.05/2)$	$= 21$ pF
2	upper $= (\varepsilon_o)(0.055/2 + 0.2 + 0.1 + 0.1 + 0.2 + 0.055/2)$	$= 6$ pF
3	left $=$ right $= (\varepsilon_o)(0.1 \times 2.3/2 + 0.35 \times (2.3 + 1)/2 + 0.325 + 0.18/2)$	$= 9.8$ pF
4	total $= 21 + 6 + 0.98 + 0.98$	$= 29$ pF

Reapplying the Monte Carlo technique to the conductor for the same grid points as before but this time with ε_{r_1} set equal to unity will lead to a new value for total charge. This new value together with the old value can be used to find characteristic impedance and effective dielectric constant. Computer program 3.2 MONTE, utilizes the probabilistic method for solving the microstrip mixed dielectric problem. This program requires the use of a random number generator routine to simulate the random walk that a particle must take when released from a node on the grid. Lines 160 to 240 represent the random number generator. Lines 250 to 1320 represent the probabilistic routine releasing particles for the random walk and then calculating probabilities of reaching a conductor. Lines 1070 to 1110 allow the probability of a move for the dielectric edge to be computed. Since the computer can produce only pseudo random numbers, i.e. a series of numbers whose statistical properties are akin to those of truly random numbers, different seed values must be used each time a particle is launched from a new node. This ensures that each random walk is different from the previous one (lines 690 to 710). Lines 1430 to the end form the numerical routine necessary to evaluate Gauss's equation for charge. Lines 1460 to 1520 represent the integration routine.

A more refined routine using Simpson's rule or some alternative method would provide less numerical error at the expense of having to select the number of ordinates correctly. The MONTE program when run in compiled form requires about 90 s per point for a confidence level based on 200 trials.

```
][ FORMATTED LISTING
FILE: PROGRAM 3.2 MONTE
PAGE-1


  10   REM
  20   REM --- MONTE CARLO ---
  30   REM
  40   REM MONTE CARLO RANDOM
  50   REM WALK CASED MICROSTRIP
  60   REM SOLN. GUASS'S EQUATION
  70   REM IS USED TO ESTIMATE
  80   REM LINE CAPACITANCE
  90   REM
 100   DIM V(12,4),E1(4),E2(4),E3(12),E4(12)
 110   LET X9 = 2
 120   LET A = 24298
 130   LET C = 9991
 140   LET AM = 199017
 150   GOTO 280
 160   REM  : RANDOM NO. GEN
 170   LET A1 = A * X9 + C
 180   LET A4 = A1 / AM
 190   LET M2 =  INT (A4)
```

```
200    LET A5 = A1 - M2 * AM
210    LET R = A5 / AM
220    LET X9 = R * 199017
230    LET X7 = R
240    RETURN
250    REM :
260    REM    :INITALISE RANDOM NO. GENERATOR
270    REM :
280    FOR I = 1 TO 100
290        GOSUB 170
300    NEXT I
310    REM    :
320    REM    :SET DIELECTRICS
330    REM :
340    LET EO = 8.854E - 12
350    LET ER = 2.3
360    LET E = (1 + ER) / 2
370    REM :SET PROBABILITIES
380    LET D = 1 + ER + 2 * E
390    LET P1 = ER / D
400    LET P2 = 1 / D
410    LET P3 = E / D
420    REM :DEFINE BOX
430    LET YBOT = 0
440    LET YTIP = 50
450    LET XBOT = 0
460    LET XTIP = 100
470    REM :DEFINE CENTER CONDUCTOR
480    REM : X1,Y1 GIVE LOWER CONDUCTOR LOCATION
490    REM : X2,Y2 GIVE UPPER CONDUCTOR LOCATION
500    LET X1 = 48
510    LET X2 = 52
520    LET Y1 = 12
530    LET Y2 = 13
540    REM :SET UP SEED TABLE
550    DIM SEED(15,5)
560    LET C2 = 2
570    FOR I = 0 TO (X2 - X1 + 2)
580        FOR J = 0 TO (Y2 - Y1 + 2)
590            LET SEED(I,J) = C2
600            LET C2 = C2 + 1
610        NEXT J
620    NEXT I
630    REM    :
640    HOME
650    PRINT "I              J                    VOLTAGE"
660    REM    :PICK POINT OF INTEREST
670    FOR I = (X1 - 1) TO (X2 + 1)
680        FOR J = (Y1 - 1) TO (Y2 + 1)
690            REM :SELECT SEED FROM LOOKUP TABLE
700            LET SA = I - (X1 - 1):
               LET SB = J - (Y1 - 1)
710            LET X9 = SEED(SA,SB)
720            IF (I - (X1 - 1)) < 0 THEN
               GOTO 1320
730            IF (I - (X1 - 1)) = 0 THEN
               GOTO 840
740            IF (I - (X1 - 1)) > 0 THEN
               GOTO 750
750            IF (I - (X2 + 1)) < 0 THEN
               GOTO 780
760            IF (I - (X2 + 1)) = 0 THEN
               GOTO 840
770            IF (I - (X2 + 1)) > 0 THEN
               GOTO 1320
780            IF (J - (Y1 - 1)) < 0 THEN
               GOTO 1310
790            IF (J - (Y1 - 1)) = 0 THEN
               GOTO 840
800            IF (J - (Y1 - 1)) > 0 THEN
               GOTO 810
810            IF (J - (Y2 + 1)) < 0 THEN
               GOTO 840
```

```
820          IF (J - (Y2 + 1)) = 0 THEN
                  GOTO 840
830          IF (J - (Y2 + 1)) > 0 THEN
                  GOTO 1310
840          LET N1 = 0
850          LET N2 = 0
860          REM :START
870          LET X = I
880          LET Y = J
890          LET N1 = N1 + 1
900          GOSUB 170
910          REM :CHECK POSITION
920          IF Y = Y1 THEN
                  GOTO 1080
930          IF (X7 - 0.25) < = 0 THEN
                  GOTO 950
940          IF (X7 - 0.25) > 0 THEN
                  GOTO 970
950          LET X = X - 1
960          GOTO 1220
970          IF (X7 - 0.5) < = 0 THEN
                  GOTO 990
980          IF (X7 - 0.5) > 0 THEN
                  GOTO 1010
990          LET X = X + 1
1000         GOTO 1220
1010         IF (X7 - 0.75) < = 0 THEN
                  GOTO 1030
1020         IF (X7 - 0.75) > 0 THEN
                  GOTO 1050
1030         LET Y = Y - 1
1040         GOTO 1220
1050         LET Y = Y + 1
1060         GOTO 1220
1070         REM   :ON THE DIELECTRIC EDGE
1080         IF (X7 - P1) < = 0 THEN
                  GOTO 1100
1090         IF (X7 - P1) > 0 THEN
                  GOTO 1120
1100         LET Y = Y - 1
1110         GOTO 1220
1120         IF (X7 - P1 - P2) < = 0 THEN
                  GOTO 1140
1130         IF (X7 - P1 - P2) > 0 THEN
                  GOTO 1160
1140         LET Y = Y + 1
1150         GOTO 1220
1160         IF (X7 - P1 - P2 - P3) < = 0 THEN
                  GOTO 1180
1170         IF (X7 - P1 - P2 - P3) > 0 THEN
                  GOTO 1200
1180         LET X = X - 1
1190         GOTO 1220
1200         LET X = X + 1
1210         REM :A HIT ?
1220         IF ((X < = XBOT) OR (X > = XTIP) OR (Y < = YBOT) OR (Y > = YT
             IP)) THEN
                  1270
1230         IF ((X > = X1) AND (X < = X2) AND (Y > = Y1) AND (Y < = Y2))
                  GOTO 1260
1240         GOTO 900
1250         REM :ITERATION LIMIT 200
1260         LET N2 = N2 + 1
1270         IF N1 < 200 THEN
                  870
1280         REM :CALCULATE VOLTAGE
1290         LET V(I - X1 + 1,J - Y1 + 1) = N2 / N1
1300         PRINT I,J,V(I - X1 + 1,J - Y1 + 1)
1310     NEXT J
1320  NEXT I
1330  REM
1340  REM FORM ELECTRIC FIELD
1350  FOR J = 0 TO ((Y2 + 1) - (Y1 - 1))
```

```
1360        LET E1(J) = V(1,J) - V(0,J)
1370        LET E2(J) = V(X2 + 1 - X1,J) - V(X2 - X1 + 2,J)
1380   NEXT J
1390   FOR J = 0 TO (X2 - X1 + 2)
1400        LET E3(J) = V(J,Y2 - Y1) - V(J,Y2 - Y1 - 1)
1410        LET E4(J) = V(J,Y2 - Y1 + 1) - V(J,Y2 - Y1 + 2)
1420   NEXT J
1430   REM FOR GAUSS'S LAW
1440   REM NEED TO INTEGRATE
1450   REM TO GET TOTAL CHARGE.
1460   REM USE TRAPEZOIDAL RULE
1470   LET F = 0
1480   LET G = 0
1490   FOR J = 1 TO (X2 - X1 + 1)
1500        LET F = F + E3(J)
1510        LET G = G + E4(J)
1520   NEXT J
1530   LET LOWQ = EO * ER * (E3(0) / 2 + E3(11) / 2 + F)
1540   LET UPPQ = EO * (E4(0) / 2 + E4(11) / 2 + G)
1550   LET LEFTQ = EO * (E1(0) / 2 * ER + E1(1) * (ER + EO) / 2 + E1(2) + E1(3)
            / 2)
1560   LET RIGHTQ = EO * (E2(0) / 2 * ER + E2(1) * (ER + EO) / 2 + E2(2) + E2(3)
            / 2)
1570   LET Q = LOWQ + UPPQ + LEFTQ + RIGHTQ
1580   PRINT
1590   PRINT "***********************"
1600   PRINT
1610   PRINT "ESTIMATED CAP= " INT (Q * 1E14 + 0.5) / 100" PF"
1620   PRINT
1630   PRINT "***********************"
1640   END
```

END-OF-LISTING

I	J	VOLTAGE
47	11	.505
47	12	.515
47	13	.78
47	14	.01
48	11	.665
48	12	.85
48	13	.675
48	14	.875
49	11	.87
49	12	.94
49	13	.97
49	14	1
50	11	.88
50	12	.94
50	13	1
50	14	.085
51	11	.82
51	12	.965
51	13	.98
51	14	.94
52	11	.505
52	12	.86
52	13	.895
52	14	.99
53	11	.39
53	12	.87
53	13	.89
53	14	1

****

ESTIMATED CAP = 36.33 PF

****

Lines 160 to 240 may be replaced by a microcomputer library routine for the generation of random numbers. However, the statistical properties of these generators should be tested. Some simple test schemes can easily be devised. The first is to check the periodicity of the sequence produced by the pseudo random generator. The criterion used in the test program 3.3 RANDOM CHECK 1 is to detect the occurrence of three consecutive numbers of selected accuracy generated at the beginning of a particular sequence. A second test program, program 3.4 RANDOM CHECK 2, determines the mean, standard deviation and distribution of uniformly distributed random numbers lying in the range zero to one.

```
][ FORMATTED LISTING
FILE: PROGRAM 3.3 RANDOM CHECK1
PAGE-1


 10   REM
 20   REM    -- RANDOM CHECK1 --
 30   REM
 40   REM    THIS PROGRAM CHECKS
 50   REM    THE LTH. OF A PSEUDO
 60   REM    RANDOM NO. GENERATOR
 70   REM    SEQUENCE PRODUCED BY
 80   REM    SUBROUTINE 370
 90   HOME
100   PRINT "TEST OF SEQUENCE LTH."
110   PRINT "I/P SIGNIFICANT PLACES EG. 0.001"
120   INPUT R
130   LET R = 1 / R
140   LET A2 = 24298:
      LET C = 9991:
      LET AM = 199017:
      LET X9 = 2
150   PRINT "WORKING:----"
160   DIM C(3)
170   FOR I = 1 TO 3
180       GOSUB 440
190       LET Y = X9 / 199017:
          LET C(I) = Y
200       LET C(I) =  INT (C(I) * R + .5) / R
210   NEXT I
220   LET N = 3
230   LET N1 = 3
240   GOSUB 440
250   LET X =  INT (Y * R + .5) / R
260   LET N = N + 1
270   LET N1 = N1 + 1
280   IF X = C(1) THEN
          330
290   IF N1 < 10000 THEN
          240
300   PRINT "NO REPEAT YET, N= "N
310   LET N1 = 0
320   GOTO 250
330   GOSUB 390
340   IF X < > C(2) THEN
          250
350   GOSUB 390
360   IF X < > C(3) THEN
          250
370   PRINT "TEST COMPLETED"
380   GOTO 520
390   PRINT "REPEATED NO. AT N= "N
400   GOSUB 440
410   LET X =  INT (Y * R + .5) / R
420   LET N = N + 1
430   RETURN
440   LET A3 = A2 * X9 + C
```

```
450   LET A4 = A3 / AM
460   LET M2 =  INT (A4)
470   LET A5 = A3 - M2 * AM
480   LET X9 = A5 / AM
490   LET Y = X9
500   LET X9 = X9 * 199017
510   RETURN
520   END
```

END-OF-LISTING

```
]RUN
TEST OF SEQUENCE LTH.
I/P SIGNIFICANT PLACES EG. 0.001
?0.01
WORKING:----

REPEATED NO. AT N= 158
REPEATED NO. AT N= 273
REPEATED NO. AT N= 590
                .
                .
                .
                .
                .
```

```
][ FORMATTED LISTING
FILE: PROGRAM 3.4 RANDOM CHECK2
PAGE-1
```

```
10    REM
20    REM --RANDOM CHECK 2 --
30    REM
40    REM THIS PROGRAM COMPUTES
50    REM THE MEAN, S.D., AND
60    REM % OF OCCURANCE OF
70    REM UNIFORMLY DIST. RANDOM
80    REM NO.S BETWEEN 0,1
90    DIM B(10):
      HOME
100   PRINT
110   PRINT "ENTER NO. OF SAMPLES"
120   INPUT L
130   PRINT
140   PRINT "ANALYSIS OF "L" SAMPLES"
150   LET U =  SQR (12)
160   LET A = 0:
      LET B = 0:
      LET T = 0:
      LET V = 0
170   REM
180   REM SET UP RANDOM NO.
190   REM GENERATOR PARAMETERS
200   LET A2 = 24298:
      LET C = 9991:
      LET AM = 199017:
      LET X9 = 2
210   PRINT
220   PRINT "WORKING:-----"
230   FOR I = 0 TO 9
240       LET B(I) = 0
250   NEXT I
260   PRINT
270   PRINT "THEORETICAL RESULTS:---"
280   PRINT "MEAN VALUE=0"
290   PRINT "STANDARD DEVIATION=1"
300   PRINT "MAX. VALUE=+1.732"
310   PRINT "MIN. VALUE=-1.732"
320   FOR I = 1 TO L
330       GOSUB 650
340       LET A = A + X
350       LET V = V + X * X
360       IF X > T THEN
              390
370       IF X < B THEN
              410
380       GOTO 420
390       LET T = X
400       GOTO 420
410       LET B = X
420   NEXT I
430   LET F0 = (V - A * A / L) / (L - 1)
440   PRINT
450   PRINT "ACTUAL RESULTS*****"
460   PRINT "MEAN= "A / L
470   PRINT "SD= " SQR (F0)
480   PRINT "MAX= "T
490   PRINT "MIN= "B
500   PRINT "SAMPLE","PERCENT"
510   FOR I = 0 TO 9
520       PRINT I,B(I) * 100 / L
530   NEXT I
540   PRINT
550   PRINT "DO YOU WANT ANOTHER GO?"
560   PRINT "ENTER 1 IF YES ; 0 IF NO"
570   INPUT Y
580   IF Y = 0 THEN
          760
590   IF Y = 1 THEN
          100
600   GOTO 550
610   REM
620   REM RAND. NO. GENERATOR
630   REM X=UNIFORMLY DIST. RAND.
640   REM NO. BETWEEN 0,1
650   LET A3 = A2 * X9 + C
```

```
660   LET A4 = A3 / AM
670   LET M2 =   INT (A4)
680   LET A5 = A3 - M2 * AM
690   LET X9 = A5 / AM
700   LET X = X9
710   LET X9 = X9 * 199017
720   LET M =   INT (10 * X)
730   LET B(M) = B(M) + 1
740   LET X = (X - 0.5) * U
750   RETURN
760   PRINT
770   PRINT "*** END OF PROGRAM ***"
780   END


END-OF-LISTING


]RUN

ENTER NO. OF SAMPLES
?1000

ANALYSIS OF 1000 SAMPLES

WORKING:-----

THEORETICAL RESULTS:---
```

```
MEAN VALUE=0
STANDARD DEVIATION=1
MAX. VALUE=+1.732
MIN. VALUE=-1.732


ACTUAL RESULTS*****
MEAN= .0158232903
SD= .969824911
MAX= 1.72820407
MIN= -1.73083238
SAMPLE              PERCENT
0                   8.7
1                   9.9
2                   9.9
3                   10.2
4                   10.1
5                   11.3
6                   10.1
7                   10.8
8                   9.7
9                   9.3

DO YOU WANT ANOTHER GO?
ENTER 1 IF YES ; 0 IF NO
?0

*** END OF PROGRAM ***
```

3.7   METHOD OF MOMENTS

An alternative to the solution of Laplace's equation by the finite difference or Monte Carlo approaches involves the use of a technique called the method of moments[8]. This technique is really a simple form of a more general and very powerful solution method, known as the finite element method[11].

As a starting point for the discussion of the method of moments consider Coulomb's law, stated below as

$$|F| = \frac{qQ}{4\pi\,\varepsilon_0\,|R|^2} \tag{3.23}$$

This equation states that the magnitude of the force between two charges, q and Q, is inversely proportional to the square of their separation R. In vector form, the above equation can be rewritten as

$$\mathbf{F} = \frac{qQ}{4\pi\,\varepsilon_0\,|R|^2}\,\mathbf{r} \tag{3.24}$$

where $\mathbf{F}$ represents a vector quantity and $\mathbf{r}$ is a unit vector directed from q to Q. This equation is a convenient way to express the fact that if one of the charges is fixed the other would be repelled or attracted depending on charge polarity. Figure 3.15 shows the situation for charges of equal polarity. For charges of opposite polarity the force vector acts along the line joining Q and q so as to attract the charges. If charge Q becomes distributed then the resulting effect on point charge q is simply the vector summation of a series of individual point charges suitably disposed in space each having a magnitude that is appropriate to this position relative to the

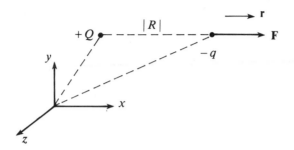

Figure 3.15   Repulsion of Charges of Equal Polarity

magnitude of the original charge distribution at the same position. The set of point charges selected are chosen so that they are equivalent to the original distributed charge distribution. Writing the above statement in the form of an equation gives the resultant force vector for a system of point charges as

$$\mathbf{F} = \sum_{\substack{\text{all point} \\ \text{charges in} \\ \text{the system}}} \frac{qQ_i}{4\pi \, \varepsilon_0 \, |R_i|^2} \mathbf{r}_i$$

where subscript i denotes the ith point charge.

When the charge distribution is continuous over a line, surface or volume, the discrete summation sign of the last equation can be replaced by a continuous integral which exists over the region where the charge exists.

$$\mathbf{F} = \int_v \frac{q\rho}{4\pi \, \varepsilon_0 \, |R|^2} \mathbf{r} \, dv \qquad\qquad (3.25)$$

In this equation the ith charge Q_i is replaced by a charge distribution ρ. Here, ρ can be made to exist in one, two or three dimensions expressed in terms of cartesian coordinates

$$\rho = \rho(x, y, z)$$

In this case, the integral in the governing force equation exists over a volume V.

Already it has been asserted that any electrical charge, when placed in proximity to a second electrical charge, will exert a force on the second charge. As a result of this force, it is possible to define a quantity called electric field intensity designated by the letter $\mathbf{E}$. This quantity has the given dimensions of force per unit charge.

If an infinitesimally small charge q is introduced into the field existing between static charges that exist in a system, then the electric field intensity is defined as

$$\mathbf{E} = \lim_{q \to 0} \frac{\mathbf{F}}{q} = \frac{Q}{4\pi \, \varepsilon_0 \, |R|^2} \mathbf{r}$$

or extended to give, for a three-dimensional charge distribution (x, y, z), a general

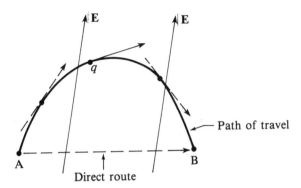

Figure 3.16   Movement of a Point Charge in the Presence of an Electric Field

expression for the electric field intensity vector

$$E = \int_v \frac{\rho(x, y, z)}{4\pi \, \varepsilon_0 |R|^2} r \, dv \qquad \text{units V m}^{-1} \text{ or N C}^{-1}$$

Consider now what happens when a point charge is allowed to move through a static electric field. Figure 3.16 illustrates the situation.

The amount of work required to produce the movement of charge q from position A to position B along some arbitrary path can be expressed as

$$\text{Work done A to B} = -q \int_A^B E \cdot dl$$

where '·' indicates the scalar dot product between two vectors (e.g. $X \cdot Y = |X||Y| \cos \theta$ where θ is the angle between them).

It is known that

$$E(x, y, z) = -\nabla \phi(x, y, z)$$

where $\nabla \phi$ is a scalar quantity, hence

$$\int_A^B E \cdot dl = -\int_A^B \nabla \phi \, dl = -\phi(B) + \phi(A)$$

hence $\phi(B) - \phi(A) = -\displaystyle\int_A^B E \cdot dl$

If point A is removed to infinity where the potential is taken by definition to be zero then the potential at point B is simply

$$\phi(B) = -\int E \cdot dl \tag{3.26}$$

Substituting the electric field intensity vector for a point charge in the above equation

$$\phi(B) = \frac{Q}{4\pi \, \varepsilon_o |R|}$$

Finally, extending this train of thought to a general distributed charge $\rho(x, y, z)$, then the potential at any point of interest will be the summation of all point charges approximating the distribution,

$$\phi(x, y, z) = \int_v \frac{\rho(x, y, z)}{4\pi \, \varepsilon_o |R|} \, dv \qquad (3.27)$$

This equation is actually the integral form of Poisson's equation

$$\nabla^2 \phi(x, y, z) = -\rho(x, y, z)$$

Suppose $\phi(x, y, z)$, the potential throughout the system under consideration, is known and the charge distribution $\rho(x, y, z)$ is unknown, then the method of moments can be used to solve for $\rho(x, y, z)$. So how does a solution proceed? One way is to assume a form of a solution for $\rho(x, y, z)$ over the volume of interest as

$$\rho(x, y, z) = \sum_{i=1}^{N} K_i f_i(\rho) \qquad (3.28)$$

Here, K_i are as yet unknown constants and $f_i(\rho)$ are as yet unknown functions. This equation indicates that the total charge distribution can be found by summing the individual charge contributions of N incremental subvolumes forming the region under consideration. Also, an unknown distribution of charge $\rho(x, y, z)$ is replaced by two sets of unknowns K_i and $f_i(\rho)$. This appears to have complicated the problem at hand.

However, the process continues by selecting a prior function $f_i(\rho)$. Consider figure 3.17 in which a simple charge distribution is shown. In the simplest representation the actual charge distribution can be considered to consist of a series of impulses suitably weighted to match the estimated magnitude of the actual charge distribution at specific points along the x-axis from which an approximate form for the actual charge distribution can be constructed. The situation is shown in figure 3.18. It should be noted that if the weighting coefficients are in error there will exist error between the actual and reconstructed charge distributions.

It should also be noted that the spacing between sample points need not be uniform. Non-uniformly spaced samples allow more definition in regions where the rate of change of charge is suspected to be large. The next type of approximation found is the step, sample and hold or piecewise constant approximation. Again, non-uniform spacing can be used. This time, the use of step functions means that depending on the actual shape of the charge distribution under consideration, parts of the curve will be overestimated while other parts will be underestimated. This means that the resulting approximation obtained for the actual charge distribution may not be very accurate. However, provided the amount of charge in each step is correct then, when total charge is calculated, this will be equivalent to averaging all the charge in the structure, i.e. finding the integral under the charge distribution

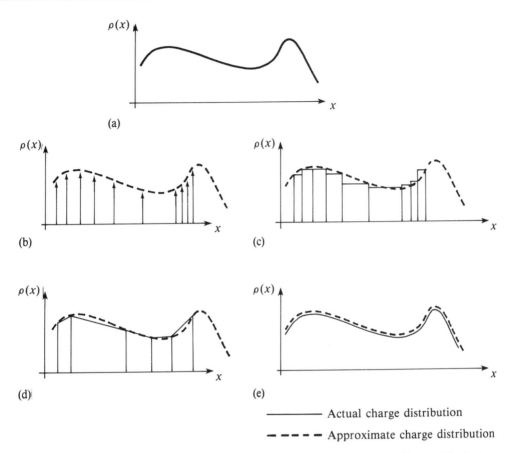

Actual charge distribution

Approximate charge distribution

Figure 3.17   Charge Distribution Approximations (a) Continuous Charge Distribution (b) Impulse Approximation (c) Step Approximation (d) Piecewise Linear (e) Nth Order Polynomial

curve, that is, low pass filtering. This has an important practical ramification in that even though localized charge distribution approximations may be poor, the total estimated capacitance of the structure calculated from the averaged or filtered charge distribution will be good.

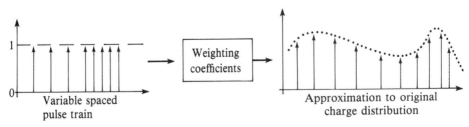

Figure 3.18   Generation of Charge Distribution using Weighted Impulses

Higher order approximations such as piecewise linear or higher order polynomials are frequently used in the finite element technique. Approximation to charge distribution obtained through these higher order approximations will usually be better than those obtained from lower order ones. Of course, these advantages are obtained at the expense of extra programming.

The same arguments used above, for simple one- and two-dimensional field distributions, apply to charge distributions that exist within a volume. In this case, piecewise linear approximations form diamond faceted surfaces approximate to the actual charge distribution surface in three dimensions.

Substituting equation (3.28) into equation (3.27) gives

$$\phi_j = \phi(x_j, y_j, z_j)$$

$$= \sum_{i=1}^{N} \frac{1}{4\pi\varepsilon_0} \int_v \frac{f_i(\rho)}{|R_{ij}|} \, dV \, \mathbf{K}_i \qquad\qquad (3.29)$$

The simplest approximation to an actual charge distribution that can be handled by a digital computer is the impulse. Consider the charge distribution to exist due to some potential on a flat rectangular conductor. The conductor is subdivided into a large number of small subunits. In a physical sense the meaning of the impulse is that each subunit comprising the conductor under investigation has an impulse of unit magnitude associated with it and its charge is assumed to be placed at the center of each subunit. When suitably weighted, see for example figure 3.19, the required impulse approximation to the actual charge distribution results. Figure 3.19 shows a hypothetical situation for a rectangular conductor comprising nine subsections each of area ΔA. Let the term in equation (3.29)

$$\frac{1}{4\pi\,\varepsilon_0} \int_v \frac{1}{|R_{ij}|} \, dV$$

be written as C_{ij}.

Also, since an impulse function of unit magnitude is to be used then equation

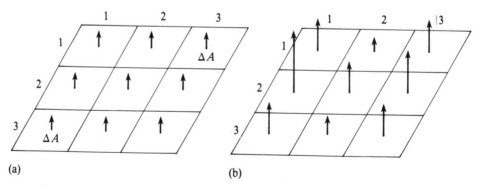

(a)   (b)

Figure 3.19   Impulse Approximation to surface Charge Distribution (a) Unweighted Charge Distribution $C_{11} = C_{12} = \ldots = C_{23} = 1$ (b) Weighted Charge Distribution $C_{11} = C_{12} = \ldots = C_{23} = $ weight

(3.29) can be represented as

$$\phi_j = \sum_{k=1}^{N} C_{ij}\mathbf{K}_i$$

or in matrix form

$$\begin{bmatrix} \phi_1 \\ \phi_2 \\ \vdots \\ \phi_N \end{bmatrix} = \begin{bmatrix} C_{11} & C_{12} & \dots & C_{1N} \\ C_{21} & C_{22} & \dots & C_{2N} \\ \vdots & \vdots & & \vdots \\ C_{N1} & C_{N2} & \dots & C_{NN} \end{bmatrix} \begin{bmatrix} K_1 \\ K_2 \\ \vdots \\ K_N \end{bmatrix}$$

$$\phi = \mathbf{CK} \tag{3.30}$$

Now, provided the C_{ij} terms can be evaluated then $\mathbf{K}$, the column vector representing the unknown set of weighting coefficients, can be found since premultiplying equation (3.30) by the inverted matrix $\mathbf{C}^{-1}$ gives the desired coefficient set

$$\mathbf{K} = \mathbf{C}^{-1}\phi$$

Returning once again to equation (3.29) for a two-dimensional rectangular conductor

$$\frac{1}{4\pi\,\varepsilon_0} \int_s \frac{f_i(\rho)}{|R_{ij}|} \, ds$$

With an impulse function used for $f_i(\rho)$ this reduces to

$$\frac{\Delta S_i}{4\pi\,\varepsilon_0 |R_{ij}|}$$

where ΔS_i is an incremental surface at the center of which the charge is concentrated.

Here $|R_{ij}|$ represents the distance in three-dimensional space from the center of subsection i on the conductor surface point (x_i, y_i, z_i) to the point (x_j, y_j, z_j) where potential is to be calculated. By simple geometry, $|R_{ij}|$ is found to be

$$|R_{ij}| = [(x_j - x_i)^2 + (y_j - y_i)^2 + (z_j - z_i)^2]^{\frac{1}{2}}$$

A simple structure that can be analyzed by the method of moments is an air-spaced capacitor. In its simplest form the capacitor consists of two parallel rectangular conductors constructed from perfectly conducting material and separated by air which acts as the capacitor dielectric. Simple analysis of this type of capacitor yields the capacitance C as

$$C = \frac{\varepsilon_0 ab}{d} \text{ Farads}$$

The quantities a, b, d are defined in figure 3.20.

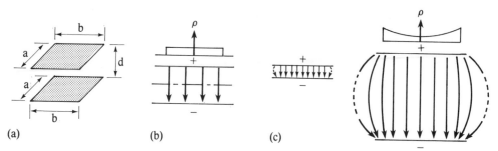

Figure 3.20   Rectangular Parallel Plate Capacitor (a) Parallel Plate Capacitor Notation (b) Electric Field Profile Neglecting Fringing Fields (c) Narrow Spaced Capacitor Displays Small Fringe Field, while Wide Spaced Capacitor Exhibits Large Fringe Field

This expression assumes a uniform distribution of charge over each plate and neglects fringing of electric fields into space. By applying the method of moments to the problem, the change in capacitance due to fringing of electrostatic fields when compared with that calculated from the above equation can be accounted for, and a better approximation to the total capacitance of the structure can be made. It is worthy of note that an approximation to the actual distribution of charge over each plate is implicit in the solution.

In order to reduce the number of calculations necessary to complete a solution, assume that one plate of the capacitor is held at a potential of $+1$ volt while the other is at -1 volt. Due to geometrical symmetry of the capacitor the charge distribution on the lower plate of the capacitor will be of opposite polarity to that on the top plate. This enables a reduction of the number of variables in the governing matrix equation. For N subunits made up of n units of area ΔA on the top conductor and n units on the lower conductor, the matrix equation can be written as

$$
\begin{bmatrix}
C_{11} & C_{12} & \dots & C_{1N} \\
C_{21} & C_{22} & \dots & C_{2N} \\
\cdot & \cdot & & \cdot \\
\cdot & \cdot & & \cdot \\
\cdot & \cdot & & \cdot \\
C_{N1} & C_{N2} & \dots & C_{NN}
\end{bmatrix}
\begin{bmatrix}
K_1 \\
K_2 \\
\vdots \\
K_n \\
\vdots \\
K_{2n}
\end{bmatrix}
=
\begin{bmatrix}
+1 \\
+1 \\
\vdots \\
+1 \\
-1 \\
\vdots \\
-1
\end{bmatrix}
\begin{array}{l}
\uparrow \text{upper conductor} \\
\\
\\
\\
\downarrow \text{lower conductor}
\end{array}
$$

This matrix contains $2N$ unknowns, however, inspection shows that the symmetry mentioned above gives a reduction in the matrix dimensions.

For the symmetrical capacitor the charge on cell 1 will be the negative of that on cell $n + 1$, similarly the charge on cell $2n$ is equal to minus the charge on cell n. Therefore, there are only n unknowns, this reduces the governing matrix equation to that given below.

$$
\begin{bmatrix}
C_{11} - C_{1n+1} & C_{12} - C_{1n+2} & \cdots & C_{1n} - C_{12n} \\
C_{21} - C_{2n+1} & C_{22} - C_{2n+2} & \cdots & C_{2n} - C_{22n} \\
\vdots & \vdots & & \vdots \\
C_{n1} - C_{nn+1} & C_{n2} - C_{nn+2} & \cdots & C_{nn} - C_{n2n}
\end{bmatrix}
\begin{bmatrix}
K_1 \\ K_2 \\ \vdots \\ K_n
\end{bmatrix}
=
\begin{bmatrix}
1 \\ 1 \\ \vdots \\ 1
\end{bmatrix}
$$

Each element in the modified C matrix as cited above represents the potential at cell i due to the charge in cell j on the upper plate together with the contribution from cell $n + j$ on the lower plate.

Returning once again to the modified C matrix it should be realized that since the elements of this matrix are to be found from an impulse approximation then this matrix is solved directly by matrix inversion since the potential vector elements are all unity. This means that the summation of the row elements of the inverted matrix C^{-1} will represent the weighting coefficient (charge) for each cell. With this information the capacitance of the structure can be found since there is ± 1 volt applied to each conductor so that

$$
C = \frac{Q}{2V}
$$

where

$$
Q = \sum_{i=1}^{N} K_i \, \Delta A_i
$$

If the cell area is selected to be unity then the total change in the structure becomes

$$
Q = \sum_{i=1}^{N} K_i = \sum_{j=1}^{N} \sum_{i=1}^{N} C^{-1}{}_{ij}
$$

from which capacitance is forthcoming.

There is one problem that does arise when computing the modified C matrix for terms along the leading diagonal. These terms represent the potential in a cell due to the charge in the same cell. The equation

$$
\frac{1}{4\pi \, \varepsilon_0} \int\!\!\int \frac{ds}{[(x_i - x_j)^2 + (y_i - y_j)^2]^{\frac{1}{2}}}
$$

must be solved. The problem arises when $i = j$ since at this point a singularity occurs. Harrington [8] showed with the use of mathematical function tables provided by Dwight [10] that this expression reduces to

$$
C_{jj} = \frac{0.8814}{\pi \, \varepsilon_0}
$$

for cells with unity area. Due to the symmetry imposed on the modified C matrix, this value must be corrected to include the effect of the cell directly below it.

The results developed in this section have been incorporated into a computer program. Program 3.5 CAPMAT solves the symmetrical capacitor equations amenable to capacitors having subsections with unit area. The program reads the dimensions of a plate that has been divided into *X* by *Y* cells each having unit area, the plate separation is then read. The modified **C** matrix is then calculated and inverted by a modified Gauss–Jordan technique. The total charge in the structure is then computed.

```
][ FORMATTED LISTING
FILE: PROGRAM 3.5 CAPMAT
PAGE-1


   10  REM
   20  REM   **** CAPMAT ****
   30  REM
   40  REM   COMPUTES CAP. OF AIR
   50  REM   SQUARE CAP. FRINGE
   60  REM   FIELDS ARE INCLUDED
   70  REM   METHOD OF MOMENTS
   80  REM   IS USED FOR CALCULATION
   90  REM   MAX NO. CELLS IS 6X6
  100  HOME
  110  PRINT
  120  DIM X(36),Y(36),Z(1296)
  130  DIM Q(36),A(36,36),B(36,36)
  140  PRINT "INPUT CAP. GEOMETRY"
  150  PRINT "LENGTH, BREATH AND SPACING"
  160  PRINT "ARE ASSUMED EQUAL"
  170  PRINT "AS ARE CELL SIZES"
  180  PRINT "NORMALISED UNITS USED"
  190  PRINT "THROUGHOUT"
  200  INPUT M
  210  LET N = M:
       LET D = M
  220  LET D2 = D ^ 2
  230  LET PI = 3.141593
  240  LET EP = 8.8553E - 12
  250  REM   .2225 ISAN EMPIRICAL CORRECTION
  260  REM FOR LOWER CELL
  270  LET P2 = 3.5256 - .2225
  280  LET NN = N * M
  290  LET N2 = NN * NN
  300  LET PE = 2 * PI * EP * M
  310  HOME
  320  PRINT "WORKING:---"
  330  FOR J = 1 TO N
  340      FOR I = 1 TO M
  350          LET I1 = I + (J - 1) * M
  360          LET X(I1) = J
  370          LET Y(I1) = I
  380      NEXT I
  390  NEXT J
  400  FOR J = 1 TO NN
  410      FOR I = 1 TO NN
  420          LET I1 = (J - 1) * NN + I
  430          IF (I - J) < > 0 THEN
               460
  440          LET Z(I1) = P2
  450          GOTO 480
  460          LET P3 = (X(I) - X(J)) * (X(I) - X(J)) + (Y(I) - Y(J)) * (Y(I) -
               Y(J))
  470          LET Z(I1) = 1 / SQR (P3) - 1 / SQR (P3 + D2)
  480      NEXT I
  490  NEXT J
  500  FOR J = 1 TO NN
  510      FOR I = J TO N2 STEP NN
  520      NEXT I
```

```
530    NEXT J
540    FOR J = 1 TO NN
550        FOR I = 1 TO NN
560            LET I1 = (J - 1) * NN + I
570            LET A(J,I) = Z(I1)
580        NEXT I
590        LET B(J,J) = 1
600    NEXT J
610    FOR J = 1 TO NN
620        FOR I = J TO NN
630            IF A(I,J) < > 0 THEN
               670
640        NEXT I
650        PRINT "SINGULAR MATRIX"
660        GOTO 980
670        FOR K = 1 TO NN
680            LET S = A(J,K)
690            LET A(J,K) = A(I,K)
700            LET A(I,K) = S
710            LET S = B(J,K)
720            LET B(J,K) = B(I,K)
730            LET B(I,K) = S
740        NEXT K
750        LET T = 1 / A(J,J)
760        FOR K = 1 TO NN
770            LET A(J,K) = T * A(J,K)
780            LET B(J,K) = T * B(J,K)
790        NEXT K
800        FOR L = 1 TO NN
810            IF L = J THEN
               870
820            LET T =  - A(L,J)
830            FOR K = 1 TO NN
840                LET A(L,K) = A(L,K) + T * A(J,K)
850                LET B(L,K) = B(L,K) + T * B(J,K)
860            NEXT K
870        NEXT L
880    NEXT J
890    FOR I = 1 TO NN
900        FOR J = 1 TO NN
910        NEXT J
920    NEXT I
930    FOR J = 1 TO NN
940        FOR I = 1 TO NN
950            LET I1 = (J - 1) * NN + I:
               LET Z(I1) = B(J,I)
960        NEXT I
970    NEXT J
980    FOR I = 1 TO NN
990        LET Q(I) = 0
1000       FOR J = 1 TO NN
1010           LET J1 = (J - 1) * NN + I
1020           LET Q(I) = Q(I) + Z(J1)
1030       NEXT J
1040       LET Q(I) = Q(I) * PE
1050   NEXT I
1060   PRINT
1070   PRINT "*********************"
1080   PRINT
1090   PRINT "INDIVIDUAL CELL CONTRIBUTIONS (pF)"
1100   PRINT
1110   FOR I = 1 TO NN
1120       PRINT  INT (Q(I) * 1E12 + .5)
1130   NEXT I
1140   PRINT
1150   LET Q3 = 0
1160   FOR J = 1 TO NN
1170       LET Q3 = Q3 + Q(J)
1180   NEXT J
1190   LET Q3 = Q3 / (M * M)
1200   PRINT "TOTAL CAP. OF STRUCTURE IS " INT (Q3 * 1E12 + .5)" (pF)"
1210   PRINT
1220   PRINT "*********************"
1230   PRINT
```

```
1240   PRINT "IF YOU REQUIRE ANOTHER"
1250   PRINT "GO TYPE RUN"
1260   PRINT
1270   PRINT "**** END OF PROGRAM ****"
```

```
END-OF-LISTING
```

```
]RUN
```

```
INPUT CAP. GEOMETRY
LENGTH, BREATH AND SPACING
ARE ASSUMED EQUAL
AS ARE CELL SIZES
NORMALISED UNITS USED
THROUGHOUT
?2
```

```
WORKING:---
```

```
**********************
```

```
INDIVIDUAL CELL CONTRIBUTIONS (pF)
```

```
24
24
24
24
```

```
TOTAL CAP. OF STRUCTURE IS 24 (pF)
```

```
**********************
```

```
IF YOU REQUIRE ANOTHER
GO TYPE RUN
```

```
**** END OF PROGRAM ****
```

A modified version of this technique would enable the characteristic impedance of sections of stripline or microstrip to be computed. The program would be run in three stages (i) with a dielectric present, (ii) without a dielectric, finally (iii) the capacitance values calculated in (i) and (ii) can be used to find characteristic impedance under static conditions.

3.8   FURTHER READING

1   Ramo, S., Whinnery, J. R. and van Duzer, T., *Fields and Waves in Communication Electronics,* Wiley, 1967.
Chapter 3 of this book is concerned with the solution of static field problems. Here the method of finite difference solution of Laplace and Poisson equations together with the relaxation method are introduced.

2   Lavenda, B.H., 'Brownian Motion', *Scientific American,* **252**(2), Feb. 1985, 56–67. Looks at the development of probabilistic potential theory and its application to a number of problems in a non-technical way.

3   Bevensee, R. M., 'Probabilistic Potential Theory Applied to Electrical Engineering Problems', *Proceedings of the I.E.E.E.* **61**, (4), April 1973, 423–37.

Royer, G. M., 'A Monte Carlo Procedure for Potential Theory Problems' *I.E.E.E. Transactions Microwave Theory and Techniques,* **19**(10), Oct. 1971, pp. 813–18.
Gives applications and limitations of probabilistic techniques applied to electrical problems including lossy transmission lines.

4 Fidler, J. K. and Nightingale, C., *Computer Aided Circuit Design,* Nelson, 1978.
Section 6.3.3 of this book provides an introduction to statistical analysis including Monte Carlo analysis. A technique for the production of random number generators is discussed.

5 Crawford, T., 'Applesoft Random Function', *Call Apple,* **4** (2), Feb. 1981, 24–7.
This article discusses some of the shortcomings of the intrinsic pseudo random number generator of a particular microcomputer.

6 Wensley, J. H. and Parker, F. W., 'The Solution of Electric Field Problems Using a Digital Computer', *Electrical Energy,* **1** (1), 1956.
Here the digital computer as a powerful number cruncher is discussed together with algorithms that make it particularly amenable for the solution of bounded electrostatic problems of the type discussed in this chapter.

7 Naylor, T. H., *Computer Simulation Techniques*, Wiley, 1966.
Hammersley, J. M. and Handscomb, D. C., *Monte Carlo Methods,* Methuen, 1964.
These books although not written for engineers provide very interesting reading for those concerned with simulation methods that involve the use of Monte Carlo modeling.

8 Harrington, R. F., *Field Computation by Moment Methods*, Macmillan, 1968.
This is a core text for anyone interested in investigating the method of moment solution of field problems. It contains many diverse problems which are systematically investigated and solved. Mixed dielectric problems are considered, these are of special interest to those working on non-TEM lines.

9 Adams, A. T., *Electromagnetics for Engineers,* Ronald Press, 1971.
This is the practical person's version of many classic electromagnetic theory textbooks. Problems are tackled in a realistic fashion and answers obtained.

10 Dwight, H. B., *Tables of Integrals and Other Mathematical Data,* Macmillan, 1957.
These tables are most useful when analytical solutions to electromagnetic problems are being sought.

11 Silvester, P. P. and Ferrari, R. L., *Finite Elements for Electrical Engineers,* Cambridge University Press, 1983.
Most books on the finite element technique for the solution of engineering problems tend to encompass many disciplines. This text is specifically designed for electrical engineering students and contains a number of sample problems. A short computer program written in Fortran is given in chapter 1 in order to illustrate the application of the finite element technique to an actual example.

12 Bajpai, A. C., Mustoe, L. R. and Walker, D., *Engineering Mathematics,* Wiley, 1974.
Anyone programming a Digital Computer to solve mathematical puzzles must be aware

of numerical approximations to a variety of standard mathematical operations. This book provides an elementary and well defined starting point containing Fortran programs for the solution of some numerical problems.

13 Numerical Algorithms Group, *Fortran Library Manual, Mark 10*, **4**, NAG Central Office, Mayfield House, 256 Banbury Road, Oxford, England.

This volume contains section FO1 which deals exclusively with matrix operations including inversion of systems of large sparse matrices.

4

Lumped and Distributed Impedance
Matching Structures and Techniques

One of the major jobs facing a designer of transmission line circuits is that of impedance matching. In chapter 1, it was illustrated how a mismatched transmission line (a line terminated in an impedance other than its characteristic impedance) would reflect some of the incident wave back along the line toward the signal generator. The interaction of the forward waves and reflected waves gave rise to a resultant waveform with nodes and antinodes at fixed points along the length of the transmission line. In this way, standing waves were formed. For lines carrying high power, these standing waves can produce peak current and voltage levels of almost twice those that exist when the line is matched. The peak voltage and current values can result in the destruction of the line insulation or dielectric support material within the line when power levels are large.

Sometimes the wave reflected from a mismatched termination will produce frequency pulling of the signal generator exciting the line, causing the generator to operate at a slightly different frequency to that intended. In a radio receiving system, mismatch of the antenna can cause loss of valuable energy to the r.f. amplifiers resulting in reduced system performance. Often a carefully controlled mismatch can improve overall system performance. This is the case with low noise amplifiers where the active devices employed have characteristics such that their terminal impedance values for the conditions of maximum power gain and minimum noise generation do not coincide. A slightly mismatched device will result therefore in an amplifier with improved noise characteristics operating at reduced gain.

What then is required to accomplish a well controlled impedance match? First, a methodical scheme for determining the degree of mismatch between source and load impedances is needed. Secondly, a method whereby the degree of mismatch can be varied in a known manner under the complete control of the designer. The methods developed in chapter 1 to assess the degree of mismatch in terms of VSWR do not tell the complete story. In order to consider how mismatch can be selectively controlled, it is instructive to have a working knowledge of the Smith Chart.

4.1   MANUAL MATCHING; THE SMITH CHART

Essentially, what is required for an effective matching exercise is an effective way of handling the expression governing the impedance transformation that occurs along a length of transmission line with known termination (equation (1.30)). The difficulty with this equation is that the evaluation of hyperbolic functions of complex numbers is not an easy task especially if hand calculation is to be used. In 1939, P. H. Smith, an engineer with the Bell Telephone Laboratories, developed a method for the graphical presentation of transmission line data [1]. The break-through made by Smith was that for passive loads (loads having a reflection co-efficient of one or less) the graphical presentation was bounded. This was unlike many of the graphical charts available at that time. The Smith Chart remains the most versatile and common matching chart used today. Figure 4.1 shows a commercially available Smith Chart. At first sight this appears to be a complicated web of unrelated information. To unravel the Smith Chart, it is necessary to go through its logical development.

Essentially, the Smith Chart is a plot of the voltage reflection governed by equation (1.37). Rewriting this equation

$$\Gamma = \frac{Z_T - Z_o}{Z_T + Z_o}$$

normalizing the load impedance to Z_o gives

$$\Gamma = \frac{Z_T}{Z_o} - 1 \left/ \frac{Z_T}{Z_o} + 1 \right. \tag{4.1}$$

Now let

$$\Gamma = u + jv \quad \text{and} \quad \frac{Z_T}{Z_o} = Z_n = R_n + jX_n$$

where subscript n denotes normalization with respect to the characteristic impedance of the line. Remember that reflection coefficient and impedance will in general be complex quantities.

Substituting into equation (4.1) gives

$$u + jv = \frac{(R_n - 1) + jX_n}{(R_n + 1) + jX_n}$$

Equating real and imaginary parts gives

$$\left.\begin{array}{l} R_n(u - 1) - X_n v = -(u + 1) \\ \text{and} \\ R_n v + X_n(u - 1) = -v \end{array}\right\} \tag{4.2}$$

Rearranging the first expression in terms of the reactive element gives

$$X_n = \frac{R_n(u - 1) + (u + 1)}{v}$$

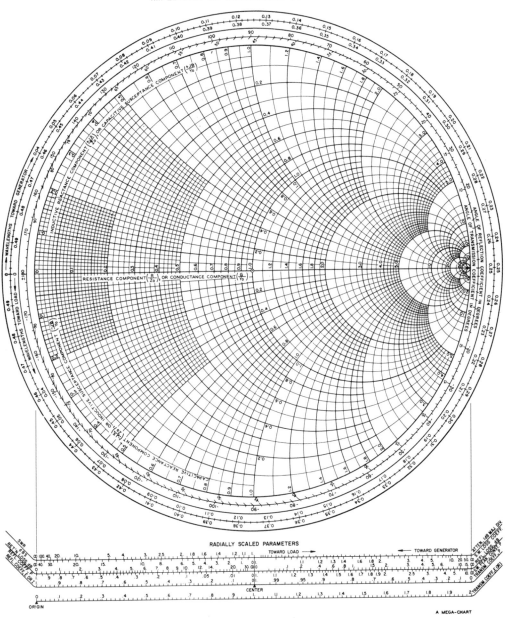

IMPEDANCE OR ADMITTANCE COORDINATES

Figure 4.1   A commercially available form of the Smith Chart with normalized impedance
coordinates (Copyrighted by Kay Electric Company, Pine Brook, N.J. and
reproduced with their permission)

Substituting this into the second expression results in

$$R_n v + \frac{[R_n(u-1) + (u+1)](u-1)}{v} = -v$$

After expanding and gathering terms in u^2 and V^2

$$v^2(R_n + 1) - 2uR_n + u^2(R_n + 1) = 1 - R_n$$

Dividing this by $R_n + 1$ gives

$$v^2 - \frac{2uR_n}{1 + R_n} + u^2 = \frac{1 - R_n}{1 + R_n}$$

Now

$$\left(u - \frac{R_n}{1 + R_n}\right)^2$$

gives

$$u^2 + \frac{R_n{}^2}{(1 + R_n)^2} - \frac{2uR_n}{1 + R_n} = 0$$

Therefore

$$v^2 + \left(u - \frac{R_n}{1 + R_n}\right)^2 = \frac{R_n{}^2}{(1 + R_n)^2} + \frac{(1 - R_n)(1 + R_n)}{(1 + R_n)(1 + R_n)}$$

and, finally, the desired relationship

$$v^2 + \left(u - \frac{R_n}{1 + R_n}\right)^2 = \frac{1}{(1 + R_n)^2} \tag{4.3}$$

When plotted on the u–v plane in cartesian coordinates, equation (4.2) represents a circle with center

$$u = \frac{R_n}{R_n + 1} \qquad V = 0$$

and radius

$$\frac{1}{(R_n + 1)}$$

For different values of R_n, a family of circles can be generated. These circles represent normalized resistance loci, some of which are shown in figure 4.2.

From equation (4.1) it can be seen that all the circles of constant normalized resistance must lie along the u-axis and must pass through the point (1,0).

If the development for the resistance loci is repeated for the normalized reactive element X_n by eliminating R_n from equation (4.2), then equation (4.4) will result

$$(u - 1)^2 + \left(v - \frac{1}{X_n}\right)^2 = \left(\frac{1}{X_n}\right)^2 \tag{4.4}$$

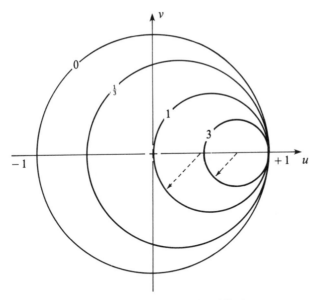

Figure 4.2   Circles of Constant Normalized Resistance

Examination of equation (4.4) reveals once again a series of circles, this time centered at

$$u = 1 \qquad v = \frac{1}{X_n}$$

and having radius

$$\frac{1}{X_n}$$

This time the circles have their centers located along the vertical asymptote placed at $u = 1$ (see figure (4.3)).

In figure 4.3, inductive reactances are placed above the horizontal line and capacitive elements denoted by negative reactances are placed below the same line.

It is possible also to overlay onto the Smith Chart constant VSWR contours as shown in figure 4.4.

Equation (4.1) implies for a matched system the reflection coefficient will be zero and VSWR will be unity. When plotted on a Smith Chart, a VSWR of unity is a point at the center of the chart. Remember that the Smith Chart is really a plot of the reflection coefficient. For a reflection coefficient of one the VSWR will be infinite and will overlay the contour for zero resistance. Notice how, near to the edges of the Smith Chart, the rate of change of VSWR begins to increase rapidly.

Now that the basics of the Smith Chart have been established it is instructive to look for landmarks on a commercially available chart. This is best done using figure 4.1 in conjunction with figures 4.3 and 4.4. In figures 4.3 and 4.4 the point

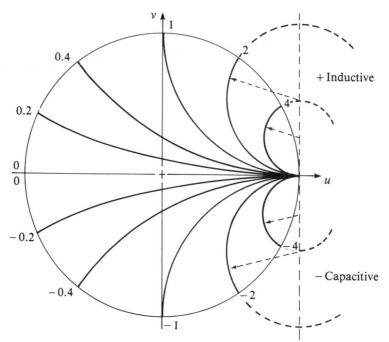

Figure 4.3 Circles of Constant Normalized Reactance

$u = 1$, $v = 0$, the normalized resistance and reactance terms, are large and in figure 4.1 the contours appear crowded. This is the position for an open circuit, i.e. infinite impedance. At position $u = 0$, $v = 1$, a value of $0 + j1$ is encountered. This represents a perfect inductive reactance. Similarly, $u = 0$, $v = 1$ represents a perfect capacitive reactance. Position $u = -1$, $v = 0$ is seen to intersect the zero resistance and zero reactance contours, i.e. this is the short circuit position. The center of the Smith Chart $u = 0$, $v = 0$ represents a normalized impedance of $1 + j0$ which, under normal circumstances, represents the transmission line characteristic impedance, assumed to be real. Around the Smith Chart periphery the angle of reflection

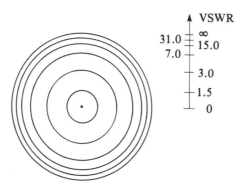

Figure 4.4 Circles of Constant VSWR

coefficient in degrees is marked. This ranges from $\pm 180°$ at the short circuit to zero degrees at the open circuit position. In this way, reflection coefficient can be plotted directly on the chart in polar coordinates.

In terms of the general sending end impedance equation (1.30), it is interesting to note that one revolution on the Smith Chart is equivalent to moving a distance of one-half guide wavelength along the line. The sending end impedance is, therefore, cyclic as was illustrated in figure 1.11. The Smith Chart can be utilized to find the impedance along a transmission either away from or toward the generator with the minimum of computation. A graphics program, program 4.1 SMITH, enables impedance points to be plotted on a computer generated Smith Chart. This program, although designed specifically for a BBC model B microcomputer can, by modification to the graphics routine, be run on most microcomputers that will support graphics.

```
][ FORMATTED LISTING
FILE: PROGRAM 4.1    Smith
PAGE-1


  10 REM
  20 REM SMITH CHART GENERATION
  30 REM    IMPEDANCE CHART
  40 REM
  50 MODE 1
  60 CLS
  70 N=60
  80 H=410
  90 K=700
 100 R=400
 110 G=1
 120 FLAG=0
 130 U=1E20
 140 D=2*PI/N
 150 MOVEK-R,H
 160 PRINTTAB(1,5)"                           "
 170 PLOT21,K+R,H
 180 PRINTTAB(0,1)"**SMITH CHART--IMPEDANCE COORDINATES**"
 190 PRINTTAB(7,12)"+j"
 200 PRINTTAB(6,19)"S/C"
 210 PRINTTAB(7,26)"-j"
 220 PRINTTAB(35,19)"0/C"
 230 MOVEK-12,H+10:VDU5:PRINT"+":VDU4
 240 GOTO1040
 250 FLAG=0
 260 PRINTTAB(0,6)"        "
 270 PRINTTAB(1,2)"ENTER NORMALISED RESISTANCE VALUE"
 280 PRINTTAB(1,3)"IF THIS VALUE > 1 THEN ENTER 0 ELSE 1"
 290 INPUT TEST
 300 PRINTTAB(1,2)"                                        "
 310 PRINTTAB(1,3)"                                        "
 320 PRINTTAB(2,4)"               "
 330 IF TEST = 0 THEN 400
 340 PRINT TAB(1,2)"ENTER DENOMINATOR VALUE"
```

```
350 PRINTTAB(1,3)"RESISTANCE LOCI"
360 INPUT DEN
370 U=DEN+1
380 G=DEN/U
390 GOTO1040
400 PRINTTAB(1,2)"ENTER NUMERATOR VALUE"
410 PRINTTAB(1,3)"RESISTANCE LOCI"
420 INPUT DEN
430 DEN=1/DEN
440 U=DEN+1
450 G=DEN/U
460 GOTO1040
470 PRINTTAB(1,3)"                                              "
480 PRINTTAB(0,4)"                                              "
490 PRINTTAB(1,2)"ENTER NORMALISED INDUCTIVE REACTANCES"
500 INPUTB:A=0
510 IFB=0THENB=.0001
520 MOVEK+R,H
530 GOTO1340
540 PRINTTAB(1,3)"                                              "
550 PRINTTAB(0,4)"                                              "
560 PRINTTAB(1,2)"ENTER NORMALISED CAPACITIVE REACTANCES"
570 INPUTB:A=0:FLAG=1
580 IFB=0THENB=.0001
590 MOVEK+R,H
600 GOTO1340
610 REM
620 REM PLOT IMPEDANCE
630 REM
640 PRINTTAB(0,4)"                                              "
650 PRINTTAB(1,2)"DO YOU WISH TO PLOT                           "
660 PRINTTAB(1,3)"AN IMPEDANCE POINT                            "
670 PRINTTAB(1,4)"IF YES ENTER 1 ELSE 0                         "
680 INPUT TEST
690 PRINTTAB(0,2)"                                              "
700 PRINTTAB(0,3)"                                              "
710 PRINTTAB(0,4)"                                              "
720 PRINTTAB(0,5)"                                              "
730 PRINTTAB(0,6)"           "
740 IF TEST=0THEN 250
750 PRINTTAB(1,2)"ENTER RESISTANCE VALUE"
760 INPUTA
770 IFA=0THENA=1E-10
780 PRINTTAB(1,2)"ENTER REACTANCE VALUE (OHMS)"
790 INPUTB
800 IFB=0THENB=1E-10
810 PRINTTAB(1,2)"ENTER NORMALISATION RESISTANCE (OHMS)"
820 INPUTZO
830 IFZO=0THEN810
840 PRINTTAB(1,2)"                                              "
850 PRINTTAB(1,3)"                                              "
860 PRINTTAB(1,4)"                                              "
870 REM MAP IMPEDANCE ONTO CHART
880 A=A/ZO:B=B/ZO
890 REAL=(A+1)*(A-1)+B*B
```

```
 900 IM=B*((A+1)-(A-1))
 910 N=(A+1)*(A+1)+B*B
 920 REAL=REAL/N*R
 930 IM=IM/N*R
 940 MOVEREAL+K-10,IM+H+10:VDU5:PRINT"*":VDU4
 950 PRINTTAB(1,2)"ALL IMPEDANCE POINTS DONE ?"
 960 PRINTTAB(0,3)"    "
 970 PRINTTAB(1,3)"IF YES ENTER 0 ELSE 1"
 980 INPUTTEST
 990 PRINTTAB(0,2)"                                            "
1000 PRINTTAB(0,3)"                                            "
1010 PRINTTAB(0,4)"                                            "
1020 PRINTTAB(0,5)"         "
1030 IFTEST=0THEN250ELSE750
1040 REM CIRCLE GENERATION
1050 MOVE K+R/U,H+G*R
1060 FOR M=0 TO 2*PI+D STEP D
1070    Y=H+G*R*COS(M)
1080    X=K+G*R*SIN(M)+R/U
1090    PLOT5,X,Y
1100    NEXTM
1110 IF U=1E20 THEN 270
1120 PRINTTAB(1,2)"                                        "
1130 PRINTTAB(1,3)"                                   "
1140 PRINTTAB(1,4)"                   "
1150 PRINTTAB(1,2)"ALL RESISTANCE LOCI DONE ?"
1160 PRINTTAB(1,3)"IF YES ENTER 0 ELSE 1"
1170 INPUT TEST
1180 IF TEST=0 THEN 470 ELSE 270
1190 REM POSITIVE REACTANCE CIRCLES
1200 FOR M =PI TO   PI+AS STEP D/8
1210    Y=H+R/A+R/A*COS(M)
1220    X=K+R+R/A*SIN(M)
1230    PLOT21,X,Y
1240    NEXTM
1250 PRINTTAB(1,2)"
1260 PRINTTAB(0,3)"
1270 PRINTTAB(1,2)"ALL INDUCTIVE LOCI DONE ?"
1280 PRINTTAB(1,3)"IF YES ENTER 0 ELSE 1"
1290 INPUTTEST
1300 IFTEST=0 AND FLAG=0 THEN 540 ELSE 470
1310 IFTEST=0 AND FLAG=1 THEN 610 ELSE 540
1320 GOTO470
1330 REM BI-LINEAR TRANSFORM
1340 REAL=(A+1)*(A-1)+B*B
1350 IM=B*((A+1)-(A-1))
1360 N=(A+1)*(A+1)+B*B
1370 REAL=REAL/N*R
1380 IM=IM/N*R
1390 A=B
1400 X=SQR((R-REAL)^2+IM^2)
1410 Y=R/A
1420 Z=R/A
1430 AS=ACS((Z^2+Y^2-X^2)/2/Z/Y)
1440 IF FLAG=0 THEN 1200 ELSE 1460
```

```
1450 REM NEGATIVE REACTANCE CIRCLES
1460 FOR M = 2*PI TO 2*PI-AS STEP -D/8
1470    Y=H-R/A+ABS(R/A)*COS(M)
1480    X=K+R+ABS(R/A)*SIN(M)
1490    PLOT21,X,Y
1500    NEXTM
1510 PRINTTAB(0,2)"
1520 PRINTTAB(0,3)"
1530 PRINTTAB(1,2)"ALL CAPACITIVE LOCI DONE ?"
1540 PRINTTAB(1,3)"IF YES ENTER 0 ELSE 1"
1550 INPUTTEST
1560 IF TEST=0 THEN 610 ELSE 540
>
```

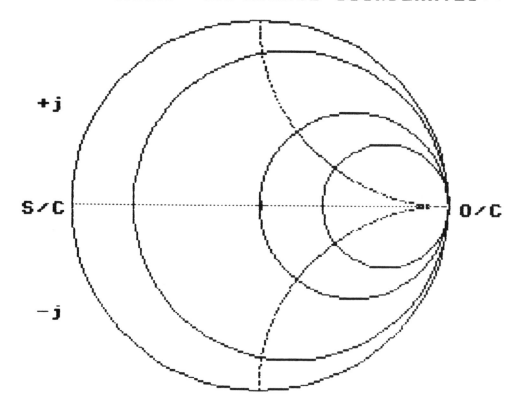

✦✦SMITH CHART--IMPEDANCE COORDINATES✦✦

To move a given impedance toward the load, movement around the Smith Chart is in an anticlockwise direction. For transformation towards the generator move clockwise. The corresponding scales on the Smith Chart are calibrated in terms of a normalized guide wavelength ranging from zero to one-half wavelength, i.e. one complete revolution.

In order to give commercial Smith Charts general appeal, the load or source impedance of interest are not located directly on the chart. Instead, a normalized impedance is plotted. The normalized impedance is simply the impedance of interest divided by the characteristic impedance of the system into which it is desired to work.

* * *

Example 4.1

Given a transmission line with characteristic impedance of 50 ohms, plot the following quantities on a Smith Chart.

(a) $Z = 15 + j25$

(b) The conjugate of Z, Z^*

(c) The admittance $Y = \dfrac{1}{Z}$

(d) The admittance equivalent to Z^*

Solution

(a) $Z = 15 + j25$

First, normalize to the characteristic impedance of the line (50 ohms in this case)

$$z_n = \frac{15 + 25j}{50} = 0.3 + 0.5j$$

The point on the Smith Chart represented by this normalized impedance is the intersection of the circle of constant resistance, value 0.3, and the constant inductance circle, value 0.5. This is shown as point z_n in figure 4.5.

(b) The conjugate of Z is

$$Z^* = 15 - 25j$$

This impedance represents a resistive component with a 25 ohm capacitive reactance associated with it. To plot this point on the Smith Chart, first normalize to the characteristic impedance of the transmission line

$$z_n^* = 0.3 - 0.5j$$

Notice how this lies on the same circle of constant resistance as z_n, the imaginary component now lies on the locus of constant capacitive reactance, value $-0.5j$, this locates z_n^*. Looking at the relationship between z and z_n^* on the Smith Chart, it can be seen that when it is necessary to plot a conjugate impedance, all that is required is to reflect the normalized reactive component of the impedance about the zero reactance locus (the horizontal axis) while maintaining the same locus of constant resistance.

NAME	TITLE	DWG. NO.
SMITH CHART FORM 82-BSPR (9-66)	KAY ELECTRIC COMPANY, PINE BROOK, N.J., © 1966. PRINTED IN U.S.A.	DATE

IMPEDANCE OR ADMITTANCE COORDINATES

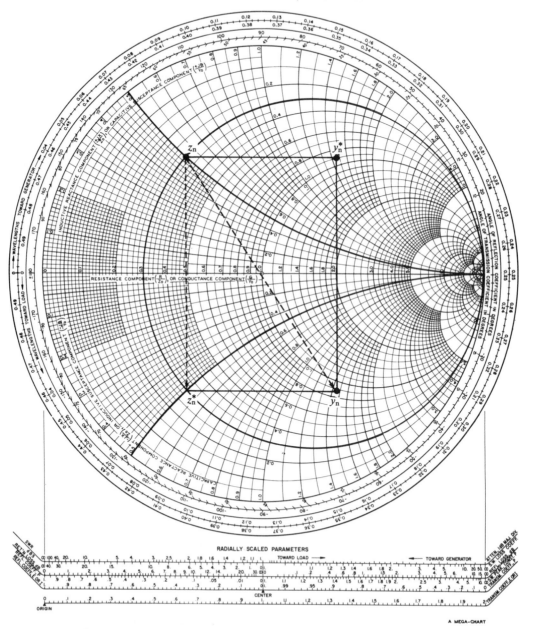

Figure 4.5   Smith Chart for Example 4.1

(c) The admittance Y

$$Y = \frac{1}{Z} = \frac{1}{15 + 25j} = \frac{15 - 25j}{15 + 25j} = (17.7 - 29.4j) \text{ mS}$$

This admittance must now be normalized to the characteristic admittance of the line, Y_o where

$$Y_o = \frac{1}{Z_o}$$

Hence $y_n = \dfrac{Y}{Y_o} = YZ_o$

$$\therefore \quad y_n = 50(17.7 - 29.4j) \text{ mS}$$
$$= (0.885 - j1.47) \text{ S}$$

Again, visual inspection of the Smith Chart shows that the point of normalized admittance is actually a reflection of z_n along the line passing through z_n and the center of the Smith Chart. This is a useful result since it means the Smith Chart can be used for graphical manipulation of complex numbers. In this way, reciprocals of complex numbers are easily found by plotting on a Smith Chart.

(d) Finally, to plot y_n^* all that is needed is to reflect y_n about the line of infinite reactance. Inspection of figure 4.5 shows z_n, z_n^*, y_n and y_n^* form a rectangle. This will always be the case when a complex term is normalized to some given number. The area of the rectangle formed will depend on the number used for normalization and on the complex term involved.

*   *   *

4.1.1 The Admittance Chart

In the above example, it was seen that a conversion of an impedance to an admittance involved a rotation of $180°$ on the Smith Chart, this is equivalent to a physical distance of one-quarter wavelength. If every point on a normalized impedance Smith Chart is mapped onto a new chart obtained by applying a $180°$ rotation, then the normalized admittance chart is formed. Here, lines of constant conductance and susceptance are used. Figure 4.6 shows the important points on the admittance chart; figure 4.7 represents a commercially available form of the admittance chart, this line with Standard Smith Chart coordinates.

Repeating example 4.1, but this time plotting points on a commercial admittance chart results in the location of points as shown in figure 4.8. Here, series to parallel conversions have been made to accommodate the scales in figure 4.8.

In practical design situations, it is often the case where the design will alternate between an impedance chart and admittance chart. Usually, series circuit elements are discussed on the impedance chart, while shunt components are discussed in

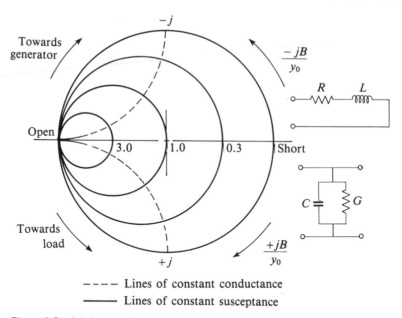

Figure 4.6   Admittance Form of the Smith Chart, normalized admittance coordinates

terms of conductance and susceptance on the admittance chart. To make life even more difficult it is quite usual to use the Smith Chart in its impedance form with overlaid portions of the admittance chart to facilitate the design. This implies that when designing on a Smith Chart one should be prepared to alternate between impedance and admittance representation. Example 4.10 in the next section gives an example of this type of design exercise. Switching from one set of coordinates to another can easily cause problems and errors are likely to occur, so be careful.

So far in this section the Smith Chart has been used to demonstrate the plotting of impedance or admittance values. The next type of problem to be examined is the one where the Smith Chart comes into its own. This is the calculation of sending end impedance. In other words, for the graphical evaluation of equation (1.30), this can be carried out with comparative ease provided that a good compass and rule are at hand. A graphical evaluation is particularly useful if tables of complex hyperbolic functions or a computer is not available. It is also useful in its own right since it gives a feel for the solution to the problem that otherwise may have been overlooked. The basic problem involves transforming a load impedance normalized to the transmission line characteristic impedance along a known electrical length of line. The next example illustrates the procedure.

* * *

Example 4.2

A transmission line with characteristic impedance of 100 ohms has a load impedance of $200 - j100$ ohms. The line is known to be 0.2 guide wavelengths long at the design

NAME	TITLE	DWG. NO.
SMITH CHART FORM 82-ESPR (9-66)	KAY ELECTRIC COMPANY, PINE BROOK, N.J. © 1966 PRINTED IN U.S.A.	DATE

IMPEDANCE OR ADMITTANCE COORDINATES

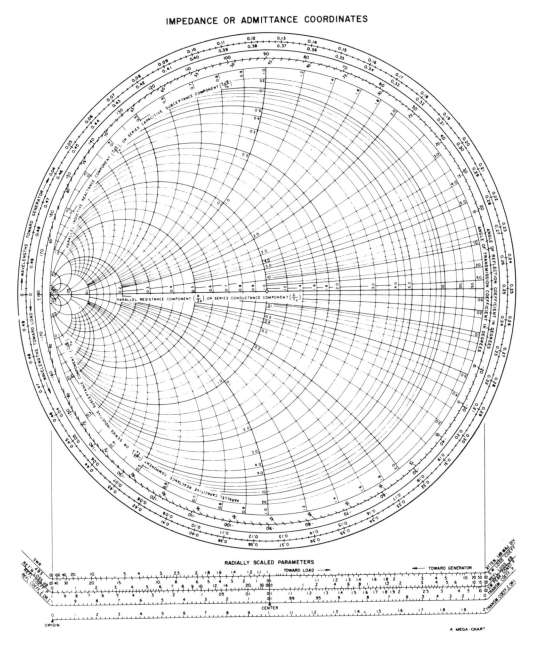

Figure 4.7   A commercially available form of the Admittance Chart (copyrighted by Kay Electric Company, Pine Brook, N.J. and reproduced with their permission)

NAME	TITLE	DWG. NO
SMITH CHART FORM 82-ESPR(9-66)	KAY ELECTRIC COMPANY, PINE BROOK, N.J. © 1966 PRINTED IN U.S.A.	DATE

IMPEDANCE OR ADMITTANCE COORDINATES

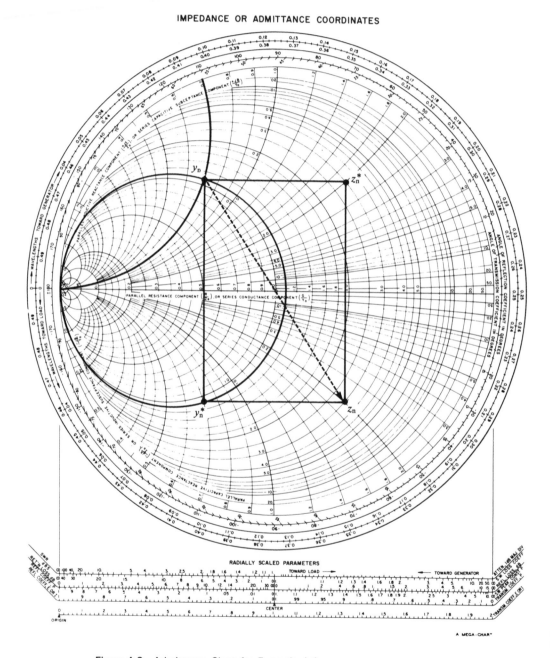

RADIALLY SCALED PARAMETERS

A MEGA-CHART

Figure 4.8   Admittance Chart for Example 4.1

frequency. Find the impedance that will be seen looking into the end of the line. Assume the line has negligible loss.

Solution

(a) Normalize the load impedance to the characteristic impedance of the line.

$$z_n = \frac{200 - j100}{100} = 2 - j1$$

This is located on the Smith Chart in figure 4.9 as point A.

(b) Extend a straight line from the center of the chart through point A to the periphery point B.

(c) Since the transformation under examination involves moving away from the load impedance along the line, i.e. toward the generator, then the outer scale (wavelengths towards the generator) is examined. The radial line constructed in part (b) intersects this scale at 0.287 λg.

(d) The final transformation of interest in this problem occurs on moving 0.2 λg further towards the generator, i.e. in a clockwise direction. Since no loss is involved, point A will move along a circle of constant VSWR that passes through it. The normalized sending end impedance will then be located on this VSWR circle along the radial line intersection, point C. Point C must lie at $(0.287 + 0.2)$ λg

(e) Point D, the normalized sending end impedance, can now be read from the chart as

$$0.39 - j0.7$$

or denormalized to give a sending end impedance of

$$(39 - j70) \text{ ohms}$$

It is useful to note that the value of VSWR for the circle constructed through point A can be obtained directly from the scale on the lower left-hand side of the chart. This is achieved by dropping a perpendicular line from the leftmost edge of the circle to the standing wave ratio scale (point E). Here, a VSWR of 2.4 (7.6 db) is obtained.

An important point to remember in any design is the direction in which the transformation is to occur, i.e. toward the load (anticlockwise) or toward the generator (clockwise).

Consider once again example 4.2, but this time with loss introduced into the line. Assume 10 dB loss per guide wavelength. To find the sending end impedance for the lossy line, the design proceeds exactly as before but with the addition of several extra steps.

Continue to step (e) as before.

(f) 0.2 λg of line is equivalent to 0.2 × 10 dB or 2 dB attenuation. Since the load is to be transferred toward the generator, any loss encountered will reduce the

NAME	TITLE	DWG. NO
SMITH CHART FORM 82-BSPR(9-66)	KAY ELECTRIC COMPANY, PINE BROOK, N.J., ©1966 PRINTED IN U.S.A.	DATE

IMPEDANCE OR ADMITTANCE COORDINATES

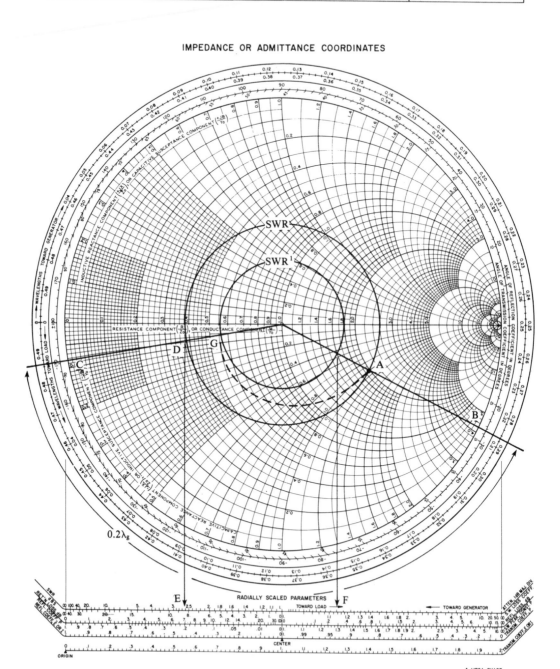

Figure 4.9   Smith Chart for Example 4.2

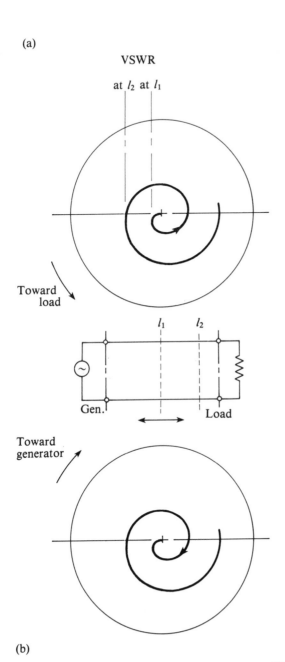

Figure 4.10 Effect of Attenuation on Smith Chart VSWR (a) Generator to Load (Toward Load) (b) Load to Generator (Toward Generator)

amount of incident as well as reflected power in the line so that a reduction in the radius of the VSWR circle will occur. The transmission loss scale, in one dB steps, is used for the calculation. Transmission loss is the loss in signal power as a function of distance along the line. When measured at different points along the line, this is assumed constant. Returning to the problem in hand, in order to finish the lossy line calculation it is necessary to move toward the center of the Smith Chart by 2 dB or two divisions on the transmission loss scale (point F) in figure 4.7. Next, extend a vertical line to the line of zero reactance.

(g) Where the vertical line intersects the line of zero reactance, construct the new SWR circle (this has value 1.7 or 4.6 dB).

(h) To finish, select the point where the radial line passing through C, D intersects the SWR[1] circle, point G. This gives the new value for sending end impedance as seen looking into a line with loss. The normalized value of this impedance is

$$0.55 - j0.05$$

or denormalized

$$(55 - j5) \text{ ohms}$$

From this example, it is evident that calculations performed on lines having loss are, in general, less accurate than those performed for lossless lines. This is due mainly to the coarseness of the transmission loss scale. Another point worthy of note is that normally only the end points, i.e. between specified positions on the line are required for the calculation to be valid.

When attenuation as a function of line length is plotted on the Smith Chart, it takes the form of a spiral, figure 4.10. So that for lossy line calculations, instead of moving along circles of constant VSWR, it is necessary to move along a spiral. The pitch of the spiral depends on the position on the chart and on the line attenuation per guide wavelength. Using only two points on the spiral equivalent to specified positions along the line length reduces the complexity of drawing a spiral to that of locating two points so that provided the transmission loss scale is available this becomes an almost trivial exercise.

* * *

To further illustrate the use of the Smith Chart, consider the solution of a problem which often arises in the areas of UHF and microwave measurement. Stated simply, the problem is to estimate the value of a load impedance from a measurement taken at some point distant from the unknown load. The technique used to solve this type of problem graphically is demonstrated in the next example.

* * *

Example 4.3

At a test frequency of 450 MHz, a 50 ohm transmission line has been used to obtain certain information about an unknown load impedance. With a short circuit placed at the end of the line it was noted that a measured minimum in the resulting VSWR pattern was observed at a distance of 12 centimeters from the end of the line that had a short circuit attached. When the short circuit was replaced by the unknown impedance, the VSWR moved to 28 centimeters from the end of the line at which the load had been placed. Under these conditions, the VSWR was measured to be 2.5. Using the experimental information obtained, calculate with the aid of a Smith Chart the value of the unknown load impedance. It can be assumed that the line used is of high quality so that attenuation can be ignored. Also, assume the line to be air-spaced so that phase velocity in the line is approximately equal to the speed of light in a vacuum, 3.0×10^{10} cm/s. The actual practicalities of the VSWR measurement are of no concern for this example.

Solution

(a) Select a VSWR value of 2.5 from the standing wave voltage ratio scale, point A in figure 4.11.

(b) Construct the VSWR circle.

(c) Convert the physical shift between voltage minima obtained for short circuit and unknown load conditions to electrical length

$$\lambda_g = \lambda = \frac{c}{f} = \frac{3 \times 10^{10}}{450 \times 10^6} = 66.7 \text{ cm}$$

distance moved away from load $= (28 - 12)$ cm
$$= 0.24 \, \lambda$$

(d) The short circuit used to locate the first voltage minimum represents point B on the impedance chart.

(e) Move a distance equal to the shift in minima calculated in step (c) away from the load (toward the generator), point C.

(f) Read off the normalized impedance point D equivalent to the unknown load impedance and denormalize to get the desired impedance

$$Z_{LOAD} = 50(2.5 + j0.3) = (125 + j15) \text{ ohms}$$

(g) If required, loss could have been included in the evaluation in the same way as it was accounted for in the previous example.

* * *

Armed with the examples in this section together with the other examples to be described in the rest of this chapter, a broad class of impedance matching and transmission line problems become immediately soluble.

The Smith Chart is of fundamental importance to those concerned with the effective design of transmission line circuits and should be mastered at an early stage. Unlike most automated attempts at solutions to transmission line problems,

NAME	TITLE		DWG. NO.
SMITH CHART FORM 82-BSPR(9-66)	KAY ELECTRIC COMPANY, PINE BROOK, N.J., ©1966. PRINTED IN U.S.A.		DATE

IMPEDANCE OR ADMITTANCE COORDINATES

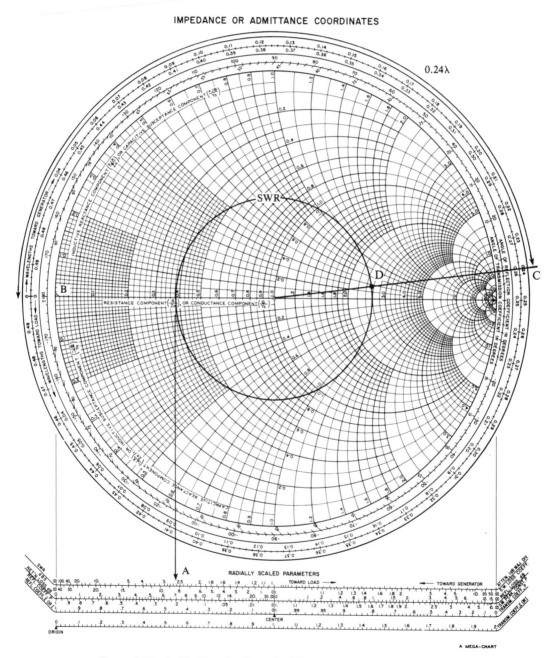

Figure 4.11   Smith Chart for Example 4.3

the Smith Chart enables an experienced designer to get to grips with a particular problem and to explore a number of alternative solutions. In this way, a ball park solution can be found which can then be refined by computer-aided methods.

4.2   THE COMPRESSED SMITH CHART

The usual form of presentation of the Smith Chart is an impedance or admittance plot displaying loci of constant resistance, reactance or loci of constant conductance, susceptance. Normally, when passive components such as lumped elements or transmission line sections are encountered, the chart is presented in such a way that the maximum reflection coefficient is unity. In this form, the chart is compactly represented within a bounding circle.

When the load is active, e.g. a semiconductor such as a tunnel or Gunn diode or perhaps an FET, then the possibility of a reflection coefficient of magnitude greater than unity exists. To accommodate these possibilities, an open form of the Smith Chart called the compressed Smith Chart is used. In a compressed Smith Chart, the standard chart described in the last section forms only part of the overall chart. In the open form, the compressed chart is of infinite extent. In view of this, it is standard practice to bound the chart. This is done by imposing a limit on the maximum value of reflection coefficient that will exist in a design. A value of 3.16 is normally selected for the reflection coefficient as this allows a significant portion of the negative resistance contours that will be encountered in practice to be plotted (figure 4.12). For active loads producing negative resistance, the region outside the unit circle is the area of interest. In this region, the definition of VSWR is redundant and it is customary to work in terms of impedance or reflection coefficient only.

The main application of the compressed Smith Chart is in the design of oscillators and negative resistance amplifiers. As with most designs of the type discussed in this text, impedance matching is the dominant issue. When designing oscillator or amplifier circuitry around an active device, it is possible to use one-port or two-port matching networks depending on the active device selected. A transistor, be it a GaAs FET or a bipolar, can be considered as a one-, two- or even a three-port system dependent on the terminal configuration selected (see, for instance, section 4.6). For transistor type circuit elements standard design techniques employing positive and negative feedback to form oscillators and amplifiers are widely used. At high r.f. frequencies the one-port, two-terminal network, and one-port device are of considerable importance in the design of reflection amplifiers and self-excited oscillators. The one-port device usually takes the form of a microwave diode such as the Gunn or Impatt.

In general, the negative conductance of active elements either of the single or multiple type behave in a similar fashion. The dependence of device negative conductance or reactance on r.f. drive level across the terminals of an active element is shown in quantitative terms in figure 4.13.

At low values of r.f. voltage drive across the active element, the magnitude of

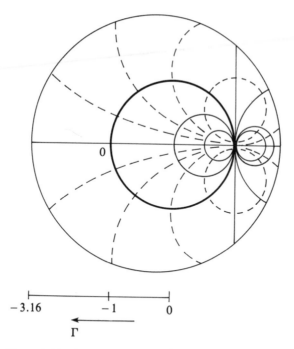

Figure 4.12   Compressed Smith Chart

negative conductance or resistance will be a maximum. This is known as the small signal condition. As r.f. voltage drive is increased, the negative resistance or conductance available at the device port of interest reduces in magnitude. The device is said to be operating in large signal mode. If it is assumed that load and circuit reactance have been selected to be resonant then, if the real port load impedance presented by the circuit to the device has a magnitude greater than the small signal

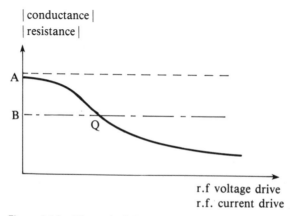

Figure 4.13   Effect of r.f. Drive Level on Device Negative Conductance or Resistance

magnitude of the device resistance or conductance then amplification can occur, point A in figure 4.13 indicates the situation. If the circuit resistance presented to the active load represents a smaller magnitude than the small signal device resistance or conductance (point B, figure 4.13), then the device resistance curve and the circuit load line will intersect, point Q.

Under these circumstances when first switched on (r.f. drive level small) the net resistance in the circuit will be negative and any noise present at the device terminals will be amplified. This will continue until the r.f. drive level in the system is such that the net resistance or conductance in the circuit is positive and the r.f. drive across the device terminals is attenuated. With the drive level across the device reduced the net resistance or conductance in the system once again becomes negative so that amplification occurs. In this way the r.f. drive across the device is held fixed about point Q and the oscillation sustained. The resonant circuit formed by the device and load reactances filter the r.f. drive signal so that sinusoidal oscillations, are produced. In order to maintain reliable start-up conditions, device negative resistance or conductance should be approximately 20 percent greater than the real part of the circuit load impedance.

The argument developed above can be concisely stated in the form of a number of simple equations. First, for start up

$$R_{\text{device small signal}} > 1.2\,R_{\text{load}}$$

Second, for resonance

$$R_{\text{device}} + R_{\text{load}} = 0$$
$$X_{\text{device}} + X_{\text{load}} = 0$$

If a relatively high-quality factor is assumed for the resonant circuit then circuit losses must be small. Therefore, a series-to-parallel circuit transformation can be determined.

$$Y = G + jB = \frac{1}{Z} = \frac{1}{R + jX}$$

hence

$$G + jB = \frac{R}{X^2} + \frac{1}{jX}$$

here R^2 is assumed to be small. Thus $R \propto G$ and $B \propto \dfrac{1}{X}$

As a result of this duality, a second set of conditions for self-excited oscillation similar to those for the series resonant circuit can be stated for the case of parallel resonance. These are

$$G_{\text{device small signal}} = 1.2\,G_{\text{load}}$$

for resonance

$$G_{\text{load}} = |-G_{\text{device}}|$$

and

$$B_{\text{load}} = | - B_{\text{device}} |$$

Examination of these equations reveals that the design of an oscillator circuit is simply another form of impedance matching problem which can be aided considerably by the compressed Smith Chart.

$$* \quad * \quad *$$

Example 4.4

Design an oscillator, given that the small signal reflection coefficient for the one-port active device to be used is $1.3 \angle 135°$.

Solution

Consider a series resonant circuit (figure 4.14). Restating the series resonant conditions necessary for oscillation

$$R_D + R_L = 0$$
$$X_D + X_L = 0$$

and redefining these in terms of reflection coefficients gives

$$\Gamma_L = \frac{Z_L - Z_o}{Z_L + Z_o} = \frac{R_L + jX_L - Z_o}{R_L + jX_L + Z_o}$$

also

$$\Gamma_D = \frac{Z_D - Z_o}{Z_D + Z_o} = \frac{R_D + jX_D - Z_o}{R_D + jX_D + Z_o}$$

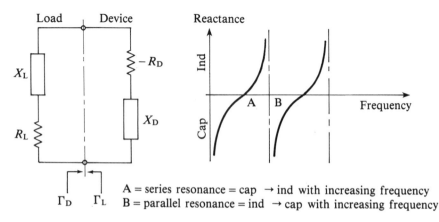

A = series resonance = cap → ind with increasing frequency
B = parallel resonance = ind → cap with increasing frequency

Figure 4.14   Series Resonant Oscillator Circuit and Resonant Impedance Characteristics

but from the resonance conductance stated above

$$\Gamma_L = \frac{-R_D - jX_D - Z_o}{-R_D - jX_D + Z_o}$$

therefore

$$\Gamma_L \Gamma_D = \frac{R_D + jX_D + Z_o}{R_D + jX_D - Z_o} \cdot \left(\frac{-1}{-1}\right) \cdot \frac{R_D + jX_D - Z_o}{R_D + jX_D + Z_o} = 1$$

This form for the resonant condition is very convenient for use with the compressed Smith Chart.

Returning to the problem under consideration

$$\Gamma_D = 1.3 \angle 135°$$

∴   for a series resonant oscillator

$$\Gamma_D \Gamma_L = 1 \angle 0° = [1.3(k)] \angle [135 + (-135)]°$$

Hence

$$k = 0.769$$

This represents the small signal drive condition. For an oscillator, remember that normal operation will occur under large signal drive condition so that

$$R_D > 1.2\, R_L$$

to ensure reliable start up. Hence Γ_D will become $1.04 \angle 135°$ assuming that device reactance is independent of r.f. drive level

∴ $\Gamma_L = 0.96 \angle 135°$

Both Γ_D and Γ_L are shown on the compressed Smith Chart (figure 4.15). Once Γ_L has been established then any suitable matching structure can be synthesized to interface the device to the load. The choice of matching structure will normally be based on an appraisal of system requirements, e.g. size, electrical loss, etc. Simple stub matching circuitry using quarter wavelength or half wavelength lines will probably suffice for most simple situations encountered in practice.

* * *

This example could have been carried out equally well for an oscillator displaying parallel resonance. Here the impedance inversion properties of the Smith Chart (valid for the compressed chart) make this an easy task. The specified reflection coefficient of the example is first converted to a normalized admittance and then moved to the line of zero susceptance. Finally, matching of the conjugate device impedance to the load is achieved as before by synthesis of a suitable matching circuit.

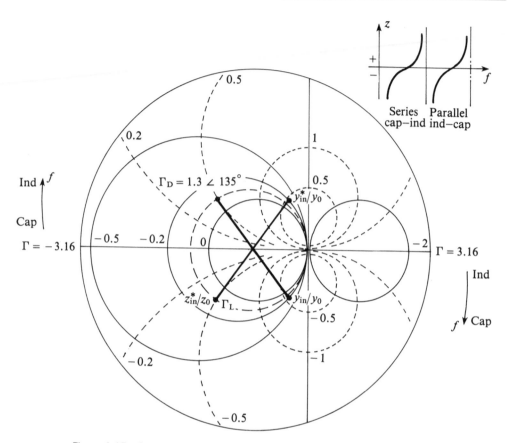

Figure 4.15   Compressed Smith Chart for Oscillator Design Example

4.3   LUMPED MATCHING CIRCUITS

At low frequencies, say below 400 MHz, some special case lumped circuits can prove to be useful. For these circuits, it is worth providing an alternative to the Smith Chart for matching purposes. The Smith Chart really comes into its own when distributed matching circuits are to be designed. Lumped matching circuits can consist of reactive or of resistive elements or a mixture of both.

When purely resistive circuits are used, impedance matching is obtained at the expense of introducing a reduction in power transmitted to the load from the generator. This reduction in power is termed attenuation. The next section will discuss a class of attenuator circuits capable of reducing power levels in a controlled manner without shifting impedance levels. In this section, a simple attenuator circuit will be developed that can match unequal impedances. Some purely reactive circuits will be analyzed to illustrate the design principles (section 4.3.2).

4.3.1 Resistive L Section Matching Circuit

This circuit consists of two resistors and represents the simplest type of matching section (figure 4.16).

The section is to be used to match between characteristic resistances R_{01} and R_{02}. For input match

$$R_{01} = R_1 + \frac{R_2 R_{02}}{R_2 + R_{02}}$$

$$= \frac{R_1 R_2 + R_1 R_{02} + R_2 R_{02}}{R_2 + R_{02}}$$

For output match

$$R_{02} = \frac{R_2 (R_1 + R_{01})}{R_1 + R_2 + R_{01}}$$

$$= \frac{R_1 R_2 + R_2 R_{01}}{R_1 + R_2 + R_{01}}$$

Hence

$$R_2 R_{01} + R_{01} R_{02} - R_1 R_2 - R_1 R_{02} - R_2 R_{02} = 0$$

$$- R_2 R_{01} + R_{01} R_{02} - R_1 R_2 + R_1 R_{02} + R_2 R_{02} = 0$$

Adding

$$R_{01} R_{02} = R_1 R_2$$

Substituting

$$R_2 (R_{01} - R_{02}) = R_1 R_{02} = \frac{R_{02}{}^2 R_{01}}{R_2}$$

$$\therefore \quad R_1 = [R_{01}(R_{01} - R_{02})]^{\frac{1}{2}}$$

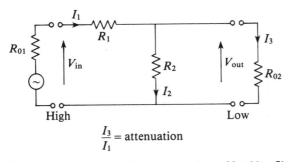

$$\frac{I_3}{I_1} = \text{attenuation}$$

Figure 4.16   Resistive L Section Impedance Matching Circuit

and

$$R_2 = \left(\frac{R_{01}R_{02}^2}{R_{01} - R_{02}}\right)^{1/2}$$

where R_{01} and R_{02} are specified by the nature of the problem.

If the circuit is designed according to these last equations then none of the incident wave will be reflected when the circuit is inserted between two dissimilar characteristic resistances. The attenuation produced by the circuit is calculated based on the current distribution within the system. If the input voltage to the circuit is V_{in} then by Kirchoff's laws

$$I_1 = V_{in}\bigg/\left[R_1 + \left(\frac{R_2R_{02}}{R_2 + R_{02}}\right)\right]$$

giving

$$\frac{V_{out}}{V_{in}} = \frac{R_2R_{02}}{(R_2 + R_{02})R_1 + R_2R_{02}}$$

hence

$$\text{attenuation} = 20\log_{10}\left[\frac{R_2R_{02}}{R_{02}(R_1 + R_2) + R_1R_2}\right] \text{ dB}$$

* * *

Example 4.5

Design an asymmetrical resistive matching network that can match a 50 ohm section of transmission line to one having a 75 ohm characteristic impedance. Calculate the attenuation that occurs on insertion of the designed circuit.

Solution

R_{02} = lower of the impedances = 50 ohm

R_{01} = larger of the impedances = 75 ohm

hence

$R_1 = [75(75 - 50)]^{1/2} = 43.3$ ohm

$R_2 = \left(\frac{75 \times 50}{75 \times 50}\right)^{1/2} = 5.5.$ ohm

$$\text{attenuation} = 20\log_{10}\left[\frac{5.5 \times 50.0}{50.0(43.3 + 5.5) + (43.3)(5.5)}\right] \approx 20 \text{ dB}$$

Therefore, only one-hundredth of the power originally transmitted reaches the second section of line.

* * *

4.3.2 Reactive Matching Circuits

In this section L, PI and TEE lumped sections will be used to form matching circuits. The basis for these circuits is that for any series circuit consisting of a series resistance R_s and reactance X_s there can be found an equivalent circuit comprising a parallel resistance R_p and reactance X_p. Figure 4.17 shows the situation.

From figure 4.17 it can be deduced for the series circuit that

$$Z = R_s + jX_s$$

$$\therefore \quad |Z| = (R_s^2 + X_s^2)^{\frac{1}{2}} \tag{4.5}$$

For the parallel circuit

$$Z = \frac{jX_pR_p}{R_p + jX_p}$$

$$\therefore \quad |Z| = \frac{X_pR_p}{(R_p^2 + X_p^2)^{\frac{1}{2}}} \tag{4.6}$$

Equations (4.5) and (4.6) are equivalent so that

$$\frac{X_pR_p}{(R_p^2 + X_p^2)^{\frac{1}{2}}} = (R_s^2 + X_s^2)^{\frac{1}{2}} \tag{4.7}$$

In order to simplify the above expression it is necessary to define quality factor Q for both the series and parallel networks in figure 4.17.

A useful definition for quality factor is stated as

$$Q = 2\pi f \, \frac{\text{stored energy in a network}}{\text{energy lost per second}}$$

When related to the series LR circuit in figure 4.17 this becomes

$$Q = \frac{X_s}{R_s} \tag{4.8}$$

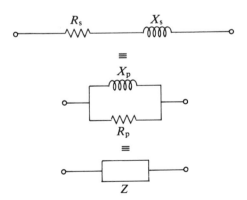

Figure 4.17   Series and Parallel LR Circuits

and for the parallel LR circuit

$$Q = \frac{R_p}{X_p} \tag{4.9}$$

Equations (4.8) and (4.9) are combined to form

$$\frac{R_p}{R_s} = Q^2 + 1 \tag{4.10}$$

Equation (4.10) is very significant since it indicates that any two resistances R_p, R_s can be matched by carefully controlling the Q of a matching network.

*   *   *

Example 4.6

Transform a 50 ohm resistive load to a 600 ohm resistance by adjusting the Q of a matching circuit. Design an L section circuit to implement the transformation at a frequency of 400 MHz.

Solution

$$\frac{R_p}{R_s} - 1 = \frac{600}{50} - 1 = Q^2$$

$$\therefore \quad Q = (11)^{\frac{1}{2}} = 3.317$$

Now, since $X_s = QR_s$, then $X_s = 50 \times 3.317 = 166$ ohms where the series circuit $R_s = 50$ ohms, also

$$X_p = \frac{R_p}{Q} = \frac{600}{3.317} = 181 \text{ ohms}$$

where for the parallel circuit $R_p = 600$ ohms.

    For the case of an L section matching network, either of the circuits shown in figure 4.18 can be used with equal facility.

    The choice of network (a) or network (b), as shown in figure 4.18, is normally a practical one based on calculated component values. For most applications circuit (b) is preferred from a systems point of view since the shunt capacitance in this circuit acts to reject any harmonics in the excitation signal. Circuit (a) is useful where a d.c. path to ground is needed, this is provided by the shunt inductor. For this reason circuit (a) is used if active devices requiring bias are present in the circuit. From figure 4.18 for circuit (b)

$$X_L = 3.317 \times 50 = 166 \ \Omega = X_s$$

$$\therefore \quad L = \frac{X_L}{2\pi f} = \frac{166}{2\pi 400 \times 10^6} = 66 \text{ nH}$$

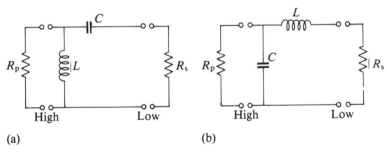

Figure 4.18 Equivalent L Networks

Similarly

$$X_c = \frac{600}{3.317} = 181 \ \Omega = X_p$$

$$\therefore \ C = \frac{1}{2\pi f X_c} = \frac{1}{2\pi 400 \times 10^6 \times 181} = 2.2 \ \text{pF}$$

Repeating the design for circuit (a) gives

$$X_s = X_c = 166 \ \Omega$$
$$\therefore \ C = 2.4 \ \text{pF}$$
$$X_p = X_L = 181 \ \Omega$$
$$\therefore \ L = 72 \ \text{nH}$$

The major problem with L–C matching circuits is that circuit Q is determined only by the ratio of the input and output circuit impedances. This is undesirable since the circuit will be able to match unequal impedances only. It is useful also to be able to maintain circuit Q at a value between 10 and 20. A low circuit Q implies that any harmonics present in the input signal are not readily suppressed. A very high Q may lead to increased circuit losses due to the large currents that can flow in the circuit at resonance. Where high Q networks or where large impedance transformation are required, often a good alternative is to cascade two or more low Q sections.

A TEE circuit can be used to overcome some of the problems stated for the L circuit. Figure 4.19 shows a generalized TEE circuit. This can be dissected into two L circuits terminated by a hypothetical resistance R_H. This dissection is very convenient since the rules that apply to the simple L circuit can be applied to this problem. The only point to note is that the hypothetical resistance must have a higher resistance than either R_1 or R_2, i.e. the resistances to be matched. This restriction follows as a result of the discussion on L sections which showed that

$$\frac{R_H}{R_{1,2}} = Q^2 + 1$$

It should be noted that the parallel combination of $X_3{}^1$ and $X_3{}^{11}$ form the shunt reactance X_3. To proceed with the design, it is assumed that the circuit Q, resistance R_1 and resistance R_2 are all known.

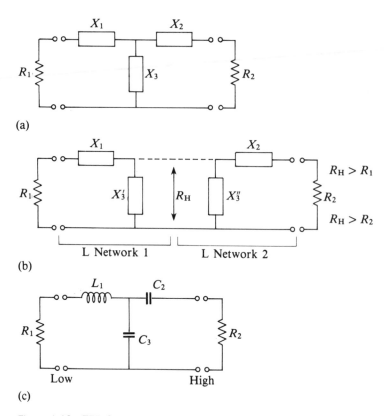

(a)

(b)

(c)

Figure 4.19   TEE Section Matching Circuit (a) General Circuit (b) Dissected PI Section
(c) Nominal Circuit

Now for the first L section, proceed as follows bearing in mind the arguments used for the L section.

$$\frac{R_H}{R_1} = Q_1^2 + 1$$

(c.f. equation (4.10)). R_1 is known and a value for Q_1 is assumed. From this the hypothetical resistance R_H is found

$$R_H = R_1(Q_1^2 + 1) \qquad\qquad (4.11)$$

At this point, if R_H is found to be less than R_2, select a higher value for Q_1 and start again. Now

$$X_3 = \frac{R_H}{Q_1} \qquad\qquad (4.12)$$

(c.f. equation (4.9)). Also

$$X_1 = R_1 Q_1 \qquad\qquad (4.13)$$

(c.f. equation (4.8)). For the second L-section

$$\frac{R_H}{R_2} = (Q_2{}^2 + 1).$$

Since R_H and R_2 are known then a value for Q_2, the Q of the second L section, is forthcoming

$$Q_2{}^2 = \frac{R_H}{R_2} - 1 \tag{4.14}$$

Now that Q_2 is known, the shunt reactance of the second section can be found

$$X_3{}^{11} = \frac{R_H}{Q_2} \tag{4.15}$$

To complete the second section, the series L section component X_2 is found

$$X_2 = Q_2 R_2 \tag{4.16}$$

Finally, the shunt reactance for the composite section can be computed as the parallel combination of the two intermediate reactances used for convenience in the preceding development

$$X_3 = \frac{X_3{}^1 X_3{}^{11}}{X_3{}^1 + X_3{}^{11}} \tag{4.17}$$

* * *

Example 4.7

Repeat example 4.6 but this time use the TEE section given in figure 4.19 to form the matching network.

Solution

Assume a value of 10 for Q_1

$$R_1 = 50 \ \Omega, \ R_2 = 600 \ \Omega, \ f = 400 \ \text{MHz}$$

Using equations (4.11) to (4.17) gives

(a) $R_H = 50(100 + 1) = 5050$ ohms
   is $R_H > R_1$ or R_2? Yes, then proceed.

(b) $X_3{}^1 = \dfrac{5050}{10} = 505$ ohms

(c) $X_1 = 50 \times 10 = 500$ ohms

(d) $Q_2 = 2.723$

(e) $X_3{}^{11} = \dfrac{5050}{2.723} = 1854$ ohms

(f) $X_2 = 2.723 \times 600 = 1634$ ohms

(g) $X_3 = \dfrac{1854 \times 505}{1854 + 505} = 397$ ohms

In figure 4.19(c)
X_1 is realized by an inductor

$$\therefore \quad L_1 = \frac{X_1}{2\pi f} = \frac{500}{800 \times 10^6 \times \pi} = 199 \text{ nH}$$

X_{C_3} by a capacitor

$$\therefore \quad C_3 = \frac{1}{2\pi f X_{C_3}} = 1 \text{ pF}$$

and for X_2, $C_2 = 0.24$ pF. This completes the design.

*   *   *

Now that suitable design techniques have been developed to cope with L section and TEE section matching circuits, consider next the PI section; figure 4.20 shows a form of the circuit suitable for discussion.

Repeating the development of a design methodology for the TEE section but applied to the PI section this time, the PI section is dissected into two distinct L sections. Each of these may now be analyzed.

If the Q of the first L section Q_1 is assumed known as are the resistances to be matched, then

$$X_1 = \frac{R_1}{Q_1} \tag{4.18}$$

also

$$\frac{R_1}{R_H} = Q_1^2 + 1$$

giving

$$R_H = \frac{R_1}{Q_1^2 + 1} \tag{4.19}$$

If the value obtained for R_H is greater than either R_1 or R_2, increase Q_1 and recalculate R_H. Next

$$X_3{}^1 = R_H Q_1 \tag{4.20}$$

After this, the Q for the second L network is calculated

$$\frac{R_2}{R_H} = (Q_2^2 + 1)$$

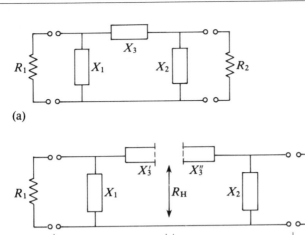

(a)

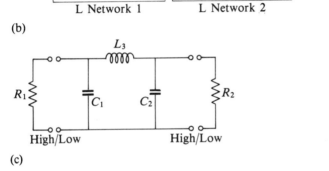

L Network 1   L Network 2

(b)

High/Low   High/Low

(c)

Figure 4.20 PI Section Matching Circuit (a) General Circuit (b) Dissected PI Section (c) Nominal Circuit

from which

$$Q_2{}^2 = \frac{R_2}{R_H} - 1 \tag{4.21}$$

The series reactance $X_3{}^{11}$ for the second section is found next

$$X_3{}^{11} = R_2 Q_2 \tag{4.22}$$

and the shunt reactance X_2 becomes

$$X_2 = \frac{R_2}{Q_2} \tag{4.23}$$

Finally, $X_3{}^1$ and $X_3{}^{11}$, the series reactances of the composite section, are combined to give X_3

$$X_3 = X_3{}^1 + X_3{}^{11} \tag{4.24}$$

* * *

Example 4.8

Repeat example 4.7 but match using the PI circuit shown in figure 4.20 part (c).

Solution

Assume $Q_1 = 10$
$R_1 = 50$ ohm, $R_2 = 600$ ohm, $f = 400$ MHz
Equations (4.19)–(4.24) yield

(a) $X_1 = \dfrac{50}{10} = 5 \ \Omega$

(b) $R_H = \dfrac{50}{(100+1)} = 0.5 \ \Omega \qquad R_H < R_1 \ \text{or} \ R_2$

(c) $X_3{}^1 = 0.5 \times 10 = 5 \ \Omega$

(d) $Q_2 = 35$

(e) $X_3{}^{11} = 21 \ \text{k}\Omega$

(f) $X_2 = 17 \ \Omega$

(g) $X_3 = 17 \ \text{k}\Omega$

(h) $C_1 = 80 \ \text{pF}$

(i) $C_2 = 23 \ \text{pF}$

(j) $L_1 = 6.8 \ \mu\text{H}$

The selection of the PI or TEE matching circuit is the designer's choice.

* * *

In order to demonstrate the utility of the equations derived in this section and also to indicate some of the side-effects associated with matching into non-resistive loads, a design example containing a reactive termination and reactive source will now be considered.

* * *

Example 4.9

An amplifier operating at 137 MHz and producing 4 watts of power is to be fitted in line with a section of 50 ohm characteristic impedance cable. A simplified circuit diagram for the arrangement is given in figure 4.21. From the manufacturer's data for the transistor, its input impedance at 137 MHz has been measured to be $(1.5 + j1.2) \ \Omega$, with a collector to emitter capacitance C_L of 60 pF when operated from a 9 volt d.c. supply. Design a suitable input and output matching network.

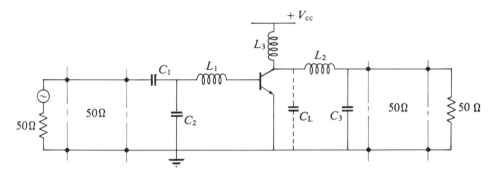

Figure 4.21 Simplified VHF Amplifier

Solution

For the output circuit 4.0 watts of power are required from a 9 volt power supply. Therefore, the output resistance of the transistor is approximately

$$R_L = \frac{V^2}{2 \times 4} = 10 \text{ ohms}$$

The Q of the output circuit is

$$Q = \left(\frac{50}{10} - 1\right)^{\frac{1}{2}} = 2$$

$$X_{L_2} = 2 \times 10 = 20 \ \Omega$$

$$X_{C_3} = 25 \ \Omega$$

Now L_3 must resonate with the collector to emitter capacitance

$$\frac{1}{2\pi f C_L} = X_{L_3}$$

hence

$$X_{L_3} = \frac{1}{2\pi 137 \times 10^6 \times 60 \times 10^{-12}} = 19.4 \ \Omega$$

also

$$L_2 = \frac{20}{2\pi 137 \times 10^6} = 23 \text{ nH}$$

$$L_3 = \frac{19.4}{2\pi 137 \times 10^6} = 23 \text{ nH}$$

$$C_3 = \frac{1}{25 \times 2\pi \times 137 \times 10^6} = 46 \text{ pF}$$

For the input network, the requirement is for 50 Ω to match $(1.5 + j1.2)$ Ω, that is a 1.5 Ω resistance in series with an inductor of reactance $1.2j$ Ω.

First match 50 Ω to 1.5 Ω. Select a Q of 10

$$R_H = 101 \times 1.5 = 152 \ \Omega$$

$$X_{C_1} = 1.5 \times 10 = 15 \ \Omega$$

$$\therefore \quad C_1 = 77 \ \text{pF}$$

$$X_{C_3} = 15.2 \ \Omega$$

$$Q_2 = 1.43$$

$$X_{C_3}{}^{11} = 106.3 \ \Omega$$

$$X_{L_1} = 71.5 \ \Omega$$

$$X_{C_2} = X_{C_3}{}^{1} + X_{C_3}{}^{11} = 121.5$$

$$\therefore \quad C_2 = 10 \ \text{pF}$$

The 1.2 Ω inductive reactance in series with the base terminal of the transistor must be absorbed into the matching network. This is achieved by subtracting the excess inductance from that already calculated. Hence

$$X_{L_1}{}^{1} = 71.5 - 1.2 = 70.3 \ \Omega$$

$$\therefore \quad L_1 = 82 \ \text{nH}$$

This completes the design.

* * *

It should be pointed out that although analytical design techniques have been used in this section to obtain the desired matching, Smith Chart techniques are valid for matching circuits which comprise lumped elements. As an example of the use of the Smith Chart in a lumped element impedance matching problem consider example 4.10.

* * *

Example 4.10

With the aid of a Smith Chart match a 50 Ω transmission line to an r.f. transistor having an input reflection coefficient measured to be 0.67 $\angle -158°$ and operating at 1 GHz. The circuit configuration shown in figure 4.21 should be used.

Solution

First plot the measured impedance on the Smith Chart, point A, figure 4.22.

NAME	TITLE	DWG. NO.
SMITH CHART FORM 82-BSPR (9-66)	KAY ELECTRIC COMPANY, PINE BROOK, N.J., © 1966. PRINTED IN U.S.A.	DATE

IMPEDANCE OR ADMITTANCE COORDINATES

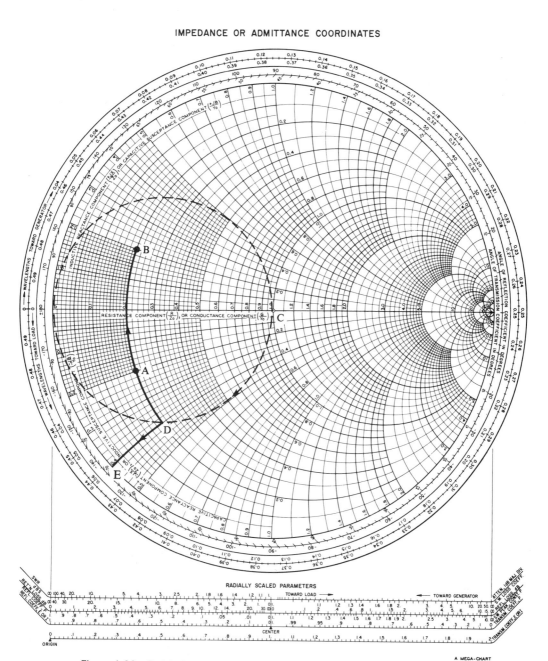

RADIALLY SCALED PARAMETERS

A MEGA-CHART

Figure 4.22   Smith Chart for Example 4.10

This corresponds to a normalized input impedance of $(0.2 - j0.2)$ or to $(10 - j10)$ Ω in a 50 Ω system. For maximum power transfer to occur, the conjugate of the input impedance is used $(0.2 + j0.2)$, point B. The problem now is to transform from point C, the center of the Smith Chart, to point B. The reflection of the unit resistance circle about a vertical line running through the center of the chart represents the circle of constant conductance, normalized value 1.0 S. In figure 4.21, it can be seen that the active device constitutes the source and the transmission line the load. For the purpose of demonstration, move along the dotted line, the circle of constant conductance, from load to source until some point D is reached. Point D is found as the point where the arc of constant conductance intersects the arc of constant resistance passing through point B. Arc CD represents the effect of a shunt capacitance and has a value of 50 mS. Arc DB represents the effect of the series inductance in the matching network in figure 4.21. The value for the required inductance is equal in magnitude to the residual capacitance resulting from a move from C to D. Found as point E in figure 4.22, in this case (0.4×50) Ω. Hence

$$X_C = \frac{1000}{50} = 20 \text{ Ω}$$

$$\therefore \quad C = 8 \text{ pF}$$

$$X_L = 20 \text{ Ω}$$

$$\therefore \quad L = 3.3 \text{ nH}$$

Note: For simplicity, capacitor C_1 has been set to a large value.

* * *

It should be appreciated that many solutions to this matching problem exist and the final topology selected for a particular job should be chosen only after several different matching networks have been thoroughly investigated in order to find that most suited to the particular design under consideration.

4.3.3 Attenuator Design

An alternative to the design of lossless matching structures is to use attenuator networks to minimize the reflections between source and load impedances. Such attenuator networks, called pads, work but should be used wisely since useful power will have to be dissipated. Apart from minimizing reflections, attenuator circuits are useful circuit elements in their own right. For example attenuators can be used for the calibration of signal strength meters in radio equipment or for providing reference levels for certain types of noise substitution measurements.

The design of a useful class of attenuator circuits, called symmetrical attenuators, can be based on TEE and PI transmission line equivalent circuits

(section 1.3). One useful property of attenuator sections is that they should introduce zero phase shift between input and output ports when inserted in circuit. This
means that the propagation constant for these circuits will be a real quantity. A
commonly used measure of attenuation is the decibel.

$$\text{Loss in decibels} = 10 \log_{10}\left(\frac{\text{power at input}}{\text{power at output}}\right) \text{dB}$$

or for a symmetrical circuit inserted between matched terminations

$$\text{Loss in decibels} = 20 \log_{10}\left(\frac{\text{input current}}{\text{output current}}\right) \text{dB}$$

or

$$20 \log_{10} N \text{ dB}$$

where $N = \exp$ (loss in nepers) (see section 1.2).

Normally when designing attenuator sections, it is usual to specify
characteristic line impedance and desired attenuation as the design parameters.

In section 1.3 an equivalence for the arms of a TEE network in terms of the
characteristic impedance of a transmission line Z_o was given. For the TEE section
attenuator shown in figure 4.23, all the circuit elements are real, so that no phase
shift will be encountered by signals passing through the network. Thus the phase
shift constant β will be zero, leaving a propagation constant of α.

Applying the equivalence given in section 1.3 for the TEE section to figure 4.23
yields

$$R_1 = R_0 \tanh\left(\frac{\alpha l}{2}\right) = R_0\left[\frac{\exp(\alpha l/2) - \exp(-\alpha l/2)}{\exp(\alpha l/2) + \exp(-\alpha l/2)}\right]$$

$$= R_0\left[\frac{\exp(\alpha l) - 1}{\exp(\alpha l) + 1}\right]$$

From the definition of the neper $N = \exp(\alpha l)$ then

$$R_1 = R_0\left(\frac{N-1}{N+1}\right) \tag{4.25}$$

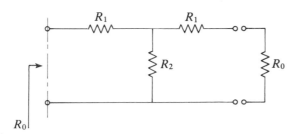

Figure 4.23   TEE Section Symmetrical Attenuator Section

In a similar fashion

$$R_2 = R_0 \frac{1}{\sinh(\alpha l)}$$

$$= \frac{2R_0}{\exp(\alpha l) - \exp(-\alpha l)}$$

$$= \frac{2R_0 \exp(\alpha l)}{\exp(2\alpha l) - 1}$$

$$R_2 = R_0 \left(\frac{2N}{N^2 - 1} \right) \tag{4.26}$$

The derived results given as equations (4.25) and (4.26) can be used to determine desired values for attenuator circuit components, for circuits of the type shown in figure 4.23. The component values calculated from equations (4.25) and (4.26) will almost certainly differ from preferred resistor values. Choosing the nearest preferred resistor value for R_1, R_2 usually gives acceptable performance. Where the experimental results of an attenuator section, constructed using preferred values, differs greatly from the theoretical design, it may be better to try the PI configuration or to use series or parallel combinations of resistors to form R_1 and R_2.

A scheme for assessing the effect on attenuation and input impedance of an attenuator circuit constructed from resistors having preferred values will be discussed later in this section. For now, return to figure 4.23 redrawn as figure 4.24 for a discussion on power dissipation within the resistive elements of a TEE section attenuator network.

The power dissipated by each resistor in the attenuator network is easily calculated from Kirchoff's laws, provided that the input power into the attenuator is known. From figure 4.24 for an input power of W watts

$$V_{in} = (R_0 W)^{\frac{1}{2}}$$

where V_{in} is the input voltage and R_0 the characteristic resistance of the section

$$I_{in} = V_{in} / \left[R_{1A} + \frac{(R_{1B} + R_0)R_2}{R_{1B} + R_2 + R_0} \right]$$

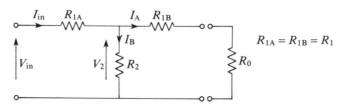

Figure 4.24   TEE Attenuator Section

this means the power dissipation in resistor $R_{1A}P_{R1}$, can be found as

$$P_{R1A} = I_{in}^2 R_{1A} \qquad (4.27)$$

Noting that

$$V_2 = V_{in} - I_{in}R_{1A}$$

then the power dissipated in resistor R_2, P_{R2} is given as

$$P_{R2} = \frac{V_2{}^2}{R_2} \qquad (4.28)$$

and finally, the power dissipated in resistor R_{1B} can be found either directly from a knowledge of the section attenuation and the input power level or by network analysis to be

$$P_{R1B} = \left(\frac{V_2}{R_{1B} + R_0}\right)^2 R_{1B} \qquad (4.29)$$

Analysis of symmetrical PI sections for attenuator applications follows along the same lines as that used for symmetric TEE sections. Figure 4.25 gives the notation for the PI section.

With reference to the PI section in section 1.3, the circuit components in figure 4.25 can be expressed as

$$R_2 = Z_o \sinh{(\alpha l)}$$

and

$$R_1 = Z_o \coth\left(\frac{\alpha l}{2}\right)$$

for pure resistive components.

Rewriting these expressions for R_1 and R_2 in terms of exponential functions yields

$$R_2 = R_0 \frac{\exp{(\alpha l)} - \exp{(-\alpha l)}}{2} = \frac{R_0}{2} \frac{\exp{(2\alpha l)} - 1}{\exp{(2\alpha l)}}$$

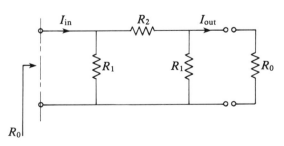

Figure 4.25   PI Attenuator Section

but remember that $\exp(\alpha l) = N$, therefore

$$R_2 = R_0 \left(\frac{N^2 - 1}{2N} \right) \tag{4.30}$$

and for R_1

$$R_1 = R_0 \frac{\exp(\alpha l/2) + \exp(-\alpha l/2)}{\exp(\alpha l/2) - \exp(-\alpha l/2)}$$

$$= R_0 \left(\frac{N + 1}{N - 1} \right) \tag{4.31}$$

The duality between PI and TEE attenuator circuits can be seen on examination of equations (4.25) and (4.31) and equations (4.26) and (4.30). In each case, each pair of equations are of a similar form. For the PI section, equations (4.30) and (4.31) provide the necessary design tools.

The power dissipated in each component in the symmetrical PI attenuator circuit can, as was done for the TEE circuit, be calculated from Kirchoff's laws. This is left as an exercise for the reader.

The choice of a TEE or a PI circuit will depend on the realizability of components. For high impedances the value of components in the shunt arms is small while for a PI circuit the series arm components values become large. When constructing attenuator circuits, resistors should be of the low inductance type, carbon film resistors are a good choice. All lead lengths should be minimized in order to reduce stray coupling. Stray coupling can be a problem in high attenuation stages where it tends to reduce the amount of actual attenuation achieved. To avoid this, high value attenuation stages are normally formed from several lower value stages in cascade.

When preferred values have to be used to approximate calculated values for PI and TEE attenuator circuits, it is useful to be able to assess their effect on circuit performance before construction proceeds. In other words, what happens to attenuation and input impedance when values other than those selected by synthesis equations (4.25), (4.26), (4.30) and (4.31) are used to construct symmetrical attenuator sections?

First investigate figure 4.23 here for input matching

$$R_0 = R_1 + \frac{R_2(R_1 + R_0)}{R_1 + R_2 + R_0}$$

$$\therefore \quad R_0 R_1 + R_0^2 + R_0 R_2 = R_1^2 + \bar{R}_1 R_0 + R_1 R_2 + R_1 R_2 + R_0 R_2$$

from this

$$R_0 = (R_1^2 + 2R_1 R_2)^{1/2} \tag{4.32}$$

also from figure 4.24

$$\frac{I_a}{I_{in}} = \frac{R_2}{R_0 + R_1 + R_2} \tag{4.33}$$

$$\therefore \quad \frac{I_{in}}{I_a} = 1 + \frac{R_1}{R_2} + \frac{(R_1{}^2 + 2R_1R_2)^{\frac{1}{2}}}{R_2} \tag{4.34}$$

after substitution for R_0.

From this the attenuation of the TEE section in decibels can be found

$$\text{Attenuation} = 20 \log_{10}\left(\frac{I_{in}}{I_a}\right)$$

Alternatively, from equation (4.33) using the fact that

$$\frac{I_{in}}{I_a} = \exp(\alpha l)$$

then

$$\exp(\alpha l) = \frac{R_0 + R_1 + R_2}{R_2}$$

this yields

$$\exp(-\alpha l) = \frac{1}{\exp(\alpha l)} = 1 \bigg/ \left(1 + \frac{R_1}{R_2} + \frac{R_0}{R_2}\right)$$

Multiplying the top and bottom of this last expression by

$$\left[\left(1 + \frac{R_1}{R_2}\right) - \left(\frac{R_0}{R_2}\right)\right]$$

gives

$$\exp(-\alpha l) = \left[\left(1 + \frac{R_1}{R_2}\right) - \left(\frac{R_0}{R_2}\right)\right] \bigg/ \left[\left(1 + \frac{R_1}{R_2}\right)^2 - \left(\frac{R_0}{R_2}\right)^2\right]$$

substitution of equation (4.32) into the above gives

$$\exp(-\alpha l) = \left(1 + \frac{R_1}{R_2} - \frac{R_0}{R_2}\right) \bigg/ 1$$

Adding this equation to equation (4.33) gives

$$\exp(\alpha l) + \exp(-\alpha l) = 2\left(1 + \frac{R_1}{R_2}\right)$$

$$\therefore \quad \cosh(\alpha l) = 1 + \frac{R_1}{R_2}$$

Hence, total attenuation αl in nepers can be found with the aid of the inverse cosh expansion in appendix B as

$$\alpha l = \cosh^{-1}\left(1 + \frac{R_1}{R_2}\right)$$

$$= \log_e\left\{\left(1 + \frac{R_1}{R_2}\right) + \left[\left(1 + \frac{R_1}{R_2}\right)^2 - 1\right]^{\frac{1}{2}}\right\} \tag{4.35}$$

This can be expressed in dB as $8.686\,\alpha l$.

The form of equation (4.35) can be shown to be valid for a PI attenuator circuit. Try this as an example and show for a PI attenuator circuit that the expression

$$\alpha l = \cosh^{-1}\left(1 + \frac{R_2}{R_1}\right)$$

is valid. The advantage in using equation (4.35) rather than equation (4.34) is that for a PI attenuator circuit the governing equation obtained by network analysis contains a quadratic equation. The solution of the quadratic can be avoided if the inverse hyperbolic cosine function is solved instead.

*   *   *

Example 4.11

Design:

(a) Using a symmetrical PI section attenuator comprising three resistors, design a pad that will produce 20 dB attenuation when placed in series with a transmission line of 50 ohm characteristic impedance.
(b) Repeat part (a) of this question but this time use a TEE section.
(c) For the TEE section attenuator pad designed in part (b), calculate the power dissipation of each resistor in the network. Assume 1 watt of input power.

Solution

Referring to figure 4.25 for a symmetrical PI section attenuator

$R_0 = 50$ ohms

$N = 10^{20/20} = 10$ since 20 dB = 20 $\log_{10}N$

(a) PI section:
from equation (4.31)

$$R_1 = 50\left(\frac{11}{9}\right) = 61 \text{ ohms}$$

equation (4.30) gives

$$R_2 = 50\left(\frac{100 - 1}{20}\right) = 248 \text{ ohms}$$

(b) TEE section:
equation (4.25) gives

$R_1 = 41$ ohms

equation (4.26) gives

$R_2 = 10$ ohms

(c) Power dissipation in the elements of the TEE circuit

$$V_{in} = (P \times R)^{\frac{1}{2}} = 7.1 \text{ volts}$$

$$I_{in} = 7.1 \Bigg/ \left[R_{in} + \frac{(41 + 50)10}{(41 + 50 + 10)} \right] = 140 \text{ mA}$$

$$V_2 = 7.1 - (0.14)(41) = 1.36$$

$$\therefore \quad P_{R_{1A}} = (0.14)^2 \, 41 = 0.803 \text{ W}$$

$$P_{R_2} = 0.185 \text{ W}$$

$$P_{R_1} = 0.009 \text{ W}$$

As a check, the sum of these individual power dissipations should add to the input power $0.803 + 0.185 + 0.009 \approx 1 \text{ W}$

Example 4.12

Assume preferred resistor values for R_1 and R_2 in part (b) of example 4.11. Calculate the input impedance and attenuation of a TEE pad constructed from the preferred components selected.

Solution

$R_1 = 41$ ohms   Preferred value 47 ohms

$R_2 = 10$ ohms   Preferred value 10 ohms

From equation (4.32)

$$R_0 = (47^2 + 2 \times 47 \times 10)^{\frac{1}{2}} = 56 \text{ ohms}$$

and equation (4.34)

$$N = 20 \log_{10}\left(1 + \frac{47}{10} + \frac{56}{10} \right) = 21.1 \text{ dB}$$

or by equation (4.35)

$$N = 8.686 \times 2.4258 = 21.1 \text{ dB}$$

*   *   *

From these calculations it is seen that the inclusion of preferred values in the design has modified the input resistance by about 12 percent and the attenuation by about 6 percent.

A computer program called ATTN, program 4.2, will allow a variety of calculations similar to those carried out in examples 4.11 and 4.12 for both symmetrical PI and symmetrical TEE attenuator pads. The program allows for the

```
][ FORMATTED LISTING
FILE: PROGRAM 4.2 ATTN
PAGE-1


 10  REM
 20  REM  --- ATTN ---
 30  REM
 40  REM   THIS PROGRAM COMPUTES
 50  REM   THE RESISTANCE VALUES
 60  REM   NECESSARY TO CONSTRUCT
 70  REM   SYMMETRICAL TEE AND PI
 80  REM   ATTENUATOR CIRCUITS
 90  REM   THE POWER DISSIPATED BY
100  REM   EACH COMPONENT IS ALSO
110  REM   CALCULATED.
120  REM   THE PROGRAM CAN ALSO
130  REM   FIND THE ATTENUATION
140  REM   AND I/P IMP. OF SYMM.
150  REM   PI AND TEE CCTS.
160  REM   WHEN THEIR COMPONENT
170  REM   VALUES ARE KNOWN.
180  REM
190  REM  RO=CHARAC. LINE RES.
200  REM  R1,R2=RESISTIVE ELEMENTS
210  REM   ALPHA=ATTN. IN DB
220  REM   POWER=I/P POWER (WATTS)
230  REM
240  REM   INPUT DATA
250  REM
260  HOME
270  PRINT
280  PRINT "DO YOU KNOW THE CCT. PARAMETERS"
290  PRINT "IF YES ENTER 1 ELSE 0"
300  INPUT A
310  IF A = 1 OR A = 0 THEN
          320ELSE235
320  IF A = 0 THEN
          340
330  IF A = 1 THEN
          770
340  PRINT "I/P CHARAC. LINE RES.(OHMS)"
350  INPUT RO
360  PRINT "ENTER DESIRED ATTN. IN DB"
370  INPUT ALPHA
380  PRINT "ENTER I/P POWER IN WATTS"
390  INPUT POWER
400  PRINT
410  PRINT "********************"
420  PRINT
430  LET N = 10 ^ (ALPHA / 20)
440  REM   TEE SECTION
450  LET K = 0
460  LET R1 = RO * (N - 1) / (N + 1)
470  LET R2 = RO * 2 * N / (N * N - 1)
480  LET V1 =   SQR (RO * POWER)
490  LET I1 = V1 / (R1 + ((R1 + RO) * R2 / (R1 + R2 + RO)))
500  LET P1 = I1 * I1 * R1
510  LET P2 = (V1 - I1 * R1) * (V1 - I1 * R1) / R2
520  LET P3 = ((V1 - I1 * R1) / (R1 + RO)) ^ 2 * R1
530  PRINT "A SYMMETRICAL TEE SECTION"
540  PRINT "WITH ATTENUATION OF "ALPHA" DB"
550  PRINT "GIVES R1 = " INT (R1 * 1000 + .5) / 1000" OHMS"
560  PRINT "AND R2 = " INT (R2 * 1000 + .5) / 1000" OHMS"
570  PRINT "POWER DISSIPATION IN R1a = " INT (P1 * 10000 + .5) / 10000" WATTS"
580  PRINT "POWER DISSIPATION IN R2 = " INT (P2 * 10000 + .5) / 10000" WATTS"
590  PRINT "POWER DISSIPATION IN R1b = " INT (P3 * 10000 + .5) / 10000" WATTS"
600  PRINT "FOR AN INPUT POWER LEVEL OF "POWER" WATTS"
610  PRINT "AND CHARAC. RESISTANCE "RO" OHMS"
620  PRINT
630  PRINT "********************"
640  PRINT
650  IF K = 1 THEN
          1110
```

```
660   REM   PI SECTION
670   LET K = 1
680   LET R1 = RO * (N + 1) / (N - 1)
690   LET R2 = RO * (N * N - 1) / 2 / N
700   LET R3 = R1 * RO / (R1 + RO)
710   LET P1 = V1 * V1 / R1
720   LET P2 = (V1 / (R2 + RO * R1 / (RO + R1))) ^ 2 * R2
730   LET P3 = (((V1 * R3) / (R2 + R3)) / R1) ^ 2 * R1
740   PRINT
750   PRINT "A SYMMETRICAL PI SECTION"
760   GOTO 540
770   PRINT
780   REM
790   PRINT "DO YOU REQUIRE TO ANALYSE"
800   PRINT "TEE OR PI SECTION"
810   PRINT "IF TEE SECTION ENTER 1 ELSE 0"
820   INPUT A
830   IF A = 1 OR A = 0 THEN
          840ELSE620
840   PRINT "INPUT RESISTORS R1,R2"
850   INPUT R1,R2
860   IF A = 0 THEN
          990
870   LET RO =  SQR (R1 * R1 + 2 * R1 * R2)
880   LET ALPHA = 20 *  LOG (1 + R1 / R2 + RO / R2) /  LOG (10)
890   PRINT
900   PRINT "*********************"
910   PRINT
920   PRINT "CHARAC. RES.RO = " INT (RO * 100 + 0.5) / 100" OHMS"
930   PRINT "TEE PAD ATTENUATION = " INT (ALPHA * 100 + .5) / 100" DB"
940   PRINT "FOR R1 = "R1" AND R2 = "R2" OHMS"
950   PRINT
960   PRINT "*********************"
970   PRINT
980   GOTO 1110
990   LET ALPHA =  LOG ((1 + R2 / R1) +  SQR ((1 + R2 / R1) ^ 2 - 1))
1000  LET ALPHA = ALPHA * 8.686
1010  LET N = 10 ^ (ALPHA / 20)
1020  LET RO = 2 * R2 * N / (N ^ 2 - 1)
1030  PRINT
1040  PRINT "*********************"
1050  PRINT
1060  PRINT "CHARAC. RES. RO= " INT (RO * 100 + .5) / 100" OHMS"
1070  PRINT "PI PAD ATTENUATION = " INT (ALPHA * 100 + .5) / 100
1080  PRINT "FOR R1= "R1" AND R2= "R2" OHMS"
1090  PRINT
1100  PRINT "*********************"
1110  PRINT
1120  PRINT "DO YOU WANT ANOTHER GO ?"
1130  PRINT "ENTER 1 IF YES ELSE ZERO"
1140  INPUT L
1150  IF L = 1 THEN
          260ELSE950
1160  PRINT
1170  PRINT "*** END OF PROGRAM ***"
1180  END


END-OF-LISTING


]RUN

DO YOU KNOW THE CCT. PARAMETERS
IF YES ENTER 1 ELSE 0
?0
I/P CHARAC. LINE RES.(OHMS)
?50
ENTER DESIRED ATTN. IN DB
?10
ENTER I/P POWER IN WATTS
?1

*********************
```

```
A SYMMETRICAL TEE SECTION
WITH ATTENUATION OF 10 DB
GIVES R1 = 25.975 OHMS
AND R2 = 35.136 OHMS
POWER DISSIPATION IN R1a = .5195 WATTS
POWER DISSIPATION IN R2 = .3286 WATTS
POWER DISSIPATION IN R1b = .0519 WATTS
FOR AN INPUT POWER LEVEL OF 1 WATTS
AND CHARAC. RESISTANCE 50 OHMS

********************

A SYMMETRICAL PI SECTION
WITH ATTENUATION OF 10 DB
GIVES R1 = 96.248 OHMS
AND R2 = 71.151 OHMS
POWER DISSIPATION IN R1a = .5195 WATTS
POWER DISSIPATION IN R2 = .3286 WATTS
POWER DISSIPATION IN R1b = .0519 WATTS
FOR AN INPUT POWER LEVEL OF 1 WATTS
AND CHARAC. RESISTANCE 50 OHMS

********************

DO YOU WANT ANOTHER GO ?
ENTER 1 IF YES ELSE ZERO
?1

DO YOU KNOW THE CCT. PARAMETERS
IF YES ENTER 1 ELSE 0
?1

DO YOU REQUIRE TO ANALYSE
TEE OR PI SECTION
IF TEE SECTION ENTER 1 ELSE 0
?0
INPUT RESISTORS R1,R2
?100
??67

********************

CHARAC. RES. RO= 50.09 OHMS
PI PAD ATTENUATION = 9.56 DB
FOR R1= 100 AND R2= 67 OHMS

********************

DO YOU WANT ANOTHER GO ?
ENTER 1 IF YES ELSE ZERO
?0

*** END OF PROGRAM ***
```

synthesis, i.e. given characteristic impedance and attenuation, find component values, and, for analysis, given component values, find characteristic impedance and attenuation.

4.4   PROPERTIES OF HALF AND QUARTER WAVE TRANSMISSION LINE SECTIONS

A number of interesting yet fundamental circuits can be constructed from sections of transmission line that are multiples of one-half or one-quarter of a guide wavelength long at a specified frequency. To aid in the discussion, equation (1.30), describing sending end impedance, is restated below in its lossless form as equation (4.36).

$$\frac{Z_S}{Z_o} = \left[\frac{Z_T}{Z_o} + \tan(\beta l)\right] \Bigg/ \left[1 + \frac{Z_T}{Z_o} \tan(\beta l)\right] \tag{4.36}$$

If βl is set equal to π radians or to $n\pi$ radians where n is an integer then equation (4.36) reduces to

$$Z_s = Z_T \tag{4.37}$$

This occurs as a result of $\tan \pi = \tan n\pi = 0$; $n = 1, 2, 3, \ldots$. Values of $\beta l = \pi$ imply that since $\beta = 2\pi/(\lambda g)$ that l must equal $\lambda g/2$. In other words, for half wavelength lines of uniform cross-section, the sending end impedance is equal to the load impedance provided the line exhibits zero loss characteristics. Inspection of the scale on the periphery of a Smith Chart indicates that one-half guide wavelength is equivalent to a $360°$ rotation on the Smith Chart so that traveling from some fixed impedance a distance of $\lambda g/2$ results in a reoccurrence of the same impedance. Half wavelength sections of transmission line are said to be electrical analogies of one-to-one transformers at a specific frequency. For this reason a $\lambda g/2$ section of uniform transmission line is called a half wavelength transformer. The main use of the half wavelength transformer is that it allows remotely located loads to be effectively moved to a more accessible position in order that additional matching can be applied to efficiently couple the load to the source.

* * *

Example 4.13

A specified load is located in the wing of an aircraft while a matched signal source specifically designed to drive the load is located in the aircraft fuselage. The input and output terminals of load and source components respectively are separated by a distance of 3.2 m. If the signal source operates at 0.45 GHz, calculate what length the transmission line, having a phase velocity of 85 percent that of the free space, must have in order to connect the load and generator while preserving the load impedance presented to the generator.

Solution

Velocity of propagation in free space $= 3 \times 10^8$ m/s

$\therefore$ Velocity of propagation in dielectric $= 0.85 \times 3 \times 10^8$

$$= 2.55 \times 10^8 \text{ m/s}$$

$$= \text{phase velocity}$$

guide wavelength $\lambda g = \dfrac{V_p}{f} = \dfrac{2.55 \times 10^8}{450 \times 10^6} = 0.57$ meters

What the question is asking for is the physical length for a half wavelength transformer. Thus

$$\frac{\lambda g}{2} = \frac{0.57}{2} = 0.285 \text{ m}$$

The distance between source and load is 3.2 meters so that the number of half wavelength sections n, needed to enable this run to be initiated, is 12, i.e.

$$12 \times 0.285 = 3.42 \text{ m} > 3.2$$

while

$$11 \times 0.285 = 3.14 \text{ m} < 3.2$$

The conclusion is that to enable the job to be completed a total cable run of 3.42 meters or 12 half wavelength sections are required. For low level power transmission where peak voltage requirements need not be considered, the characteristic impedance of the cable is not critical. This means that any standard cable can be used, e.g. 50 ohms or 75 ohms, since the properties of the half wave transformation are independent of characteristic impedance (see equation (4.37)).

If a cable run of 3.42 meters was too long due to physical cabling restrictions then a 3.2 meter length of cable of correct characteristic impedance would have to be specially constructed or else compact matching networks designed and standard characteristic impedance cable used. Overall, the proposition of multiple half wave transformer sections to connect a load to an already matched source appears the most attractive provided the extra 0.22 meters of cable can be housed.

* * *

One further point about half wave transformers is that they produce a 180 degree phase shift between voltage and current in the line. This fact is utilized in producing feed circuits for antenna arrays.

Returning once again to equation (4.36), further interesting results can be produced if βl is allowed to equal $\pi/2$ radians. At $\pi/2$ or $n\pi/2$ radians where this time n is an odd integer, the tangent function goes to infinity so that the Z_T/Z_o multiplier becomes negligible. Hence

$$\frac{Z_s}{Z_0} = \frac{Z_o}{Z_T}$$

for which

$$Z_o = (Z_s Z_T)^{\frac{1}{2}} \tag{4.38}$$

Equation (4.37) means that a low impedance can be connected to high impedance by placing a quarter wavelength section or multiple quarter wavelength sections of line with characteristic impedance equal to the geometric mean of the impedances to be matched between them. When both Z_T and Z_s are complex, one impedance must be of equal magnitude and of opposite phase to the other so that Z_0, the characteristic impedance of the quarter wave line section, will be real. In practice, this restriction is not needed since quarter wavelength transformer sections are normally employed only when resistive loads are to be connected to resistive sources. A restriction exists on the use of quarter wavelength sections. This restriction is that the characteristic equation, satisfied by equation (4.38) subject to the source and load impedance values to be matched, should be physically realizable. For example, if the design is to be implemented in a microstrip medium then thin lines (>100 ohms) or wide lines (<20 ohms) will prove to be impractical.

*   *   *

Example 4.14

A 20 ohm load impedance is required to be matched to a signal generator with 50 ohm output impedance. The load impedance is located 2 meters from the generator which is operating at 144 MHz. Design a quarter wavelength transformer section that uses coaxial line with a phase velocity of 2×10^8 m/s.

Solution

$$\lambda g = \frac{V_p}{f} = \frac{2 \times 10^8}{144 \times 10^6} = 1.39 \text{ m}$$

$$\therefore \quad \frac{\lambda g}{4} = 0.35 \text{ m}$$

Now find the characteristic impedance

$$Z_0 = (20 \times 50)^{\frac{1}{2}} = 31.6 \text{ ohms}$$

Therefore, the design could be implemented by seven quarter wavelength sections each of length 0.35 m, which gives a total length of 2.45 metres of 31.6 ohm coaxial cable.

*   *   *

The solution just obtained is not the best one, however. It would be better to use one-quarter wavelength line section followed by 2.1 meters of standard low loss 50 ohm coaxial cable. The reason for this is that, like half wavelength transformers, quarter wavelength transformers are sensitive to variations in frequency, so that the phase shift introduced along seven quarter wavelength sections due to a variation in the frequency of oscillation of the generator will be seven times more than that for a single section.

The problem of impedance mismatch as a result of frequency perturbation can be solved by applying the Smith Chart. First, consider the design of a matching circuit that will match a complex load to a resistive generator.

* * *

Example 4.15

Given an air spaced transmission line with a characteristic impedance of 75 ohms and a load impedance of $(150 - j112)$ ohms. Design a matching section that employs a quarter wavelength transformer to operate at 920 MHz.

Solution

As was pointed out previously in this section, the quarter wavelength impedance transformer matches resistive elements only. The first part of this design should eliminate the reactive part of the load impedance. This can be done in a number of ways. In this instance, assume a series section of transmission line connected to the load whose length has been selected to give a sending end impedance that is purely resistive. This resistance can then be matched to the generator resistance by a quarter wavelength section of appropriate characteristic impedance. The situation is shown in figure 4.26. The steps in the design are as follows:

(a) Normalize the load impedance to the characteristic impedance of the transmission line, 75 ohms in this case

$$Z_L = \frac{150 - j112}{75} = 2 - j1.49$$

(b) Plot this on the Smith Chart, shown in figure 4.27 as point A.
(c) Draw the circle center $(1, 0)$ on the chart that passes through point A. This is the VSWR circle and its value can be obtained from the scale at the bottom left-hand side of the chart. From the standing wave ratio scale a value of 3.3 is found. Here, lossless line has been assumed.

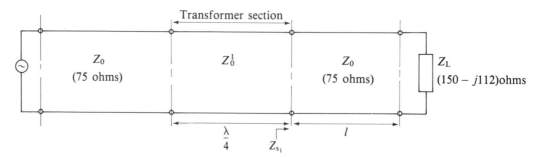

Figure 4.26   Quarter Wave Matching of a Complex Load Impedance to a Transmission Line

NAME	TITLE	DWG. NO.
SMITH CHART FORM 82-BSPR (9-66)	KAY ELECTRIC COMPANY, PINE BROOK, N.J., © 1966 PRINTED IN U.S.A.	DATE

IMPEDANCE OR ADMITTANCE COORDINATES

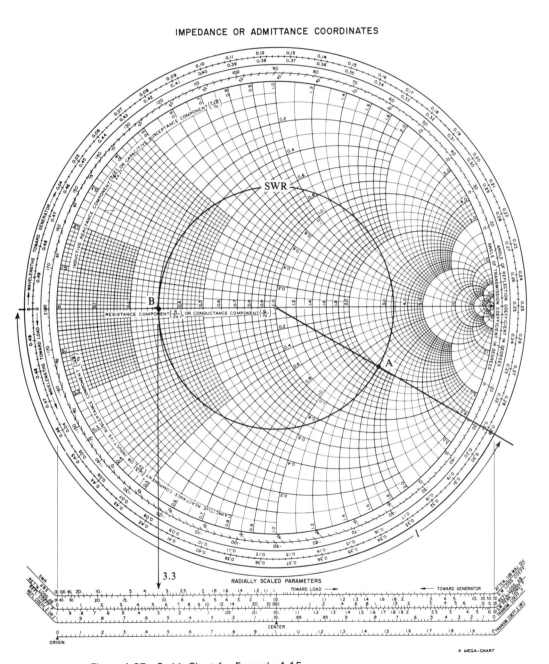

Figure 4.27   Smith Chart for Example 4.15

(d) Move away from the load located at point A toward the generator (in a clockwise direction) until the point where the VSWR circle intersects the line of zero reactance is reached, i.e. the horizontal line point B. The distance l for the section of 75 ohm transmission line required to make the complex load impedance appear real can now be found.

(e) To find this length draw a line from the chart center through point A to the chart periphery. This line intersects the wavelengths towards the generator scale at 0.291λ. By extending the horizontal line, the line of zero reactance from the center through B length l for the required section of 75 ohm line can be found. Hence the distance from the load at which the line looks resistive is

$$l = (0.291 - 0.5)\lambda = 0.21\lambda$$

(f) A quarter wavelength transformer section can now be inserted between the end of the cable length determined in part (e) and the generator. Point B lies at $0.3 = j0$ on the Smith Chart. This point represents an impedance of

$$75 \times (0.3 + j0) = (22.5 + j0) \text{ ohms}$$

The characteristic impedance for the quarter wavelength section can now be found

$$Z_0{}^1 = (75 \times 22.5)^{\frac{1}{2}} = 41.1 \text{ ohms}$$

(g) To finish the design it is necessary to calculate the lengths required for each section. Since the line is air spaced, the phase velocity will be 3×10^8 m/s

$$\therefore \quad \lambda = \frac{c}{f} = \frac{3 \times 10^8}{920 \times 10^6} = 0.326 \text{ m}$$

hence

$$l = 6.85 \text{ cm}$$

and

$$\frac{\lambda}{4} = 8.15 \text{ cm}$$

This completes the design.

*   *   *

Consider now what would happen if the circuit designed in example 4.14 was operated at a frequency removed from the design frequency. The evaluation of this is illustrated in the next design example.

*   *   *

Example 4.15 (continued)

Find qualitatively the effective impedance mismatch of a given quarter wavelength impedance matching circuit if the frequency is increased by 10 percent of the design frequency.

Solution

Here

$$l = 0.21$$

$$\lambda = 9.5 \text{ cm}$$

$$\frac{\lambda}{4} = 8.15 \text{ cm}$$

$$Z_o = 75 \text{ ohms}$$

$$Z_o^1 = 41.1 \text{ ohms}$$

$$f = 920 \text{ MHz}$$

(a) If frequency is increased by 10 percent then the line section l will appear 10 percent electrically longer.

$$l^1 = (0.21 + 0.021)\lambda = 0.23\lambda$$

This is shown on the Smith Chart in figure 4.28. Here point A represents the normalized load impedance, point B the end of arc l^1.

(b) The line through B from the center intersects the 3.3 SWR circle at point C. Point C represents the sending end impedance Z_{s_1} in figure 4.26; as a normalized impedance, this is equal to

$$0.3 + j0.12$$

or denormalized with reference to a 75 Ω characteristic impedance line

$$(22.5 + j9) \text{ ohms}$$

(c) This means the quarter wavelength section is no longer presented with a pure resistance. The quarter wavelength section itself appears to be 10 percent electrically longer at the increased frequency. So that its electrical length becomes 0.275 λ.

   The next step of the design appraisal involves the quarter wavelength section. The characteristic impedance of this section of line is different to that of the first line section so that it becomes necessary to normalize Z_{L_1} to the characteristic impedance of the quarter wavelength section, in this case 41.1 ohms

$$Z_L^1 = \frac{22.5 + j9}{41.1} = 0.547 + j0.219$$

plotted as point D and giving an SWR^1 circle of value 1.85.

(d) The radial line through point D intersects the periphery at point E, 0.05λ.

(e) Adding the new electrical length of the quarter wave transformer to this gives

$$(0.05 + 0.275)\lambda \qquad \text{point F}$$

NAME	TITLE	DWG. NO.
SMITH CHART FORM 82-BSPR(9-66)	KAY ELECTRIC COMPANY, PINE BROOK, N.J., © 1966. PRINTED IN U.S.A.	DATE

IMPEDANCE OR ADMITTANCE COORDINATES

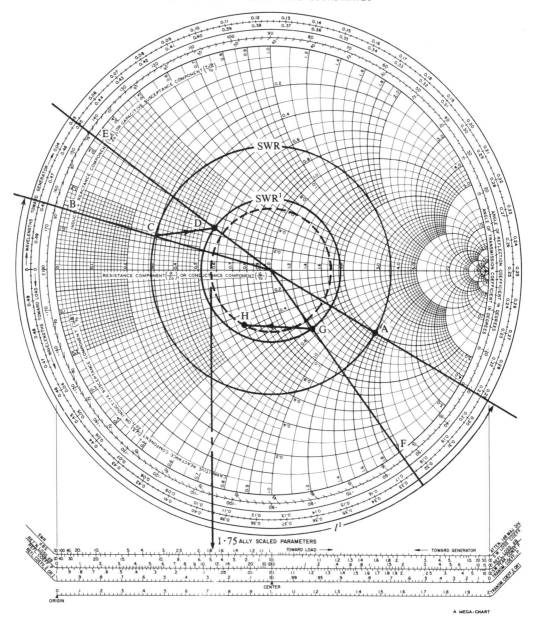

Figure 4.28   Effect of Frequency on Design Example 4.15

(f) The radial line through point F intersects the SWR[1] circle at point G

$1.25 - j0.7$

or denormalized to 41.1 ohms

$(51.4 - j28.8)$ ohms

(g) To get back to the 75 ohm system so that a comparison with the exact match obtained at the design frequency can be made, normalize point G to 75 ohms

$$\frac{51.4 - j28.8}{75} = 0.69 - j0.38 \qquad \text{point H}$$

This represents a VSWR value of 1.75. The conclusion then is that a properly matched circuit at 920 MHz (VSWR = 1) is mismatched at 1012 MHz (VSWR = 1.75). This means that at 1012 MHz only 90 percent of the power incident on the matching structure will reach the load, the rest being reflected.

* * *

After that gruelling exercise, the next example shows how quarter wavelength transformers matching resistive terminations can, with the aid of a simple design, produce a rather elegant circuit for use in the telecommunications industry.

* * *

Example 4.16

Design a suitable phasing harness constructed from coaxial cable such that two antennae each having an input impedance of 50 ohms can be connected to a power amplifier having a 50 ohm output impedance.

Solution

The geometry for the problem is shown in figure 4.29. Here, the antennae are depicted as purely resistive elements.

If l_1 and l_2 in figure 4.29 are made an odd integer times one-quarter guide wavelength long and Z_{o_1}, Z_{o_2} the characteristic impedance of the lines feeding antennae A, B are made 75 ohms then applying equation (4.38) to each line segment length l_1, l_2 gives

$$Z_{s_A} = Z_{s_B} = \frac{75 \times 75}{50} = 112.5 \text{ ohms}$$

These sending end impedances appear in parallel as shown in figure 4.29 so that the total sending end impedance is

$$Z_s = \frac{(112.5)^2}{2 \times 112.5} = 56.25 \text{ ohms}$$

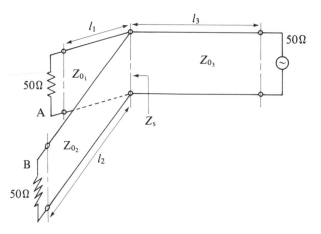

Figure 4.29   Phasing Harness Design Example

If Z_{o_3} the characteristic impedance of the connecting line is selected to be 50 ohms then a reflection coefficient of

$$|\Gamma| = \frac{56.25 - 50}{56.25 + 50} = 0.059$$

and a

$$VSWR = \frac{1 + 0.059}{1 - 0.059} = 1.125$$

is achieved.

In other words, 99.7 percent of the energy present in the incident wave will be transferred to the load.

* * *

The use of quarter wave sections in this example have enabled cheap readily available 50 and 75 ohm characteristic impedance cable (instead of special purpose 25 ohm cable) to allow two antennae of equal input impedance to be connected directly to a generator having 50 ohm output impedance.

One problem often encountered in installing a transmission line between a load and a generator is knowing exactly where the terminal locations lie electrically. Electrically the connection of a termination to a quarter wavelength transformer section may present a discontinuity due to the difference in physical dimensions of the line cross-sections. These discontinuities often make the transformer section lengths appear electrically different to their design lengths. This results in the introduction of a slight mismatch into the circuit.

Another problem exists when the line has been cut incorrectly, either too short or too long. Again this will result in a mismatch in the circuit. If either the

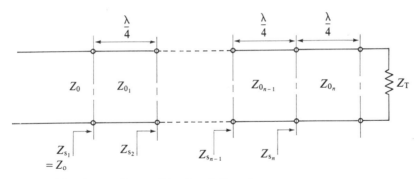

Figure 4.30   Multiple Quarter Wavelength Transformer

discontinuity can be estimated (see for example section 5.2.1) or the length error is known, then their effects can be calculated using the same technique evolved in example 4.15 for evaluating the effect of frequency variation on transmission line mismatch.

In many applications the mismatch produced by frequency excursions about the design frequency cannot be tolerated. One way round the problem is to design circuits whose VSWR varies only slightly over a broad range of frequencies. This is known as broadband design. If the load impedance is assumed to be constant over a range of frequency then the bandwidth over which matching can be held within specified limits of VSWR can be increased by using multiple quarter wavelength transformer sections in cascade (figure 4.30).

The problem with the configuration shown in figure 4.30 is how best to choose the characteristic impedance of adjacent line sections so that the best match between Z_T and Z_o is achieved.

One simple method is to make the ratio of adjacent impedances constant

$$\frac{Z_T}{Z_{O_n}} = \frac{Z_{O_n}}{Z_{S_n}} = \frac{Z_{S_n}}{Z_{O_{n-1}}} = \frac{Z_{O_{n-1}}}{Z_{S_{n-1}}} = \cdots = \frac{Z_{O_1}}{Z_{S_1}}$$

This obeys equation (4.38) since by grouping alternative pairs of impedances

$$Z_{O_n} = (Z_T Z_{S_n})^{\frac{1}{2}}$$

and

$$Z_{O_{n-1}} = (Z_{S_{n-1}} Z_{S_n})^{\frac{1}{2}}, \text{ etc.}$$

* * *

Example 4.17

Design a broadband matching circuit using two stages of transformation to match a 600 ohm load to a 50 ohm transmission line.

Solution

Two stages, i.e. $n = 2$

$$\frac{Z_T}{Z_{o_2}} = \frac{Z_{o_2}}{Z_{s_2}} = \frac{Z_{s_2}}{Z_{o_1}} = \frac{Z_{o_1}}{Z_{s_1}}$$

here $Z_T = 600$ ohms and $Z_{s_1} = Z_o = 50 \ \Omega$. Grouping the two center terms gives

$$Z_{s_2} = (Z_{o_1} Z_{o_2})^{\frac{1}{2}}$$

Grouping the outer terms

$$Z_{o_2} Z_{o_1} = Z_T Z_{s_1} = 600 \times 50 = 30\ 000 = Z_{s_2}{}^2$$

$$\therefore \quad Z_{s_2} = 173 \text{ ohms}$$

Hence, grouping alternate pairs of terms gives

$$Z_{o_2} = (Z_T Z_{s_2})^{\frac{1}{2}} = 322 \text{ ohms}$$

$$Z_{o_1} = (Z_{s_2} Z_{s_1})^{\frac{1}{2}} = 93 \text{ ohms}$$

Line sections with these characteristic impedances each of length one-quarter wavelength can be synthesized for a selected line type by using one of the computer programs given in chapter 2.

To complete the design the 600 ohm load is preceded by a quarter wavelength section of line with characteristic impedance of 322 ohms, which itself is preceded by a quarter wavelength section of impedance 93 ohms. This joins the 50 ohm feed line to finish the overall matching structure. It should be noted that a matching structure constructed in this way will be physically long, having $n\lambda/4$ sections for an n-section transformer. The bandwidth obtained for the transformer (designed by the above technique) is shown in figure 4.31. This was computed by means of computer program CASCADE which will be discussed in section 4.6.

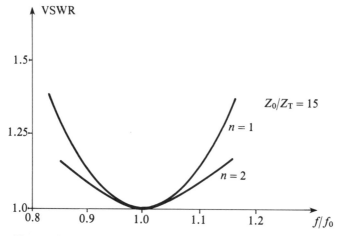

Figure 4.31   Two Section Impedance Transformer Normalized Frequency Response

* * *

In design example 4.17, it was seen that in a broadband structure a number of line sections each displaying a step change in characteristic impedance are used to connect source and load impedances. The abrupt change in line cross-section at these steps may lead to unwanted reflections. These reflections can be minimized by reducing the difference in characteristic impedance, i.e. the change in line cross-section, between adjacent steps to a minimum. Taking this idea to its limit, a matching transformer consisting of a long tapered section would represent a smooth transition from source to load impedance. This type of impedance matching circuit should provide excellent broadband performance. This is pursued further in section 5.6 where tapered lines are discussed.

4.5   TUNING STUBS

In section 1.5 a number of equations were derived to illustrate the behavior of the input impedance of arbitrary lengths of transmission line with perfect open or short circuit terminations. Figure 1.9 shows how the sending end impedance of an open circuit or short circuit line section varies as line length is changed. For line lengths of one-eighth guide wavelength or multiples thereof, it can be seen that a pure inductive or capacitive reactance can be realized depending on the type of termination used. In the previous section it was discovered that quarter wavelength sections give an impedance inversion while half wavelength sections behave as one-to-one transformers. In example 4.15, it was indicated that a fixed physical length of transmission line can be made to have a variable electrical length that is dependent on the deviation of operating frequency from the design frequency for which the physical line length is selected. Equation (1.31) describes the sending end impedance for a given length of lossless transmission line terminated with a perfectly conducting short circuit. This equation makes it possible to plot the variation of sending end impedance, not as a function of length as in figure 1.9, but as a function of frequency length being held constant (figure 4.32).

Figure 4.32 illustrates a number of important points. First, the line appears to be periodic due to the cyclic nature of the tangent function in the governing equation. At low frequencies the reactance seen looking into the line from the sending end becomes more inductive until the frequency at which the line is operated approaches $V_p/4l$. At this point, the line appears electrically equivalent to a physical length of one one-quarter wavelength. The value of the tangent function is infinity since its argument is $\pi/2$ radians. The infinite reactance state is similar to the reactance obtained for a parallel resonant circuit. At this frequency then, a parallel resonant equivalent circuit provides a suitable lumped representation for the line. It can be noted from the Smith Chart than an impedance inversion will have occurred from the short circuit position to this quarter wavelength position. As frequency is further increased, the tangent function becomes negative indicating a capacitive reactance when viewed from the sending end of the line. When the frequency of excitation becomes equal to $V_p/2l$ the tangent function goes to zero, passing from a capacitive to inductive reactance at this frequency. This situation is

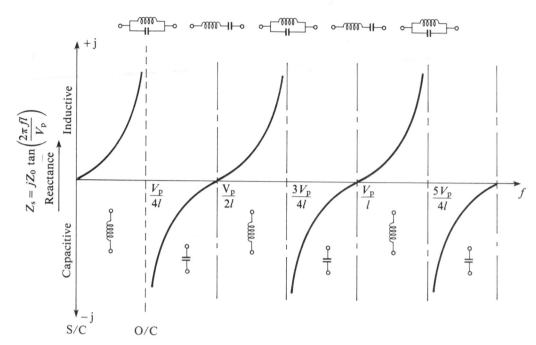

$$Z_s = jZ_0 \tan\left(\frac{2\pi fl}{V_p}\right)$$

Figure 4.32 Sending End Impedance as a Function of Frequency for a Uniform Transmission Line with Perfectly Reflecting Terminations

the type of behavior exhibited by series resonant LC circuits. Any further increase in frequency will cause a repetition of the cycle of events just discussed. Note that at the frequency at which series resonance occurs the line has an electrical length of one-half guide wavelength so that it behaves as a one-to-one transformer.

A final point about figure 4.32 is that close to the zero impedance transitions the rate of change of impedance is approximately linear since tan (small angle) ≈ small angle in radians. This means that for frequency changes over which the electrical line length varies by ±λ/20, a short circuit or open circuit line section, sometimes called a stub, can be approximated by a lumped capacitor or inductor.

Inspection of a Smith Chart will verify graphically the sequence of events discussed above. For a loss free transmission line terminated with a perfectly reflecting short circuit, movement will be confined to the locus of zero resistance. As frequency is increased the load appears to move further away from its initial position which is fixed. This is the same as moving around the chart in a clockwise direction, i.e. toward the generator. Moving clockwise around the chart periphery traces out the same impedance variation as predicted by equation (1.31).

Figure 4.32 can of course be reproduced for a line of variable length operated at a fixed frequency. This makes open and short circuit stubs powerful circuit elements for use in many impedance matching problems. One very useful matching circuit is the stub tuner. Here the application takes the form of interfacing a complex

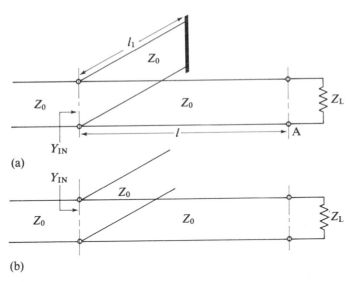

Figure 4.33   Single Stub Impedance Matching (a) Short Circuit Stub (b) Open Circuit Stub

load to a line or generator with real impedance. The advantage gained by use of this type of matching structure rather than quarter wavelength sections is that the transmission line, having the same characteristic impedance as that of the main transmission line, can be used to synthesize the matching network.

In a single stub impedance matching circuit, the stub is connected in parallel with the main transmission line at some point a distance l away from the load (figure 4.33). The stub line is connected in parallel with the main line so that it is convenient to work in terms of admittance. In the light of this, distance l is selected to yield

$$Y_{in} = 1 \pm jB$$

This gives a resistive match leaving only the susceptance component to be negated, so that a perfect match between source and load can be formed. The excess susceptance component can be compensated by suitable choice of stub length.

*   *   *

Example 4.18

Match a load impedance of $(75 - j25)$ ohms to a low loss line having characteristic impedance of 50 ohms. The design frequency is 500 MHz and it may be assumed that the phase velocity of the cable is that of free space. A single stub tuning circuit should provide the matching network.

Solution

See figure 4.34

$$f = 0.5 \text{ GHz}$$
$$V_p = 3 \times 10^{10} \text{ cm/s (air spaced)}$$
$$Z_o = 50 \text{ ohms}$$
$$Z_L = (75 - j125) \text{ ohms}$$

attenuation = 0 dB

(a) $z_L = \dfrac{75 - j125}{50} = 1.5 - j2.5$, point A on the chart.

(b) Draw SWR circle (value 5).

(c) Convert point A to an admittance, point B. When denormalized this becomes

$$(4 - j6) \text{ mS}$$

(d) Extend radial line to periphery and read 'wavelengths to generator' scale (remember moving from load to generator), point C.

$$0.05 \lambda$$

(e) Note the intersection of the SWR circle and the unit conductance circle, point D.

(f) Extend a radial line through point D, this results in point E. Note the scale reading

$$0.186 \lambda$$

(g) The distance l from load to stub along the main line can be found

$$(0.186 - 0.05)\lambda = 0.136\lambda$$

This completes the first stage of the design.

Next, a suitable stub length l_1 must be chosen to negate any residual susceptance remaining at point D, $(1 + j1.9)$.

(h) If a short circuit stub is selected then its length must be chosen to present a susceptance of $-j1.9$ to the main transmission line so that a match can be accomplished. Since the design is executed in terms of admittance, a short circuit termination is located at point F on the right-hand side of the Smith Chart.

(i)   What remains now is to move along the stub away from the short circuit towards the main line, toward the generator, until a value of normalized susceptance of -1.9 is obtained, point G (0.315λ). From this, length, l_1 can be found

$$l_1 = (0.315 - 0.25)\lambda = 0.065\lambda$$

(j) Find the required physical lengths for l_1 and l.

$$l = 0.136\lambda$$
$$l_1 = 0.065\lambda$$

NAME	TITLE	DWG. NO.
SMITH CHART FORM 82-BSPR (9-66)	KAY ELECTRIC COMPANY, PINE BROOK, N.J., © 1966 PRINTED IN U.S.A.	DATE

IMPEDANCE OR ADMITTANCE COORDINATES

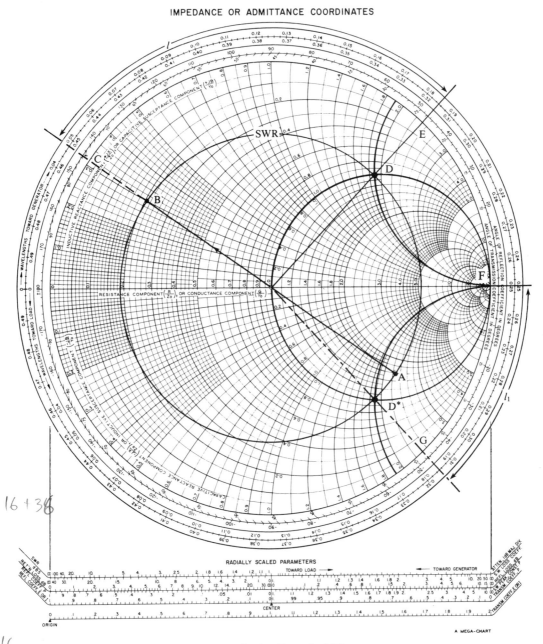

16 + 3⅙

16
36
52

7 ⅟₇

Figure 4.34   Short Circuit Stub Design Example 4.18

$$\lambda = \frac{V_p}{f} = \frac{3 \times 10^{10}}{500 \times 10^6} = 60 \text{ cm}$$

$$\therefore \quad l = 0.136 \times 60 = 8.16 \text{ cm}$$

$$l_1 = 3.9 \text{ cm}$$

and

$$Z_o = 50 \text{ ohm}$$

This completes the design.

Example 4.19

Repeat example 4.18 but this time use a stub line with an open circuit termination to provide reactive tuning of the main line in order to present a match between source and load impedances.

Solution

Steps (a) to (g) of the previous design example are valid. Step (h) refers to figure 4.35. On the admittance plane an open circuit appears at point F. Point D* represents the conjugate impedance at the end of a series section of transmission line length *l*. The susceptance at this point has to be canceled to allow an impedance match at the design frequency. Moving towards the generator a total distance of

$$| 0.25 + (0.315 - 0.25) | = 0.315 \lambda$$

so that with an open circuit stub at 500 MHz

$$l_1 = 18.9 \text{ cm}$$

*   *   *

Inspection of figures 4.34 and 4.35 reveals that both designs are virtually identical. In the second example a quarter wavelength line section, i.e. one-half rotation on the Smith Chart, is required to move the open circuit termination to the short circuit position; with the result that the short circuit design produces a tuning stub that is shorter than an open circuit stub required to provide the same matching function. The attraction here would be to choose the design yielding the shortest stub length, i.e. the short circuit stub in this case (each design should be assessed on its own merit). Normally it is advantageous to keep all transmission line sections used in matching networks as short as possible. This ensures the smallest mismatch when the operating frequency of a circuit is varied slightly about the design frequency. Once again the final choice between open or short circuit stub will be based on practical considerations. For example, if the designs discussed in examples 4.18 and 4.19 were to be implemented on a microstrip medium, the short circuit stub option may not be necessarily the first choice. The reason for this is that short circuits are

NAME	TITLE	DWG. NO.
SMITH CHART FORM 82-BSPR (9-66)	KAY ELECTRIC COMPANY, PINE BROOK, N.J., © 1966. PRINTED IN U.S.A.	DATE

IMPEDANCE OR ADMITTANCE COORDINATES

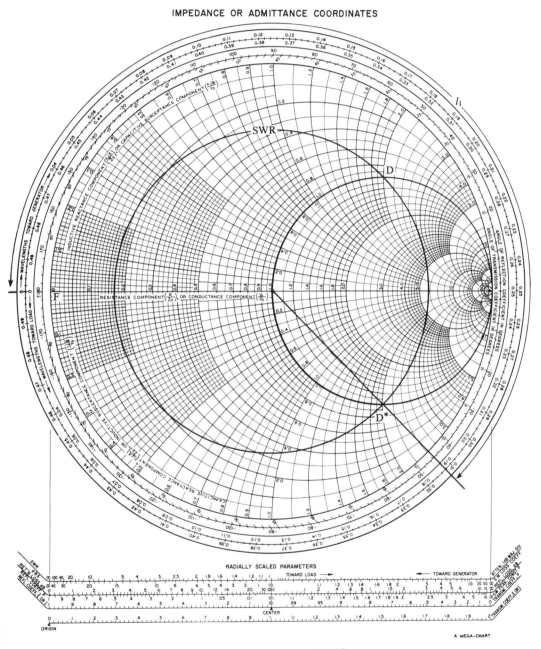

Figure 4.35   Open Circuit Stub Design Example 4.19

difficult to manufacture on microstrip. On the other hand, open circuit terminations, albeit imperfect, are easy to manufacture. This appears to suggest that the
open circuit stub would be the best option to follow. However, space limitations on
the substrate forming the microstrip line or indeed the loss associated with the long
length of line required to produce the open circuit stub may force the designer to
use the short circuit stub. Again, it must be stressed that each matching problem has
to be taken on its own merit.

For example 4.18, the short circuit stub design problem, consider the effect of
a frequency variation on matching performance.

*   *   *

Example 4.20

Calculate, using a Smith Chart, the mismatch produced by a 10 percent reduction
in frequency relative to the design frequency for the matching structure synthesized
in example 4.18.

Solution

(a) Redraw the relevant portions of the design in example 4.18 as shown in figure
4.36.

(b) A 10 percent reduction in frequency relative to the design frequency will cause
a corresponding increase in wavelength

$$\lambda \pm \Delta\lambda = \frac{V_p}{(f \pm \Delta f)}$$

This assumes that the phase velocity is independent of frequency, the line has
zero dispersion. As a result of this reduction in frequency, a fixed physical length
of line will appear electrically shorter, fewer wavelengths per unit length.
Therefore

$$l = 0.136\lambda = 8.16 \text{ cm at } 500 \text{ MHz}$$

becomes

$$l^1 = (0.136 - 0.1 \times 0.136)\lambda = 0.1224\lambda \text{ at } 550 \text{ MHz}$$

and

$$l_1 = 0.065\lambda = 3.9 \text{ cm at } 500 \text{ MHz}$$

becomes

$$l_1^{\,1} = 0.0585\lambda$$

(c) Next draw the arc length represented by l^1 starting from the load admittance,
point B. The radial line from the center to the periphery strikes the load VSWR

NAME	TITLE	DWG. NO.
SMITH CHART FORM 82-BSPR (9-66)	KAY ELECTRIC COMPANY, PINE BROOK, N.J., © 1966. PRINTED IN U.S.A.	DATE

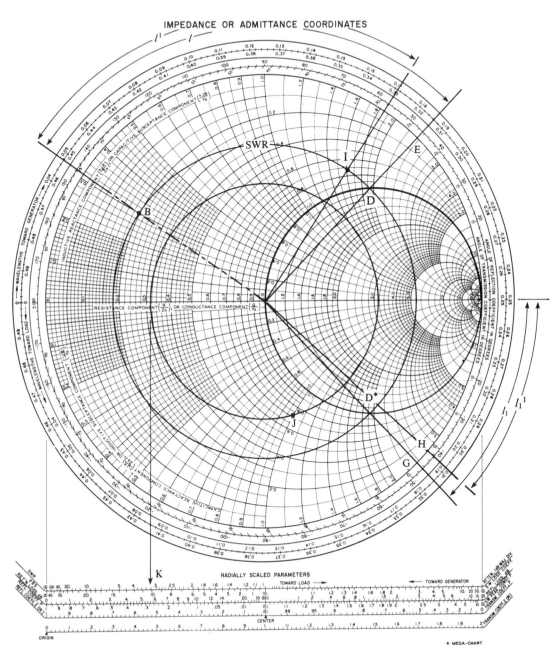

IMPEDANCE OR ADMITTANCE COORDINATES

RADIALLY SCALED PARAMETERS

A MEGA-CHART

Figure 4.36   Effect of Frequency Variation on Short Circuit Stub Design Example 4.18

circle at point I, corresponding to a normalized susceptance of

$0.75 + j1.6$

(d) Locate $l_1{}^1$ on the Smith Chart periphery and draw the radial line to this point, point H. This position corresponds to a normalized susceptance of $-j2.6$.

(e) The total normalized admittance at the junction between the stub line and series transmission line is

$0.75 + j1.6 - j2.6$

$= 0.75 - j1.0$ (point J)

Denormalized, this gives a value of

$(15 - j20)$ mS

(f) The VSWR circle passing through this point has a value of 3.0, point K. This means that at the design frequency an original VSWR value of approximately 6 was reduced to zero. However, a 10 percent frequency reduction has led to the VSWR increasing to 3.0. In terms of the amount of incident power being transmitted to the load, a 25 percent reduction occurs between the perfectly matched condition and the mismatched condition at 10 percent frequency reduction.

*   *   *

If example 4.20 is repeated for the open circuit stub design example 4.19, then step (d) results in point H becoming $-j4.8$. In step (e), point J becomes $0.75 - j3.2$ which results in the VSWR circle of step (f) having a high value of 15 so that only 24 percent of the power in the incident wave will reach the load. In this case, a 10 percent reduction in frequency has resulted in a matching structure which actually increases mismatch since without it a VSWR of about 5 is obtained.

The use of two or perhaps three stubs, placed $3\lambda/8$ apart and tuned experimentally, normally allows increased bandwidth operation. In a practical implementation of a paper design, trial and error is used to refine the design. This post manufacture trimming is almost always necessary because of the difficulty in precisely locating the length of the series line from the load to the center of the stub line.

A simple computer program, program 4.3 STUB based on program ZSEND, enables the calculations for single open or short circuit stub matching circuits such as those discussed in examples 4.18 and 4.19 to be carried out. The program can also, with straightforward modification, allow the effects of mismatch at frequencies removed from the design frequency to be evaluated.

```
][ FORMATTED LISTING
FILE: PROGRAM 4.3 STUB
PAGE-1


   10   REM
   20   REM   **** STUB ****
   30   REM
```

```
 40   REM   SYNTHESIS OF SINGLE
 50   REM   STUB MATCHING CCT.
 60   REM   BOTH LINES ARE
 70   REM   ASSUMED TO HAVE
 80   REM   IDENTICAL PARAMETERS
 90   REM
100   REM   +L GIVES MOVEMENT
110   REM   TOWARDS GENERATOR
120   REM
130   REM   F=FREQ.(GHZ)
140   REM   L=LINE LTH.(CM)
150   REM   A2=ATTENUATION (DB/CM)
160   REM   ZO=CHARC. IMP.(OHMS)
170   REM   E=REL./EFF. PERM.
180   HOME
190   PRINT "INPUT FREQ.(GHZ)"
200   INPUT F
210   PRINT "I/P LOAD RES.(OHMS)"
220   INPUT R1
230   PRINT "INPUT LOAD REACTANCE (OHMS)"
240   INPUT X1
250   PRINT
260   REM   SERIES LINE
270   PRINT "FOR SERIES LINE ENTER"
280   PRINT "INPUT CHARC. IMP.(OHMS)"
290   INPUT ZO
300   PRINT "INPUT ATTN.(DB/CM)"
310   INPUT A2
320   PRINT "INPUT REL.OR EFF. PERM."
330   INPUT E
340   LET R = R1
350   LET X = X1
360   LET F6 = ZO
370   LET W1 = 30 / (F *  SQR (E))
380   LET MA = 0
390   HOME
400   PRINT "WORKING:---"
410   PRINT
420   FOR L = 0 TO 1000 STEP .01
430       LET R2 = (R1 * R1 - ZO * ZO + X1 * X1) / ((R1 + ZO) * (R1 + ZO) + X1
          * X1)
440       LET X2 = 2 * X1 * ZO / ((R1 + ZO) * (R1 + ZO) + X1 * X1)
450       IF X2 <  > 0 THEN
              490
460       IF R2 >  = 0 THEN
              480
470       LET G =  - 3.1415927:
          GOTO 590
480       LET G = 1E - 20:
          GOTO 590
490       IF R2 <  = 0 THEN
              520
500       IF X2 = 0 THEN
              450
510       LET G =  ATN (X2 / R2):
          GOTO 590
520       IF R2 <  > 0 THEN
              560
530       IF X2 >  = 0 THEN
              550
540       LET G =  - 1.5707963:
          GOTO 590
550       LET G = 1.5707963:
          GOTO 590
560       IF R2 >  = 0 THEN
              590
570       IF X2 = 0 THEN
              450
580       LET G = 3.1415927 +  ATN (X2 / R2)
590       LET T1 = G
600       LET M1 =  SQR (R2 * R2 + X2 * X2)
610       LET A2 = A2 / 8.686
620       LET T2 = T1 - 4 * 3.1415927 * L / W1
```

```
630     LET M2 = M1 *  EXP ( - (2 * A2 * L))
640     LET D = 1 - 2 * M2 *  COS (T2) + M2 * M2
650     IF D = 0 THEN
            LET D = 1E - 20
660     LET R1 = ZO * (1 - M2 * M2) / D
670     LET X1 = ZO * 2 * M2 *  SIN (T2) / D:
        IF MA = 0 THEN
            700
680     LET F8 = X1:
        IF  SGN (F8) =  SGN (Y) AND  ABS (F8) /  ABS (Y) < 1.0125 AND  ABS (F
        8) /  ABS (Y) > .9875 THEN
            960
690     GOTO 710
700     IF (ZO * R1 / (R1 * R1 + X1 * X1)) < 1.005 AND (ZO * R1 / (R1 * R1 +
        X1 * X1)) > 0.995 THEN
            730
710     LET R1 = R:
        LET X1 = X
720 NEXT L
730 PRINT :
    PRINT "***********************":
    PRINT
740 PRINT "OPERATING FREQUENCY.(GHZ) "F
750 PRINT
760 PRINT "LOAD IMPEDANCE"
770 PRINT "   " INT (1000 * R / 1000 + .5)"  " INT (1000 * X / 1000 + .5)" j O
    HMS"
780 PRINT
790 PRINT "SERIES LINE LTH.(CMS.) " INT (L * 1000 + .5) / 1000
800 PRINT
810 PRINT "REL. OR EFF. PERM. "E
820 PRINT "ATT.(DB/CM) " INT (A2 * 1000 * 8.686 + .5) / 1000
830 PRINT "CHARAC. IMP.(OHMS) "ZO
840 PRINT
850 PRINT "AT THIS POINT THE"
860 PRINT "SENDING END IMPEDANCE IS "
870 PRINT "   " INT (1000 * R1 / 1000 + .5)"  " INT (1000 * X1 / 1000 + .5)" j
     OHMS"
880 IF MA = 0 THEN
        LET F3 = R1
890 IF MA = 0 THEN
        LET F4 = X1
900 PRINT
910 REM  IDEAL SHORT CCT. LINE
920 LET Y =  - X1
930 LET R1 = 0:
    LET X1 = 0:
    LET R = R1:
    LET X = X1
940 LET MA = 1
950 GOTO 420
960 PRINT
970 IF MA = 2 THEN
        PRINT "OPEN CIRCUIT STUB LTH.(CMS.) " INT (L * 1000 + .5) / 1000
980 IF MA = 2 THEN
        1000
990 PRINT "SHORT CIRCUIT STUB LTH.(CMS.) " INT (L * 1000 + .5) / 1000
1000 PRINT
1010 PRINT "IMPEDANCE LOOKING INTO STUB"
1020 PRINT "IS " INT (R1 * 1000 + .5) / 1000"  " INT (X1 * 1000 + .5) / 1000"
     j OHMS"
1030 IF MA = 2 THEN
        1080
1040 REM  IDEAL OPEN CCT. LINE
1050 LET R1 = 1E10:
     LET X1 = 1E10
1060 LET R = R1:
     LET X = X1:
     LET MA = 2
1070 GOTO 420
1080 LET F5 = ( - F3 * F4 + F4 * F4) / F3
1090 PRINT
1100 PRINT "FINAL INPUT RESISTANCE IS " INT (F5 * 1000 + .5) / 1000" OHMS"
```

```
1110   LET F5 =  ABS ((F5 - F6) / (F5 + F6))
1120   LET F5 = (1 + F5) / (1 - F5)
1130   PRINT
1140   PRINT "VSWR FOR THIS DESIGN IS " INT (F5 * 1000 + .5) / 1000
1150   PRINT :
       PRINT "**********************"
1160   PRINT
1170   PRINT "FINISHED? IF NO ENTER 1"
1180   INPUT T
1190   IF T = 1 THEN
            180
1200   PRINT
1210   PRINT "**** END OF PROGRAM ****"
1220   END
```

END-OF-LISTING

```
]RUN
INPUT FREQ.(GHZ)
?1
I/P LOAD RES.(OHMS)
?200
INPUT LOAD REACTANCE (OHMS)
?-300

FOR SERIES LINE ENTER
INPUT CHARC. IMP.(OHMS)
?150
INPUT ATTN.(DB/CM)
?0
INPUT REL.OR EFF. PERM.
?1
WORKING:---


**********************

OPERATING FREQUENCY.(GHZ) 1

LOAD IMPEDANCE
   200  -300 j OHMS

SERIES LINE LTH.(CMS.) 3.8

REL. OR EFF. PERM. 1
ATT.(DB/CM) 0
CHARAC. IMP.(OHMS) 150

AT THIS POINT THE
SENDING END IMPEDANCE IS
   37  -65 j OHMS


SHORT CIRCUIT STUB LTH.(CMS.) 1.93

IMPEDANCE LOOKING INTO STUB
IS 0  64.166 j OHMS

OPEN CIRCUIT STUB LTH.(CMS.) 9.43

IMPEDANCE LOOKING INTO STUB
IS 0  64.166 j OHMS

FINAL INPUT RESISTANCE IS 178.615 OHMS

VSWR FOR THIS DESIGN IS 1.191

**********************

FINISHED? IF NO ENTER 1
?0

**** END OF PROGRAM ****
```

4.6   CASCADED LINE APPRAISAL

Very often when dealing with a series of transmission line sections and lumped
elements placed in cascade the ability to perform an analysis of overall behavior is
essential. This analysis normally takes the form of computing sending end
impedance as a function of frequency. From this the variation of circuit VSWR with
frequency can be found. For simple systems containing only a few elements, a direct
analysis approach is the simplest to implement. With direct analysis, the impedance
seen looking into a circuit comprising of line segments, stubs and lumped elements
is computed on a per element basis. The input impedance of each section is
calculated starting with the element furthest away from the generator. The input
impedance of the section nearest to the load acts as the load impedance for the next
section closest to the generator. The transmission line equation represents series line
elements directly. For stub elements, the terminal impedance is converted to
admittance. Lumped elements and the interconnection of elements are calculated by
network analysis. The technique of direct analysis is simple to visualize and to
implement. A simple program that allows calculations of this type is program 4.4
CASCADE. In this program series transmission lines, open and short circuit shunt
lines together with series and shunt resistances, capacitors and inductances are
catered for.

```
][ FORMATTED LISTING
FILE: PROGRAM 4.4 CASCADE
PAGE-1

    1    REM
    5    REM    **** CASCADE ****
    7    REM
   10    DIM CO(100),RO(100),L1(100),C3(100)
   20    DIM R(100),L(100),Z0(100),E(100),A(100)
   30    DIM C1(100),L2(100),C(100)
   40    REM   100 ELEMENTS CAN BE USED
   50    REM   TO INCREASE CHANGE DIM
   60    HOME
   70    PRINT
   80    PRINT "NUMBER OF ELEMENTS REQUIRED"
   90    PRINT "TO TERMINATE I/P 0"
  100    INPUT N1
  110    IF N1 = 0 THEN
            1580
  120    PRINT "LOAD RESISTANCE"
  130    INPUT R9
  140    PRINT
  150    PRINT "CODING IS AS FOLLOWS"
  160    PRINT "TRANSMISSION LINE   = 1"
  170    PRINT "OPEN CIRCUIT STUB   = 2"
  180    PRINT "SHORT CIRCUIT STUB = 3"
  190    PRINT "SHUNT RESISTANCE    = 4"
  200    PRINT "SHUNT CAPACITANCE   = 5"
  210    PRINT "SERIES INDUCTANCE   = 6"
  220    PRINT "SERIES RESISTANCE   = 7"
  230    PRINT "SERIES CAPACITANCE = 8"
  240    PRINT "SHUNT INDUCTANCE    = 9"
  250    FOR K = 1 TO N1
  260        INPUT CO(K)
  270    NEXT K
  280    FOR J = 1 TO N1
```

```
290      IF CO(J) = 1 THEN
            680
300      IF CO(J) = 2 THEN
            630
310      IF CO(J) = 3 THEN
            600
320      IF CO(J) = 4 THEN
            570
330      IF CO(J) = 5 THEN
            530
340      IF CO(J) = 6 THEN
            490
350      IF CO(J) = 7 THEN
            460
360      IF CO(J) = 8 THEN
            420
370      IF CO(J) = 9 THEN
            380
380      PRINT "SHUNT INDUCTANCE (nH)"
390      INPUT L
400      LET L2(J) = L * 1E - 9
410      GOTO 700
420      PRINT "SERIES CAPACITANCE (pF)"
430      INPUT C
440      LET C1(J) = C * 1E - 12
450      GOTO 700
460      PRINT "SERIES RESISTANCE (OHMS)"
470      INPUT RO(J)
480      GOTO 700
490      PRINT "SERIES INDUCTANCE (nH)"
500      INPUT L
510      LET L1(J) = L * 1E - 9
520      GOTO 700
530      PRINT "SHUNT CAPACITANCE (pF)"
540      INPUT C
550      LET C3(J) = C * 1E - 12
560      GOTO 700
570      PRINT "SHUNT RESISTANCE (OHMS)"
580      INPUT R(J)
590      GOTO 700
600      PRINT "S/C STUB : ZO,LTH.,EFF OR ER,ATT.(dB/CM)"
610      INPUT ZO(J),L(J),E(J),A(J)
620      GOTO 700
630      PRINT "O/C STUB:ZO,LTH,EFF OR ER,ATT(dB/CM),END CAP.(pF)"
640      INPUT ZO(J),L(J),E(J),A(J),C(J)
650      LET C(J) = C(J) * 1E - 12
660      IF C(J) = 0 THEN
                 LET C(J) = 1E - 15
670      GOTO 700
680      PRINT "LINE: ZO,LTH,EFF OR ER,ATT(dB/CM)"
690      INPUT ZO(J),L(J),E(J),A(J)
700  NEXT J
710  LET R1 = R9
720  LET X1 = 0
730  PRINT "FREQUENCY (GHZ)"
740  INPUT F
750  LET W = 2 * 3.141579 * F * 1E9
760  PRINT
770  PRINT
780  PRINT "****** RESULTS ******"
790  PRINT
800  PRINT "SENDING END IMPEDANCE"
810  PRINT "LOOKING INTO THE :--"
820  PRINT
830  FOR I = 1 TO N1
840      IF CO(I) = 1 THEN
            1530
850      IF CO(I) = 2 THEN
            1390
860      IF CO(I) = 3 THEN
            1320
870      IF CO(I) = 4 THEN
            1240
```

```
880        IF CO(I) = 5 THEN
           1150
890        IF CO(I) = 6 THEN
           1100
900        IF CO(I) = 7 THEN
           1060
910        IF CO(I) = 8 THEN
           1010
920        PRINT "SHUNT INDUCTANCE"
930        LET G1 = R1 / (R1 * R1 + X1 * X1)
940        LET B1 =  - (X1 / (R1 * R1 + X1 * X1))
950        LET B2 = (1 / (W * L2(I))`
960        LET B1 = B1 + B2
970        LET R1 = G1 / (G1 * G1 + B1 * B1)
980        LET X1 = (B1 / (G1 * G1 + B1 * B1))
990        PRINT "R= "R1" X= "X1
1000       GOTO 1560
1010       PRINT "SERIES CAPACITANCE"
1020       LET X2 =  - (1 / (W * C1(I)))
1030       LET X1 = X1 + X2
1040       PRINT "R= "R1" X= "X1
1050       GOTO 1560
1060       PRINT "SERIES RESISTANCE"
1070       LET R1 = R1 + RO(I)
1080       PRINT "R= "R1" X= "X1
1090       GOTO 1560
1100       PRINT "SERIES INDUCTANCE"
1110       LET X2 = W * L1(I)
1120       LET X1 = X1 + X2
1130       PRINT " R = ",R1" X= "X1
1140       GOTO 1560
1150       PRINT "SHUNT CAPACITANCE"
1160       LET G1 = R1 / (R1 * R1 + X1 * X1)
1170       LET B1 =  - (X1 / (R1 * R1 + X1 * X1))
1180       LET B2 = W * C3(I)
1190       LET B1 = B1 + B2
1200       LET R1 = G1 / (G1 * G1 + B1 * B1)
1210       LET X1 =  - (B1 / (G1 * G1 + B1 * B1))
1220       PRINT "R= ",R1" X="X1
1230       GOTO 1560
1240       PRINT "SHUNT RESISTANCE"
1250       LET G1 = R1 / (R1 * R1 + X1 * X1)
1260       LET B1 =  - (X1 / (X1 * X1 + R1 * R1))
1270       LET G1 = G1 + 1 / R(I)
1280       LET R1 = G1 / (G1 * G1 + B1 * B1)
1290       LET X1 =  - (B1 / (G1 * G1 + B1 * B1))
1300       PRINT "R= "R1" X= "X1
1310       GOTO 1560
1320       PRINT "SHORT CIRCUIT STUB"
1330       LET G2 = R1 / (R1 * R1 + X1 * X1)
1340       LET B2 =  - (X1 / (R1 * R1 + X1 * X1))
1350       LET X1 = 0
1360       LET R1 = 0
1370       GOSUB 1620
1380       GOTO 1450
1390       PRINT "OPEN CIRCUIT SHUNT STUB"
1400       LET G2 = R1 / (R1 * R1 + X1 * X1)
1410       LET B2 =  - (X1 / (R1 * R1 + X1 * X1))
1420       LET X1 =  - (1 / (W * C(I)))
1430       LET R1 = 0
1440       GOSUB 1620
1450       LET G3 = R1 / (R1 * R1 + X1 * X1)
1460       LET B3 =  - (X1 / (R1 * R1 + X1 * X1))
1470       LET G1 = G2 + G3
1480       LET B1 = B2 + B3
1490       LET R1 = G1 / (G1 * G1 + B1 * B1)
1500       LET X1 =  - (B1 / (G1 * G1 + B1 * B1))
1510       PRINT " R= "R1" X= "X1
1520       GOTO 1560
1530       PRINT "SERIES TRANSMISSION LINE"
1540       GOSUB 1620
1550       PRINT "R= "R1" X= "X1
1560  NEXT I
```

```
1570    GOTO 70
1580    PRINT
1590    PRINT "**** END OF PROGRAM ****"
1600    END
1610    REM
1620    REM   LINE TRANSFORMATION
1630    REM
1640    LET W1 = 30 / (F *  SQR (E(I)))
1650    PRINT "WAVE= "W1
1660    LET R2 = (R1 * R1 - ZO(I) * ZO(I) + X1 * X1) / ((R1 + ZO(I)) ^ 2 + X1 * X
        1)
1670    LET X2 = 2 * X1 * ZO(I) / ((R1 + ZO(I)) ^ 2 + X1 * X1)
1680    REM   ATN QUADRANT CHECK
1690    IF X2 <  > 0 THEN
             1730
1700    IF R2 >  = 0 THEN
             1720
1710    LET G =  - 3.1415927:
             GOTO 1830
1720    LET G = 1E - 20:
             GOTO 1830
1730    IF R2 < 0 THEN
             1760
1740    IF X2 = 0 THEN
             1690
1750    LET G =  ATN (X2 / R2):
             GOTO 1830
1760    IF R2 <  > 0 THEN
             1800
1770    IF X2 >  = 0 THEN
             1790
1780    LET G =  - 1.5707963:
             GOTO 1830
1790    LET G = 1.5707963:
             GOTO 1830
1800    IF R2 >  = 0 THEN
             1830
1810    IF X2 = 0 THEN
             1690
1820    LET G = 3.1415927 +  ATN (X2 / R2)
1830    LET T1 = G
1840    LET M1 =  SQR (R2 * R2 + X2 * X2)
1850    LET A2 = A(I) / 8.686
1860    LET T2 = T1 - 4 * 3.1415927 * L(I) / W1
1870    LET M2 = M1 *  EXP ( - (2 * A2 * L(I)))
1880    LET D = 1 - 2 * M2 *  COS (T2) + M2 * M2
1890    LET R1 = ZO(I) * (1 - M2 * M2) / D
1900    LET X1 = ZO(I) * 2 * M2 *  SIN (T2) / D
1910    RETURN
```

END-OF-LISTING

]RUN

```
NUMBER OF ELEMENTS REQUIRED
TO TERMINATE I/P 0
?4
LOAD RESISTANCE
?50

CODING IS AS FOLLOWS
TRANSMISSION LINE   = 1
OPEN CIRCUIT STUB   = 2
SHORT CIRCUIT STUB  = 3
SHUNT RESISTANCE    = 4
SHUNT CAPACITANCE   = 5
SERIES INDUCTANCE   = 6
SERIES RESISTANCE   = 7
SERIES CAPACITANCE  = 8
SHUNT INDUCTANCE    = 9
?1
?6
```

```
?7
?8
LINE: ZO,LTH,EFF OR ER,ATT(dB/CM)
?50
??2
??1
??0.1
SERIES INDUCTANCE (nH)
?1
SERIES RESISTANCE (OHMS)
?10
SERIES CAPACITANCE (pF)
?2
FREQUENCY (GHZ)
?1


****** RESULTS ******

SENDING END IMPEDANCE
LOOKING INTO THE :--

SERIES TRANSMISSION LINE
WAVE= 30
R= 50 X= 0
SERIES INDUCTANCE
 R =            50 X= 6.283158
SERIES RESISTANCE
R= 60 X= 6.283158
SERIES CAPACITANCE
R= 60 X= -73.2946594

NUMBER OF ELEMENTS REQUIRED
TO TERMINATE I/P 0
?0

**** END OF PROGRAM ****
```

4.6.1 *ABCD* Matrices

For large systems or for analysis routines that must be called many thousands of times, which might be the case where a circuit analysis routine is embedded in an optimization program, an efficient scheme for representing circuit elements is required. If program CASCADE were to be used, the execution time for the program would be longer than if the circuit elements were coded in the form of *ABCD* matrices. The *ABCD* representation (sometimes called chain or transmission parameter representation) of a circuit element is based on the concept of a two-port network. Figure 4.37 shows the basic two-port network together with the voltage and current configuration commonly used.

In terms of figure 4.37, the circuit parameters are defined by equations (4.39) and (4.40) as

$$V_1 = A V_2 - B I_2 \tag{4.39}$$

$$I_1 = C V_2 - D I_2 \tag{4.40}$$

or in matrix form as

$$\begin{bmatrix} V_1 \\ I_1 \end{bmatrix} = \begin{bmatrix} A & B \\ C & D \end{bmatrix} \begin{bmatrix} V_2 \\ -I_2 \end{bmatrix}$$

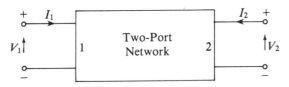

Figure 4.37   Two-Port Configuration

Notice how current I_2 has been defined as flowing into the two-port network. It should also be noted that the choice of V_1, V_2 and I_1, I_2 in this order is not a unique choice to define the network. In fact, six possible pairs of choices exist each leading to a different 2×2 matrix. Stated briefly these are:

(1) Open circuit impedance parameters Z
(2) Short circuit admittance parameters Y
(3) $ABCD$, chain or transmission parameters T
(4) Inverse transmission parameters B
(5) Hybrid parameters H
(6) Inverse hybrid parameters G

Any one set can be transformed to any other set.

By redefining the terminal quantities of interest, another type of 2×2 matrix, the S parameter matrix, can be defined (see section 4.6.3). In the unlikely event that a completely general computer analysis program is needed that can accept data in the form of a random mixture of matrix representations and then produce results in the form of a selected matrix type, a large amount of programming will be required. If, however, the $ABCD$ matrix T is selected as an intermediate conversion step between a matrix of one type and a matrix of another type, then only 12 out of a possible 36 transformations need be programmed.

$$[Z, Y, B, H, G, S, T] \rightarrow [T] \rightarrow [Z, Y, B, H, G, S, T]$$

This results in a more efficient program code.

In a real analysis situation, each circuit element is considered to be a single $ABCD$ matrix placed in series with the previous matrix (figure 4.38). The resulting connection for two cascaded matrices can be represented by a single two-port with parameters A, B, C, D where

$$A = A_1A_2 + B_1C_2$$

$$B = A_1B_2 + B_1D_2$$

$$C = A_2C_1 + C_2D_1$$

$$D = B_2C_1 + D_1D_2$$

In other words n cascaded two-ports defined as $ABCD$ matrices can be represented as a single two-port after repeated 2×2 matrix multiplication. This is very attractive since a computer can perform this multiplication efficiently at little programming

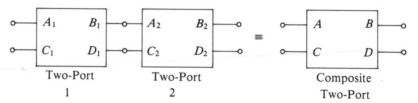

Figure 4.38   Cascaded Two-Ports

expense. The utility of the *ABCD* transformation to reduce the amount of programming necessary for large analysis programs receiving various data types, together with the fact that all commonly used basic networks have an *ABCD* representation in conjunction with the ease of interconnection of elements, makes the *ABCD* matrix a natural choice for circuit analysis programs.

4.6.2 *ABCD* Circuit Element Representation

For transmission line work, five basic circuits require examination, these are:

(1) Series (impedance) element
(2) Shunt (admittance) element
(3) TEE element
(4) PI element
(5) General transmission line.

Considering each circuit in turn.

   Series (impedance) element (figure 4.39)
   From the defining equations for the *ABCD* representation of the two-port applied to the series element

$$V_1 = V_2 + I_2 Z$$

and

$$I_1 = -I_2$$

hence $A = 1$, $B = Z$, $C = 0$, $D = 1$.

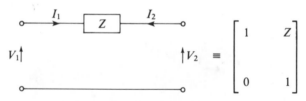

Figure 4.39   Series (Impedance) Element

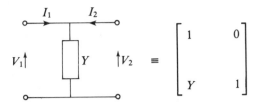

Figure 4.40   Shunt (Admittance) Element

Shunt (admittance) element (figure 4.40)

Here

$$V_1 = V_2$$

and

$$I_1 + I_2 = YV_1 = YV_2$$

$$\therefore \quad I_1 = YV_2 - I_2$$

hence $A = 1$, $B = 0$, $C = Y$, $D = 1$

The admittance matrix can be obtained directly from the impedance matrix if it is noted that $Y = 1/Z$ on transposing the impedance matrix.

Asymmetrical TEE element (figure 4.41)

From Kirchoff's voltage law for the I_1 loop

$$V_1 = I_1Z_1 + I_1Z_3 + I_2Z_3$$

From this equation

$$I_1 = \frac{V_1 - I_2Z_3}{Z_1 + Z_3}$$

For the I_2 loop

$$V_2 = I_2Z_2 + I_2Z_3 + I_1Z_3$$

Combining the last two equations gives

$$V_2(Z_1 + Z_3) = I_2(Z_1Z_2 + Z_2Z_3 + Z_3Z_1) + Z_3V_1$$

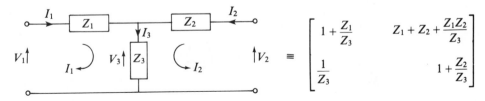

Figure 4.41   Asymmetrical TEE Element

Hence

$$V_1 = V_2\left(1 + \frac{Z_1}{Z_3}\right) - I_2\left(Z_1 + Z_2 + \frac{Z_1 Z_2}{Z_3}\right)$$

Comparing this equation with the equations defining the *ABCD* matrix

$$A = 1 + \frac{Z_1}{Z_3}$$

$$B = Z_1 + Z_2 + \frac{Z_1 Z_2}{Z_3}$$

In order to find the *C* and *D* elements of the matrix, rearrange the governing equation for the I_2 loop so that

$$I_1 Z_3 = V_2 - I_2(Z_2 + Z_3)$$

from this

$$I_1 = \frac{V_2}{Z_3} - I_2\left(1 + \frac{Z_2}{Z_3}\right)$$

thus

$$C = \frac{1}{Z_3}$$

$$D = \left(1 + \frac{Z_2}{Z_3}\right)$$

For a symmetrical TEE network the *A, B, C, D* coefficients derived above can be simplified. In the symmetrical network $Z_1 = Z_2 = Z$, the reduced *ABCD* matrix is stated below.

$$\begin{bmatrix} 1 + \dfrac{Z}{Z_3} & 2Z + \dfrac{Z^2}{Z_3} \\[2mm] \dfrac{1}{Z_3} & 1 + \dfrac{Z}{Z_3} \end{bmatrix}$$

Asymmetrical PI element (figure 4.42)
The *ABCD* coefficients can be deduced for this network either directly by

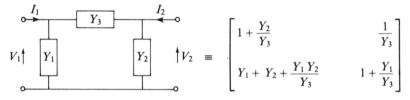

Figure 4.42   Assymetrical PI Element

Kirchoff's laws or indirectly by matrix transposition of the asymmetrical TEE network. Let

$$Y_1 = \frac{1}{Z_1}, \quad Y_2 = \frac{1}{Z_2} \quad \text{and} \quad Y_3 = \frac{1}{Z_3}$$

this yields the desired result

$$
\begin{bmatrix}
1 + \dfrac{Y_2}{Y_3} & \dfrac{1}{Y_3} \\[3mm]
Y_1 + Y_2 + \dfrac{Y_1 Y_2}{Y_3} & 1 + \dfrac{Y_1}{Y_3}
\end{bmatrix}
$$

General Transmission Line (figure 4.43)

In section 1.3, a symmetrical TEE circuit was shown to be equivalent to a lossy transmission line provided the series elements Z_1 and Z_2 of figure 4.41 are made equal to

$$Z_o \tanh\left(\frac{\psi l}{2}\right)$$

and the shunt element Z_3 is assigned a value

$$\frac{Z_o}{\sinh(\psi l)}$$

substitution of these values into the *ABCD* matrix for the symmetrical TEE circuit yields

$$A = D = 1 + \tanh\left(\frac{\psi l}{2}\right) \sinh(\psi l)$$

$$= 1 + \frac{2 \sinh(\psi l/2)}{\cosh(\psi l/2)} \sinh\left(\frac{\psi l}{2}\right) \cosh\left(\frac{\psi l}{2}\right)$$

$$= 1 + 2 \sinh^2\left(\frac{\psi l}{2}\right)$$

$$= \cosh(\psi l)$$

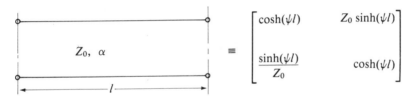

Figure 4.43 General Transmission Line

element C is simply

$$\frac{1}{Z_3} = \frac{1}{Z_o} \sinh (\psi l)$$

In section 1.3 it was shown that for a PI equivalent circuit

$$Z_3 = \frac{1}{Y_3} = Z_o \sinh (\psi l)$$

The B element of the $ABCD$ matrix representation for the transmission line element is equivalent to the Z_3 element of the symmetrical PI circuit, therefore

$$B = Z_o \sinh (\psi l)$$

This means that all the elements of the $ABCD$ matrix representation of the general transmission line have been found.

For a line displaying zero loss, the generalized transmission line matrix can be rewritten by replacing the hyperbolic functions with their trigonometric equivalents. So that the lossless lie matrix becomes

$$\begin{bmatrix} \cos (\beta l) & jZ_o \sin (\beta l) \\ j\dfrac{\sin (\beta l)}{Z_o} & \cos (\beta l) \end{bmatrix}$$

Numerical evaluation of the $ABCD$ matrix elements derived above is valid at a single frequency only, if they are concerned with elements other than lumped resistors. This means that if circuit performance over a range of frequencies is to be evaluated then the coefficients of each $ABCD$ matrix must be recomputed at each frequency. In effect, the frequency response of a circuit will be computed only at discrete frequencies across the range of interest.

In networks that are symmetrical, coefficient A is equal to coefficient D in the transmission matrix. If the network is reciprocal (i.e. ports one and two of the circuit element can be reversed without any change in the electrical characteristics of the network) then, for the transmission matrix describing the two-port behavior, $AD - BC = 1$. This serves as a check on the validity of the matrix elements computed for reciprocal components, especially if they turn out to be very large or very small since they may be prone to truncation effects.

* * *

Example 4.21

Show that a uniform transmission line element exhibits reciprocal properties.

Solution

For a reciprocal network

$$AD - BC = 1$$

Substituting the *ABCD* coefficients for a uniform transmission line gives

$$\cosh^2(\psi l) - \sinh^2(\psi l) = 1$$

or

$$1 + \sinh^2(\psi l) = \cosh^2(\psi l)$$

If this is true then the network is reciprocal. Rewriting the hyperbolic functions in terms of exponents

$$\tfrac{1}{4}(e^x + e^{-x})^2 = 1 + \tfrac{1}{4}(e^x - e^{-x})^2$$

where $x = \psi l$.

Expanding the squared terms and equating both sides gives

$$e^{2x} + 2 + e^{-2x} = 4 + e^{2x} - 2 + e^{-2x}$$

$$\therefore \quad 4 = 4$$

which is certainly true. Therefore, a uniform transmission line exhibits reciprocal properties.

*   *   *

When the *ABCD* matrices for each element in a particular circuit at a specific frequency have been evaluated, they can be multiplied together to form the composite *ABCD* matrix for the circuit at that frequency. Once the overall *ABCD* matrix is known then input impedance seen looking into either of the ports of the *ABCD* two-port can be found.

Consider figure 4.44, this defines the input impedance seen looking toward the load through the network from port one to port two as Z_{in_1}, while Z_{in_2} defines the impedance seen looking back through the network from port two to port one towards the generator impedance Z_1. Here, the ratio of equations (4.39) and (4.40), the defining equations for the *ABCD* representation of a two-port, yield Z_{in_1}

$$Z_{in_1} = \frac{V_1}{I_1} = \frac{AV_2 - BI_2}{CV_2 - DI_2} = \frac{AZ_2 + B}{CZ_2 + D} \tag{4.41}$$

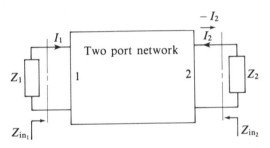

Figure 4.44   Terminated Two-Port

where

$$Z_2 = \text{load impedance} = \frac{V_2}{-I_2}$$

For $Z_{in_2} = V_2/I_2$ the defining equations (4.39) and (4.40) must be rearranged. Restating these equations

$$V_1 = AV_2 - BI_2$$

and

$$I_1 = CV_2 - DI_2$$

$$\therefore \quad I_2 = \frac{AV_2 - V_1}{B} = \frac{CV_2 - I_1}{D}$$

hence

$$V_2(AD - BC) = DV_1 - BI_1$$

similarly

$$V_2 = \frac{V_1 + BI_2}{A} = \frac{I_1 + DI_2}{C}$$

hence

$$I_2(AD - BC) = CV_1 - AI_1$$

$$\therefore \quad Z_{in_2} = \frac{V_2}{I_2} = \frac{DV_1 - BI_1}{CV_1 - AI_1} = \frac{DZ_1 + B}{CZ_1 + A} \qquad (4.42)$$

where

$$Z_1 = \frac{V_1}{-I_1}$$

Now that equations (4.41) and (4.42) have been established, the voltage transfer function V_g/V_2 for the overall network will be found. The voltage transfer function is particularly useful when filter circuits are being evaluated. The defining parameters for this type of circuit are shown in figure 4.45.

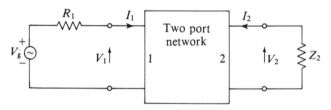

Figure 4.45   Two-Port Notation for Calculation of Voltage Transfer Function

From the input loop

$$\frac{V_g - V_1}{R_1} = I_1$$

from the output loop

$$V_2 = -I_2 Z_2$$

Taking the ratio of these expressions

$$\frac{V_g}{V_2} = \frac{I_1 R_1 + V_1}{-I_2 Z_2}$$

where

$$\frac{V_1}{-I_2 Z_2} = \frac{A V_2}{-I_2 Z_2} + \frac{B}{Z_2} = \frac{A Z_2}{Z_2} + \frac{B}{Z_2}$$

and

$$\frac{I_1 R_1}{-I_2 Z_2} = \left(\frac{C V_2}{-I_2} + D\right) \frac{R_1}{Z_2}$$

$$= C R_1 + \frac{D R_1}{Z_2}$$

Collecting these terms gives the desired transfer function

$$\frac{V_g}{V_2} = A + \frac{B}{Z_2} + C R_1 + \frac{D R_1}{Z_2}$$

Finally

$$\frac{V_2}{V_g} = \frac{Z_2}{A Z_2 + B + C R_1 Z_2 + D R_1} \tag{4.43}$$

Equations (4.41)–(4.43) facilitate an examination of overall circuit behavior. It should, of course, be noted that the coefficients A, B, C, D defining the two-port network are, in general, complex terms.

With an expression for the voltage transfer function now at hand, equation (4.43), it is possible to calculate the total loss introduced on inserting the circuit under consideration between the generator and load. This loss is specified as the transducer loss ratio P_1/P_2. Where P_1 is the power available at the input port of the overall network element given by the maximum power transfer theorem as

$$P_1 = \frac{|V_g|^2}{4 R_1}$$

and where P_2 is the power delivered to load reactance Z_2 given as

$$P_2 = \frac{|V_2|^2}{|Z_2|}$$

where V_2 can be calculated from equation (4.39).

For all the power available at port one of the two-port network to be transferred to the load impedance, the ratio P_1/P_2 will have to equal unity. Taking logarithms to the base ten of the transducer loss enables the attenuation, denoted as L, to be calculated in decibels

$$L = 10 \log_{10}\left(\frac{P_1}{P_2}\right) \text{ dB}$$

$$\therefore \quad L = 10 \log_{10}\left(\left|\frac{V_g}{V_2}\right|^2 \frac{|Z_2|}{4R_1}\right) \text{ dB} \tag{4.44}$$

From the discussion of the *ABCD* two-port matrix representation, it has been seen that this method of representation of circuit elements provides a compact and powerful means for circuit analysis. The method allows a number of circuit evaluation criteria such as loss, input impedance and transfer function to be calculated in one step. This leads to increased computational efficiency which is essential when large scale analysis routines are required. A number of large scale programs use the *ABCD* matrix representation for circuit elements in their analysis sections.

One early attempt at such a program is one called DEMON [4]. This program is written in the Fortran language and provides a number of conversion routines for the transformation of one set of two-port matrices, *Z*, *Y*, etc., to transmission matrices, reverse transformations are included also. Most types of commonly used circuit elements are catered for in the program and provision is made for a user supplied routine to be included.

4.6.3 Scattering Parameters

At low frequencies, say below 0.5 GHz, voltage and current can be measured and any of the six parameter sets mentioned in section 4.6.1 will adequately define the circuit element. At higher frequencies, when circuit elements have to be experimentally characterized, scattering parameters are used. Experimental measurement is necessary for components operating at higher frequencies since parasitics associated with the device packaging, e.g. bonding lead capacitance and inductance, or with transmission line discontinuities result in a deterioration of circuit element performance. This deterioration is often impossible to predict beforehand and must be measured for some or all of the elements used in a circuit.

Conventional methods of impedance measurement by voltage and current become unpredictable or impossible due to the lack of adequate measuring equipment at high frequencies. In order to derive experimentally a two-port *ABCD* matrix element a series of calibration pieces, precision short circuits, matched loads, etc., are required. Again, at high frequencies the precision of these calibration components cannot be maintained over frequency ranges of greater than a few percent of the center frequency due to surface loss and fringing capacitance together with repeatability of connection. Finally, if the system under investigation contains active devices such as diodes or transistors, placing an open or short circuit calibration

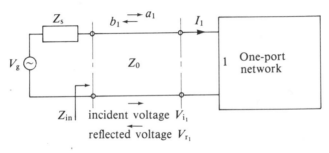

Figure 4.46   One-Port S-Parameter Notation

piece on one port of the device may result in instability leading to device oscillation. This will invalidate any characterization measurements made.

The method devised for the two-port characterization of networks at high frequencies is based on incident and reflected waves rather than voltage or current. The use of waves will lead to the definition of the scattering parameter matrix or S matrix. Figure 4.46 gives the required notation for a one-port network.

In terms of the one-port network, a signal launched by the generator V_g matched to the characteristic impedance of the line Z_0 will travel along the line as an incident wave. Part of this wave will be reflected (scattered) and the rest will be dissipated in the one-port. The scattered wave will be absorbed in the voltage generator provided an exact match between generator and line has been maintained.

Defining the normalized incident and scattered voltage waves as

$$a = \frac{V_{i_1}}{\sqrt{Z_0}}$$

and

$$b_1 = \frac{V_{r_1}}{\sqrt{Z_0}}$$

respectively, will ultimately lead to the scattering parameter or S-parameter representation of the two-port. If the dimensions of the definitions for incident and reflected waves are examined, it is found that they are defined in units of power since

$$\text{incident power} = \frac{|V_i|^2}{Z_0}$$

$$\text{and scattered power} = \frac{|V_r|^2}{Z_0}$$

The main reason for defining incident and reflected waves in this manner is that, at frequencies from about 100 MHz to 100 GHz and beyond, power measurements can be made with relative ease, and that a_i and b_i can be set to zero.

Combining the definition for the input reflection coefficient with those for incident and reflected voltage waves gives

$$\Gamma_{in} = \frac{V_{r_1}}{V_{i_1}} = \frac{b_1}{a_1} = S_{11}$$

here S_{11} is the scattering parameter for the one-port under test.

Extending the argument given above to the two-port shows the utility of S-parameter characterization (figure 4.47).

This time when a signal is launched by the signal source considered to be exactly matched to the input line, characteristic impedance Z_{o_1}, some of the incident wave a_1 will reach port two of the two-port network interposed between lines one and two. Here it will contribute to reflected wave b_2. Part of the incident wave will be reflected at port one of the network back into line one. Similarly, at port two of the two-port network, part of the a_2 wave will be transmitted to wave b_1, the rest will be reflected back into line two. If the load termination on line two is not matched exactly to this line, the process of partial reflection and partial transmission of waves a_2 and b_2 will continue until the system losses absorb any energy remaining in the waves.

For the incident and reflected waves on the output side of the network the remaining definitions are

$$a_2 = \frac{V_{i_2}}{\sqrt{Z_o}}$$

and

$$b_2 = \frac{V_{r_2}}{\sqrt{Z_o}}$$

The scattering parameters for both parts of the two-port are defined as

$$b_1 = S_{11}a_1 + S_{12}a_2 \tag{4.45}$$

$$b_2 = S_{21}a_1 + S_{22}a_2 \tag{4.46}$$

or in matrix form as

$$\begin{bmatrix} b_1 \\ b_2 \end{bmatrix} = \begin{bmatrix} S_{11} & S_{12} \\ S_{21} & S_{22} \end{bmatrix} \begin{bmatrix} a_1 \\ a_2 \end{bmatrix}$$

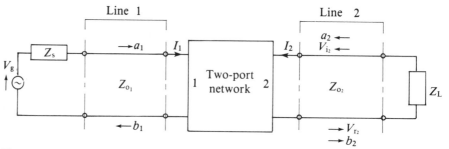

Figure 4.47   Two-Port S-Parameter Notation

What remains to be done now is to define a method for evaluating the coefficients of the S-matrix.

To find S_{11} the output port of the two-port network is terminated in a matched load. This sets a_2 equal to zero so that S_{11}, the input reflection coefficient, can be found from equation (4.45)

$$b_1 = S_{11}a_1 + S_{12}0$$

$$\therefore \quad S_{11} = \frac{b_1}{a_1}\bigg|_{a_2=0}$$

Under this condition S_{21}, the forward transmission coefficient for the network, can be found from equation (4.46)

$$b_2 = S_{21}a_1 + S_{22}0$$

$$\therefore \quad S_{21} = \frac{b_2}{a_1}\bigg|_{a_2=0}$$

It should be recognized that S_{21}, the forward transmission coefficient for the network, is a measure of attenuation for passive circuits and a measure of gain for circuits containing active devices, e.g. an amplifier.

When the input side of the circuit is matched, i.e. $a_1 = 0$, then S_{22} the output reflection coefficient is given as

$$S_{22} = \frac{b_2}{a_2}\bigg|_{a_1=0}$$

S_{12}, the reverse transmission coefficient, can also be found since

$$S_{12} = \frac{b_1}{a_2}\bigg|_{a_1=0}$$

When measuring the scattering parameters for a network it is often impractical to connect measurement equipment to the device under test directly. Measurement is usually carried out at some remote location by introducing sections of transmission line between the device under test and the measuring equipment. The phase shift introduced by the connecting line sections can be calculated and removed to recover the original S-parameters of the network under test. A two-port network under test is embedded between two sections of transmission line each with different electrical length (figure 4.48).

In figure 4.48 the S-parameters measured at plane A and plane B can be denoted by the transformed matrix $\mathbf{S}^1$. Consider what happens when the input reflection coefficient is required at port one of the two-port under test. The measured input reflection coefficient $S_{11}{}^1$ is equivalent to the reflection coefficient measured at the two-port terminals S_{11} together with a phase shift of $2\psi l_1$ since wave a_1 has to travel along the length of the embedding line l_1. For the forward transmission coefficient term S_{21} the measured term will equal the parameter measured at the two-port plus the additional phase shifts experienced by a_1 traveling along l_1 and by b_2 through line length l_2.

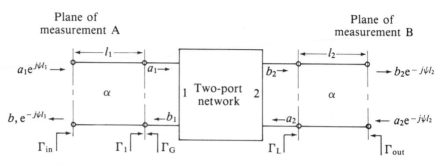

Figure 4.48   Embedded Two-Port

Summarizing what has been said in mathematical terms:

$$S_{11}{}^1 = S_{11} \exp{(-2j\psi l)}$$

$$S_{21}{}^1 = S_{21} \exp{[-j\psi(l_1 + l_2)]}$$

Similar expressions can be written for $S_{12}{}^1$ and $S_{22}{}^1$. Gathering together all the terms for the measured scattering parameters the $\mathbf{S}^1$ matrix can be written as

$$\mathbf{S}^1 = \begin{bmatrix} \exp{(-2j\psi l_1)}S_{11} & \exp{[-j\psi(l_1 + l_2)]}S_{12} \\ \exp{[-j\psi(l_1 + l_2)]}S_{21} & \exp{(-2j\psi l_2)}S_{22} \end{bmatrix}$$

or alternatively

$$\mathbf{S}^1 = \begin{bmatrix} \exp{(-j\psi l_1)} & 0 \\ 0 & \exp{(-j\psi l_2)} \end{bmatrix} \begin{bmatrix} S_{11} & S_{12} \\ S_{21} & S_{22} \end{bmatrix} \begin{bmatrix} \exp{(-j\psi l_1)} & 0 \\ 0 & \exp{(-j\psi l_2)} \end{bmatrix}$$

or as

$$\mathbf{S}^1 = \mathbf{TST}$$

From this matrix equation, the S-parameters of interest, i.e. the $\mathbf{S}$ matrix can be found by pre- and post-multiplying $\mathbf{S}^1$ by the inverse $\mathbf{T}$ matrix so that

$$\mathbf{S} = \mathbf{T}^{-1}\mathbf{S}^1\mathbf{T}^{-1}$$

The ability to measure the $\mathbf{S}^1$ matrix and then to perform the transformation of measurement planes as outlined above is extremely useful since a large class of practical transmission line circuit design problems involve a mixture of analytically or numerically modeled elements together with experimentally characterized components.

From the practical viewpoint of including de-embedded S-parameters in a matrix analysis program that employs $ABCD$ two-port circuit element representations, a set of transforms are required to map from S-parameters to $ABCD$ matrices. Table 4.1 gives the required transformation S to $ABCD$ and the inverse transform $ABCD$ to S.

A valuable extension of the S-parameter representation of circuit elements is the use of a graphical form to pictorially represent the interaction of incident and

Table 4.1   Transformation of S–ABCD–S Parameters

S	$\rightarrow$	ABCD	
$\begin{bmatrix} S_{11} & S_{12} \\ S_{21} & S_{22} \end{bmatrix} \rightarrow \dfrac{1}{(A+B+C+D)}$		$\begin{bmatrix} A+B-C-D & 2(AD-BC) \\ 2 & -A+B-C+D \end{bmatrix}$	

S	$\leftarrow$	ABCD
$\dfrac{1}{2S_{21}}\begin{bmatrix} (1+S_{11})(1-S_{22})+S_{12}S_{21} \\ (1-S_{11})(1-S_{22})-S_{12}S_{21} \end{bmatrix}$	$\begin{bmatrix} (1+S_{11})(1+S_{22})-S_{12}S_{21} \\ (1-S_{11})(1+S_{22})+S_{12}S_{21} \end{bmatrix} \leftarrow$	$\begin{bmatrix} A & B \\ C & D \end{bmatrix}$

reflected waves within a system of cascaded line segments. In this way, a systems designer can obtain a feel for circuit performance. The system under consideration is represented by means of a signal flow diagram (SFD). The technique of signal flow analysis can be used to provide analytical information about multiple reflections that can occur within cascaded line segments.

To introduce the concept of signal flow analysis, it is necessary to present the S-parameter defining equations in a graphical format. This graphical format is called the signal flow diagram. In signal flow format each node is made to represent a wave variable a_1, a_2, b_1 and b_2. Each branch is made to be an S-parameter. Finally, each node is equal to the sum of the branches that enter it.

Applying these rules to equations (4.45) and (4.46) then the signal flow diagram can be constructed (figure 4.49).

The signal flow representations for the two S-parameter governing equations can be superimposed to form the signal flow representation for the complete two-port. Examination of figure 4.49(c) shows that when incident wave a_1 meets the two-port, part of it is transmitted through the network to become part of wave b_2 while part of it is reflected to become part of wave b_1. Wave a_2 entering the second port of the two-port splits in a similar fashion, part of it joining wave b_2 and part adding to wave b_1.

The signal flow diagram representation for a two-port network excited with a generator exhibiting internal impedance and terminated in a load with reflection coefficient Γ_L is shown in figure 4.50.

To find out what is happening to the incident and reflected voltage waves within the system, consider figure 4.50(b). Here the available power from the generator reaches the two-port and undergoes a series of interactions as described for figure 4.49(c). In addition, part of the incident wave a_1 is reflected to node b_1 where the mismatch between generator and network input port and Γ_g allows some of the reflected wave to contribute to the main incident wave once again. In this way, multiple reflections occur at the input side of the two-port. At the output side is a similar mismatch between network and load results in the Γ_L branch shown in figure 4.50(b). This branch will once again accommodate multiple reflections, in addition to the ordinary operation of the two-port described in figure 4.49(c).

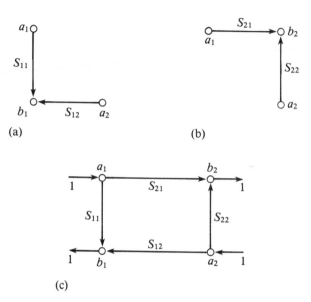

Figure 4.49 Signal Flow Representation of a Two-Port Network (a) Signal Flow Representation of Equation (4.44) (b) Signal Flow Representation of Equation (4.45) (c) Signal Flow Representation for Complete Two-Port

Application of signal flow analysis in a method known as Mason's rule enables complicated network transfer functions to be easily determined. This technique is

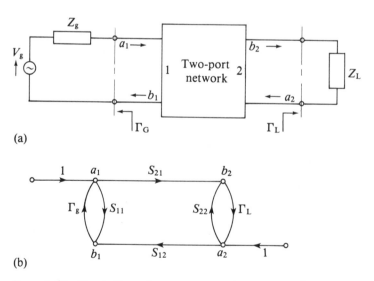

Figure 4.50 Signal Flow Representation of Excited Two-Port Network (a) Circuit Configuration (b) Signal Flow Representation

especially valuable when active devices are under investigation and quantities such as voltage gain and power gain are of interest [6]. This is beyond the scope of this text since active device behavior is not of primary concern. For this reason, an elementary analytical alternative to Mason's rule can be applied.

Figure 4.50(b), describing the embedded two-port, indicates that

$$\Gamma_{in} = \Gamma_1 \exp\left(-2\psi l_1\right) = \frac{b_1}{a_1}$$

This equation together with the governing equations for the S-parameter matrix, equations (4.45) and (4.46), and the governing equation for Γ_L, i.e.

$$\Gamma_L = \frac{a_2}{b_2}$$

gives

$$b_2 = \frac{a_2}{\Gamma_L} = S_{21}a_1 + S_{22}a_2$$

from this

$$a_1 = \frac{a_2(1 - S_{22}\Gamma_L)}{S_{21}\Gamma_L}$$

now

$$\Gamma_1 = \frac{b_1}{a_1}$$

therefore

$$\Gamma_1 = S_{11} + S_{12}\frac{a_2}{a_1}$$

substituting for a_1 in the last equation yields

$$\Gamma_1 = S_{11} + \frac{S_{12}S_{21}\Gamma_L}{(1 - S_{22}\Gamma_L)} \tag{4.47}$$

From this, the input reflection coefficient for the embedded system can be found as

$$\Gamma_{in} = \exp\left(-2\psi l_1\right)\left(S_{11} + \frac{S_{12}S_{21}\Gamma_L}{1 - S_{22}\Gamma_L}\right) = S_{11}{}^1 \tag{4.48}$$

For port two of the two-port network a similar expression can be derived, this time for the output reflection coefficient

$$\Gamma_{out} = \exp\left(-2\psi l_2\right)\left(S_{22} + \frac{S_{12}S_{21}\Gamma_G}{1 - S_{11}\Gamma_G}\right) = S_{22}{}^1 \tag{4.49}$$

When $l_1 = l_2 = 0$ and the generator and load impedance are matched to the input and output ports of the two-port, i.e. $\Gamma_L = \Gamma_G = 0$ then

$$\Gamma_{in} = S_{11} \quad \text{and} \quad \Gamma_G = S_{22}$$

as might be expected.

Equation (4.48) provides an alternative method to that provided by program CASCADE for calculating the input impedance frequency sensitivity of a series combination of transmission lines.

Figure 4.51 shows a typical low pass filter structure as it might appear if synthesized on a strip transmission line material. The filter consists of a cascade connection of lines of high impedance (the narrow bands) and lines of low impedance (the broad bands). Electrically, these can be thought of in a generalized sense as a series of transmission line segments each with a different phase shift, characteristic impedance, line loss and guide wavelength. Reflection coefficients R_{1n} and R_{2n} occur at the discontinuity between section $n-1$ and section n due to the impedance in the line $Z_{o_{n-1}}$ to Z_{o_n}. The reactive discontinuities, discussed in the next

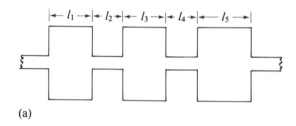

(a)

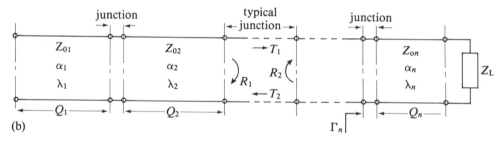

(b)

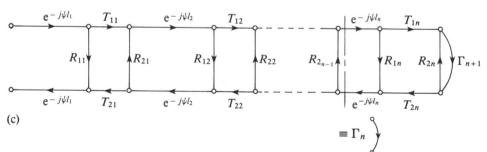

(c)

Figure 4.51   Cascaded Line Analysis Using a Signal Flow Diagram (a) Physical Construction (b) Electrical Equivalent (c) Signal Flow Representation

chapter have been neglected. The quantities R_{1n}, R_{2n} can be written as

$$R_{1n} = -R_{2n} = \frac{Z_{n+1} - Z_n}{Z_{n+1} + Z_n}$$

in the same way the transmission coefficients T_{1n} and T_{2n} are given as

$$T_{1n} = 1 + R_{1n}$$

$$T_{2n} = 1 + R_{2n} = 1 - R_{1n}$$

Figure 4.51(c) shows that each discontinuity can be lumped with the transformation in phase associated with each corresponding line section to form the termination reflection coefficient for the next cascade section. All that is required to start the analysis process going is the reflection coefficient between the final line section and the load. Rewriting equation (4.48) in terms of the cascaded line analysis notation gives

$$\Gamma_n = \exp(-2j\psi l_n)\left[R_{1n} + \frac{(1 - R_{1n})(1 + R_{1n})\Gamma_{n+1}}{1 - R_{1n}\Gamma_{n+1}}\right]$$

$$\therefore \quad \Gamma_n = \exp(-2j\psi l_n)\left[R_{1n} + \frac{(1 - R^2_{1n})\Gamma_{n+1}}{1 - R_{1n}\Gamma_{n+1}}\right] \tag{4.50}$$

Provided the reflection coefficient at the load is known, then equation (4.50) is sufficient to enable the necessary line analysis.

Knowing the lengths and electrical properties of each line segment and the reflection coefficient at the load, each preceding reflection coefficient is calculated until the overall reflection coefficient for the complete system Γ_1 is found. The discontinuity reflection at each abrupt junction between line segments can be calculated by this method, as can the amount of power reflected or absorbed by the overall structure. Program 4.5 SIGFLOW, shows how equation (4.50) can be implemented for the analysis of series connected line segments.

```
][ FORMATTED LISTING
FILE: PROGRAM 4.5 SIGFLOW
PAGE-1

 10   REM
 20   REM   *** PROGRAM SIGFLOW ***
 30   REM
 40   REM   THIS PROGRAM COMPUTES
 50   REM   THE REFLECTION COEFF.
 60   REM   SENSITIVITY FOR
 70   REM   CASCADED LINES AS A
 80   REM   FUNCTION OF FREQ.
 90   REM   MICROSTRIP LINES ARE
100   REM   ACCOMMODATED FOR.
110   REM
120   REM   ROE = REFLECTION COEFF.
130   REM   ZO  = CHARAC. IMP.
140   REM   ER  = RELATIVE DIE. CONST.
150   REM   P1  = PHYSICAL SECTION LTH.
160   REM   W1  = WAVELTH.
170   REM
180   DIM RROE(20),IROE(20),RI(20),I1RI(20)
190   DIM Z(20),W1(20),P1(20),ARRI(20)
200   DIM REFLECTA(20),ANGLE(20)
210   DIM THETAI(20),XANG(20),YANG(20)
```

```
220   LET JJ = 1
230   HOME
240   REM
250   REM   INPUT DATA
260   REM
270   PRINT :
      PRINT "I/P REL. DIE. CONST."
280   PRINT "GREATER THAN 1 IF MICROSTRIP ANALYSIS REQUIRED"
290   INPUT ER
300   PRINT "I/P REQUIRED NO.OF SECTIONS"
310   INPUT N
320   FOR J = 1 TO N
330        PRINT "WORKING FROM LOAD ; I/P IMP. OF SECTION "J
340        INPUT Z(J)
350   NEXT J
360   PRINT "I/P START , STOP , CENTER AND INCREMENT FREQ (GHZ)"
370   INPUT FS,FF,FC,DF
380   LET FRQ = FS
390   FOR L = 1 TO N
400        GOSUB 1450
410   NEXT L
420   IF FRQ > FS THEN
           530
430   LET KL = N - 1
440   FOR J = 2 TO KL
450        PRINT "STARTING AT LOAD I/P PHYSICAL LTH. OF SECTION "J" IN MMS"
460        INPUT P1(J)
470   NEXT J
480   LET P1(N) = 0
490   GOSUB 1390
500   IF JJ = 0 THEN
           250
510   PRINT "I/P REFLECTION COEFF. AT LOAD IN POLAR FORM"
520   INPUT RROE(1),IROE(1)
530   REM
540   REM   CALCULATE LINE REF. COEFFS.
550   REM
560   LET LF = N - 1
570   FOR J = 1 TO LF
580        LET K = J + 1
590        LET ARRI(K) = ((Z(J) - Z(K)) / (Z(J) + Z(K)))
600        LET RI(K) = ARRI(K)
610        LET I1RI(K) = 0
620        LET THETAI(K) = (2 * 3.14159265 * P1(K) / W1(K)) * ( - 2.0)
630   NEXT J
640   FOR J = 1 TO LF
650        LET K = J + 1
660        LET C =  COS (THETAI(K))
670        LET S =  SIN (THETAI(K))
680        LET XA = 1 - RI(K) * RI(K)
690        LET XB = IROE(J) * XA
700        LET XC = RROE(J) * XA
710        LET XD = 1 + RROE(J) * RI(K)
720        LET XE = IROE(K) * RI(K)
730        LET XH = XB * XD - XC * XE
740        LET XG = XB * XE + XD * XC
750        LET XF = XD * XD + XE * XE
760        LET XI = XH / XF
770        LET XJ = RI(K) + XG / XF
780        LET RROE(K) = (C * XJ - S * XI)
790        LET IROE(K) = (C * XI + S * XJ)
800        LET REFLECTA(K) =  SQR (RROE(K) ^ 2 + IROE(K) ^ 2)
810        LET YANG(K) = IROE(K)
820        LET XANG(K) = RROE(K)
830        LET A =  ATN (YANG(K) / XANG(K))
840        IF ((XANG(K) < 0) AND (YANG(K) > = 0)) THEN
                LET ANG = ANG + 3.14159653
850        IF ((XANG(K) < 0) AND (YANG(K) < 0)) THEN
                LET ANG = ANG - 3.14159653
860        LET ANGLE(K) = (360 / (2 * 3.14159653)) * A
870   NEXT J
880   LET RFL = REFLECTA(N) * REFLECTA(N)
890   LET AB = (1 - RFL)
900   LET DB = 10 *  LOG (AB) /  LOG (10)
```

```
 910   REM
 920   REM    PRINT RESULTS
 930   REM
 940   PRINT :
       PRINT
 950   PRINT "**** THE RESULTS WILL NOW BE PRINTED ****"
 960   PRINT
 970   PRINT "THE RELATIVE PERMITTIVITY IS "ER
 980   PRINT "THE EFFECTIVE PERMITTIVITY IS "EEF" FOR MICROSTRIP"
 990   PRINT
1000   PRINT "THE NO. OF SECTIONS ARE "N
1010   LET J = 1
1020   PRINT
1030   PRINT "SECTION NO. "J
1040   PRINT "REFLECTION COEFF. IN POLAR FORM "RROE(J)" ; "IROE(J)
1050   PRINT "LINE IMP "Z(J)" OHMS"
1060   LET J = J + 1
1070   IF J > N THEN
              1220
1080   LET LV = J - 1
1090   PRINT
1100   PRINT "SECTION NO. "J
1110   PRINT "SECTION LTH. = "P1(J)" MMS"
1120   PRINT "SECTION IMP. = "Z(J)" OHMS"
1130   PRINT "GUIDE WAVELTH. OF SECTION = "W1(J)" MMS"
1140   PRINT
1150   PRINT "CELL INDICATION "LV" -- "J
1160   PRINT
1170   PRINT "DISCONTINUITY REFL. COEFF. "ARRI(J)
1180   PRINT "I/P REFL. FACTOR IS GIVEN AS "
1190   PRINT "AMPLITUDE "REFLECTA(J)" PHASE(DEGREES)"ANGLE(J)
1200   PRINT "REAL PART "XANG(J)" IMG. PART "YANG(J)
1210   GOTO 1060
1220   PRINT :
       PRINT :
       PRINT
1230   PRINT "POWER REFLECTION RATIO "RFL
1240   PRINT "POWER ABSORPTION RATIO "AB
1250   PRINT "POWER TRANSMITTED "DB" DB"
1260   LET F8 = FRQ + FC
1270   LET F8 = F8 / FC
1280   PRINT :
       PRINT "FREQ = "FRQ" NORMALISED FREQ. = "F8
1290   LET FRQ = FRQ + DF
1300   IF FRQ < = FF GOTO 390
1310   PRINT
1320   PRINT :
       PRINT "**** COMPUTATION COMPLETE ****"
1330   PRINT :
       PRINT "TYPE IN A NO. > 0 TO REPEAT"
1340   INPUT LOOP
1350   IF LOOP < > 0 THEN
              220
1360   PRINT :
       PRINT " **** END OF PROGRAM ****"
1370   END
1380   REM
1390   REM    CHECK ROUTINE
1400   REM
1410   PRINT "IS DATA CORRECT ? IF YES I/P 1 ; IF NO I/P 0"
1420   INPUT JJ
1430   RETURN
1440   REM
1450   REM    WAVELTH. ROUTINE
1460   REM    > 1 GIVES EEFF. FOR MICROSTRIP
1470   REM
1480   LET AQ = 2 / 3.141593
1490   LET B = 377 * 3.141593 / 2 / SQR (ER) / Z(L)
1500   LET AFR2 = LOG (B - 1) + 0.39 - 0.61 / ER
1510   LET WH = AQ * (B - 1 - LOG (2 * B - 1) + ((ER - 1) / 2 / ER) * AFR2)
1520   LET EEFF = (ER + 1) / 2 + ((ER - 1) / 2) / SQR (1 + 12 / WH)
1530   LET W1(L) = 3E11 / FRQ / 1E9 / SQR (EEFF)
1540   RETURN
```

END-OF-LISTING

```
]RUN

I/P REL. DIE. CONST.
GREATER THAN 1 IF MICROSTRIP ANALYSIS REQUIRED
?1
I/P REQUIRED NO.OF SECTIONS
?3
WORKING FROM LOAD ; I/P IMP. OF SECTION 1
?50
WORKING FROM LOAD ; I/P IMP. OF SECTION 2
?48
WORKING FROM LOAD ; I/P IMP. OF SECTION 3
?50
I/P START , STOP , CENTER AND INCREMENT FREQ (GHZ)
?1,1,1,1
STARTING AT LOAD I/P PHYSICAL LTH. OF SECTION 2 IN MMS
?150
IS DATA CORRECT ? IF YES I/P 1 ; IF NO I/P 0
?1
I/P REFLECTION COEFF. AT LOAD IN POLAR FORM
?0,0


**** THE RESULTS WILL NOW BE PRINTED ****


THE RELATIVE PERMITTIVITY IS 1
THE EFFECTIVE PERMITTIVITY IS 1 FOR MICROSTRIP

THE NO. OF SECTIONS ARE 3

SECTION NO. 1
REFLECTION COEFF. IN POLAR FORM 0 ; 0
LINE IMP 50 OHMS

SECTION NO. 2
SECTION LTH. = 150 MMS
SECTION IMP. = 48 OHMS
GUIDE WAVELTH. OF SECTION = 300 MMS

CELL INDICATION 1 -- 2

DISCONTINUITY REFL. COEFF. .0204081633
I/P REFL. FACTOR IS GIVEN AS
AMPLITUDE .0204081632 PHASE(DEGREES)4.19094642E-07
REAL PART .0204081632 IMG. PART 1.49277355E-10

SECTION NO. 3
SECTION LTH. = 0 MMS
SECTION IMP. = 50 OHMS
GUIDE WAVELTH. OF SECTION = 300 MMS

CELL INDICATION 2 -- 3

DISCONTINUITY REFL. COEFF. -.0204081633
I/P REFL. FACTOR IS GIVEN AS
AMPLITUDE 1.54548213E-10 PHASE(DEGREES)-74.9930964
REAL PART -4.00177669E-11 IMG. PART 1.49277355E-10


POWER REFLECTION RATIO 2.38851503E-20
POWER ABSORPTION RATIO 1
POWER TRANSMITTED 0 DB

FREQ = 1 NORMALISED FREQ. = 2

**** COMPUTATION COMPLETE ****

TYPE IN A NO. > 0 TO REPEAT
?0

  **** END OF PROGRAM ****
```

4.7   FURTHER READING

1   Smith, P. H., *Electronic Applications of the Smith Chart in Waveguide, Circuit and Component Analysis*, McGraw-Hill, July 1969.

This book gives basic and advanced uses of the Smith Chart. Theoretical background is also given for the problems encountered. Four plastic overlay charts are supplied also.

2   Analog Instrument Co, *Smith Chart and Accessories*, PO Box 808, New Providence, NJ 07974.

This company provides a range of Smith Chart graph paper and specialized slide rules for Smith Chart calculations. All these charts are sold under copyright agreement with Kay Electric Company, Pine Brook, N.J.

3   Weinberg, L. *Network Analysis and Synthesis*, McGraw-Hill, 1962.

This book is written for first-year graduates in engineering and as such tends to be fairly mathematical in content. However, it is an excellent book and chapter 1 together with the beginning of chapter 2 should be consulted for further information regarding two-port and *n*-port networks and introductory network theory pertinent to attenuator design.

4   Perlman, B. S., 'Computer Aided Design, Simulation and Optimisation', *Advances in Microwaves*, **8**, 321–99, 1974.

This paper provides an excellent overview of the field of microwave circuit design, analysis and subsequent optimization by computer-aided techniques. Anyone interested in large scale design and optimization is strongly recommended to read this article. A large scale computer program called DEMON (Diminishing Error Method of Optimisation for Networks) is described and a source listing in the Fortran Language provided.

5   Emery, F. E. and Policky, G. J., 'Computer Aided Design of Microwave Circuits', *Texas Instruments Application Report 1.19*, 1968.

This application report discusses, in a non-mathematical way, the strategies and organization used when designing a computer-aided design program for microwave circuit design and analysis problems. Circuit sensitivity to component perturbation is briefly discussed.

6   Hewlett Packard, 'S-Parameter Design', *Hewlett Packard Application Note 154*, April 1972.

This application note describes how amplifiers can be constructed provided the S-parameters quantifying the active devices can be found. The use of signal flow diagrams and Mason's rule are fully described.

7   Fitzpatrick, J., 'Error Models for Systems Measurement', *Microwave Journal*, **21**(5), 63–66, May 1978.

This short paper details how signal flow diagrams can be used to model errors in network analyzer systems.

8   Thomas, R. L., *A Practical Introduction to Impedance Matching*, Artech House, 1976.

Contains many design examples that exploit the Smith Chart for impedance matching using lumped circuits. Of particular interest is the use of the Carter Chart as an alternative to the Smith Chart. A description of manual broadband matching techniques is also given.

9   Liao, S. Y., *Microwave Devices and Circuits*, 2nd Edn, Prentice-Hall, 1985.
This book should provide interesting reading for those who want to incorporate active devices into designs that require the use of impedance matching techniques of the type described in this chapter. Almost all of the solid state devices currently employed in the microwave design industry are included in this text.

5

Transmission Line Circuits

The previous chapters itemized in some detail the terminology of transmission line analysis and design. This, together with the analysis and synthesis techniques described for lines having standard and non-standard geometries leads to the design and synthesis of a number of useful circuit elements. The design and synthesis of these circuits is helped by computer-aided techniques in the form of the design programs already discussed in previous sections, together with those to be given in this chapter.

The circuits described in this chapter represent some of the basic building blocks used in many types of radio frequency and microwave equipment. Detailed accurate design of these components is often difficult. This is especially true when the frequency of operation is high, when parasitics such as radiation or discontinuities cannot be neglected. The design algorithms in this chapter are therefore approximate. However, they are good enough to allow a reasonably close first attempt at the design of a circuit to meet a desired specification.

5.1 LUMPED LOW PASS FILTER DESIGN

The traditional starting place for the design of a transmission line filter circuit is the examination of a lumped element filter comprising passive circuit elements. The lumped element filter topology selected is normally synthesized from tables. These tables ensure that the lumped filter has component values that enable it to approximately exhibit the electrical properties called for in a preliminary specification.

Once the lumped component values have been established these values are then transformed to the desired distributed components. This conversion will be discussed later. The emphasis in this section will be to release the designer from the use of tables by substituting for them simple computerized data generation programs.

The discussion will be confined to two well-known low pass filter responses. It will be shown how once the design prototype has been found for the low pass filter circuit, simple transformations can be applied that allow high pass and band pass prototypes to be constructed. For the purposes of this discussion, a low pass filter is defined as a frequency selective circuit having a single passband between frequencies of 0 rad/s and some cut-off frequency of ω_c rad/s. The ideal filter response shown in figure 5.1 is not realizable in practice due to the infinitely steep cut-off curve at ω_c. The ideal filter response can be approximated in a number of different ways. One common approximation is given as

$$|G(\omega)| = \frac{1}{(1 + \omega^{2n})^{\frac{1}{2}}} \qquad n = 1, 2, 3, \ldots \qquad (5.1)$$

This represents the gain of the filter as a function of frequency and was first proposed by Butterworth [1]. The Butterworth approximation gives rise to a class of filter known as maximally flat or indeed Butterworth. The class of filters defined by equation (5.1) is known as maximally flat since the derivative of the function at frequencies much less than the cut-off frequency has a small rate of change with frequency.

Another well-known approximation to the ideal filter response occurs when

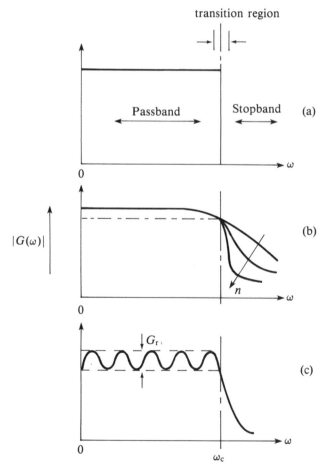

Figure 5.1   Low Pass Filter Responses (a) Ideal (b) Butterworth (c) Chebyshev

the gain function

$$|G(\omega)| = \frac{1}{[1 + \varepsilon C_n^{\,2}(\omega)]^{\frac{1}{2}}} \qquad n = 1, 2, 3, \ldots \tag{5.2}$$

is employed.

This expression, called the Chebyshev function, is somewhat more complicated to evaluate than the Butterworth function. In the low pass Chebyshev function, ε is a constant and $C_n(\omega)$ is a Chebyshev polynomial of the first kind and
degree n.

This polynomial is expressed as

$$C_n(\omega) = \cos\,[n\,\cos^{-1}(\omega)] \qquad 0 \le \omega \le 1$$

$$= \cosh\,(n\,\cosh^{-1}(\omega)) \qquad \omega > 1 \tag{5.3}$$

From equation (5.3) it can be seen that

$$C_0(\omega) = 1$$

$$C_1(\omega) = \omega$$

At this point, if higher order polynomials of the first kind are to be found then
for $0 \le \omega \le 1$, $\cos^{-1}(\omega)$ may be set equal to θ. Then

$$C_n(\omega) = \cos\,(n\theta)$$

$$\therefore \quad C_{n+1}\,(\omega) = \cos\,[(n+1)\theta]$$

$$= \cos\,(n\theta)\,\cos\,(\theta) - \sin\,(n\theta)\,\sin\,(\theta)$$

$$C_{n-1}(\omega) = \cos\,[(n-1)\theta]$$

$$= \cos\,(n\theta)\,\cos\,(\theta) + \sin\,(n\theta)\,\sin\,(\theta)$$

adding these two equations

$$C_{n+1}(\omega) + C_{n-1}(\omega) = 2\,\cos\,(n\theta)\,\cos\,(\theta) = 2\omega C_n(\omega)$$

hence

$$C_2(\omega) = 2\omega C_n(\omega) - C_0(\omega) = 2\omega^2 - 1$$

In this way a table of Chebyshev polynomials of the first kind can be constructed
(table 5.1).

A typical Chebyshev low pass filter amplitude characteristic is displayed in
figure 5.1(c). The filter response displays ripple in the passband, each ripple having
equal magnitude. For this reason Chebyshev filters are also called equiripple filters.
In the Chebyshev response, passband ripple amplitude can be traded for filter roll-
off rate for frequencies above ω_c. In general as the roll-off rate is increased for a
fixed number of sections, the in-band ripple amplitude increases. Passband ripple
can be held constant and roll-off rate adjusted by varying the number of sections
comprising the filter. Comparison of Butterworth and Chebyshev low pass filters
show the Chebyshev filter to have superior performance in the stopband. The

Table 5.1 Chebyshev Polynomials of the First Kind

Order n	Polynomial $C_n(\omega)$
0	1
1	ω
2	$2\omega^2 - 1$
3	$4\omega^3 - 3\omega$
4	$8\omega^4 - 8\omega^2 + 1$
5	$16\omega^5 - 20\omega^3 + 5\omega$
6	$32\omega^6 - 48\omega^4 + 18\omega^2 - 1$

Chebyshev filter has a greater roll-off rate than a Butterworth filter with the same number of reactive components.

The greater the number n of reactive components in a filter the more nonlinear its phase response will be. It turns out that a Butterworth filter has a more linear phase response than a comparable Chebyshev filter. For both filter types lower order filters exhibit a more linear response than do their higher order counterparts. If phase response is important for a particular application then the superiority of the Chebyshev amplitude response may be offset by an overly nonlinear phase response. When a linear phase response is of paramount importance then another class of filter, known as the Bessel filter should be considered. This type of filter has linear phase response over sections of its passband, but exhibits an inferior amplitude response when compared with Butterworth or Chebyshev low pass equivalents. Bessel filters find a variety of applications in phase shifting and time delay networks [1]. The design of Bessel filters will not be pursued any further in this text. However, the design techniques to be developed for Chebyshev and Butterworth filters can be extended to include Bessel types if required.

The attenuation characteristics $L(\omega)$ measured in dB of Butterworth and Chebyshev low pass filters can be found by taking logarithms of equations (5.1) and (5.2), i.e. the governing equations for each filter characteristic. This gives the following insertion loss equations.

Butterworth insertion loss

$$L(\omega) = -20 \log_{10}\left\{\frac{1}{[1 + (\omega/\omega_c)^{2n}]^{\frac{1}{2}}}\right\}$$

$$= 10 \log_{10}\left[1 + \left(\frac{\omega}{\omega_c}\right)^{2n}\right] \text{ dB} \qquad (5.4)$$

Chebyshev insertion loss

$$L(\omega) = -20 \log_{10}\left\{\frac{1}{[1 - \varepsilon C_n^{2}(\omega/\omega_c)]^{\frac{1}{2}}}\right\}$$

$$= 10 \log_{10}\left\{1 + \varepsilon \cos^2\left[n \cos^{-1}\left(\frac{\omega}{\omega_c}\right)\right]\right\}\Bigg|_{0 \le \omega \le \omega_c} \quad \text{dB}$$

$$= 10 \log_{10}\left\{1 + \varepsilon \cosh^2\left[n \cosh^{-1}\left(\frac{\omega}{\omega_c}\right)\right]\right\}\Bigg|_{\omega > \omega_c} \tag{5.5}$$

where

$$\varepsilon = \underset{10}{\quad}\left(\frac{\text{ripple amplitude in dB}}{10}\right)_{-1}$$

Here, frequency has been normalized to some cut-off frequency ω_c. Equations (5.4) and (5.5) are valuable for computing the attenuation characteristics of Butterworth and Chebyshev low pass filters. Computer program 5.1 LPATNN will compute attenuation tables for both filter types with known number of sections across the normalized frequency range 0 to 1 in 0.1 steps. For the Chebyshev response, the ripple in the passband G_r given in decibels is required as input data. For Butterworth filters this should be set to zero. Using the results from this program a graph of attenuation against normalized frequency for a Chebyshev filter having 2 dB ripple in the passband and consisting of five elements is plotted in figure 5.2.

```
]] FORMATTED LISTING
FILE: PROGRAM 5.1 LPATNN
PAGE-1


 10   REM
 20   REM    **** LPATNN ****
 30   REM
 40   REM    THIS PROGRAM CALCULATES
 50   REM    ATTENUATION CHARAC.
 60   REM    FOR Nth ORDER PASSIVE
 70   REM    BUTTERWORTH AND
 80   REM    CHEBYSHEV FILTERS
 90   REM
100   DIM LB(100),LC(100)
110   REM   DATA ENTRY
120   HOME
130   PRINT "ENTER ORDER OF FILTER"
140   INPUT N
150   PRINT "ENTER PASSBAND RIPPLE IN DB"
160   PRINT "IF ZERO BUTTERWORTH RESPONSE ASSUMED"
170   INPUT G1
180   HOME
190   PRINT "WORKING:------"
200   LET E = 10 ^ (G1 / 10) - 1
210   LET M = 1
220   REM   IN BAND RESPONSE
230   IF G1 = 0 THEN
      290
240   FOR I = 0 TO 1 STEP 0.1
250       LET F =  -  ATN (I /  SQR ( - I * I + 1)) + 1.570796326
260       LET LC(M) = 10 *  LOG (1 + E *  COS (N * F) *  COS (N * F)) /  LOG (1
      0)
270       LET M = M + 1
280   NEXT I
290   LET M = 1
300   FOR I = 0 TO 1 STEP .1
310       LET LB(M) = 10 *  LOG (1 + I ^ (2 * N)) /  LOG (10)
320       LET M = M + 1
330   NEXT I
340   REM   OUT OF BAND RESP.
```

```
350   REM   STEP DEFINES RESOLUTION
360   IF G1 = 0 THEN
            LET M = 1:
            GOTO 450
370   FOR I = 1 TO 10 STEP 0.5
380       LET P = N *  LOG (I +  SQR (I * I - 1))
390       LET Q = E * 0.25 * ((( EXP (P) +  EXP ( - 1 * P)) ^ 2))
400       LET LC(M) = 10 *  LOG (Q) /  LOG (10)
410       IF LC(M) < = 0 THEN
                LET LC(M) = 0
420       LET M = M + 1
430   NEXT I
440   LET M = 11
450   FOR I = 1 TO 10 STEP 0.5
460       LET LB(M) = 10 *  LOG (1 + I ^ (2 * N)) /  LOG (10)
470       LET M = M + 1
480   NEXT I
490   IF G1 = 0 THEN
            670
500   PRINT
510   PRINT "********************"
520   PRINT
530   PRINT "REMEMBER TO CHECK INCREMENTS USED"
540   PRINT
550   PRINT "STEP                    ATTN.(DB)"
560   PRINT "                 CHEBY.      BUTTERWORTH"
570   PRINT
580   LET Q = 0
590   FOR I = 0 TO M - 1
600       PRINT Q, INT (LC(I) * 100 + .5) / 100, INT (LB(I) * 100 + .5) / 100
610       LET Q = Q + 1
620   NEXT I
630   PRINT
640   PRINT "*********************"
650   PRINT
660   GOTO 790
670   PRINT
680   PRINT "*******************"
690   PRINT
700   PRINT "STEP                      ATTN.(DB)"
710   PRINT "CHECK INCREMENTS USED     BUTTERWORTH"
720   PRINT
730   FOR I = 0 TO M - 1
740       PRINT I,LB(I)
750   NEXT I
760   PRINT
770   PRINT "********************"
780   PRINT
790   PRINT
800   PRINT "DO YOU WANT ANOTHER GO ?"
810   PRINT "ENTER 1 IF YES"
820   INPUT A
830   IF A = 1 THEN
            120
840   PRINT
850   PRINT "**** END OF PROGRAM ****
860   END


END-OF-LISTING


]RUN
ENTER ORDER OF FILTER
?4
ENTER PASSBAND RIPPLE IN DB
IF ZERO BUTTERWORTH RESPONSE ASSUMED
?1
WORKING:------

********************

REMEMBER TO CHECK INCREMENTS USED
```

STEP	CHEBY.	ATTN.(DB) BUTTERWORTH
0	0	0
1	1	0
2	.86	0
3	.51	0
4	.13	0
5	.01	0
6	.27	.02
7	.73	.07
8	1	.24
9	.73	.67
10	.06	1.55
11	0	3.01
12	21.55	14.25
13	33.87	24.1
14	42.55	31.84
15	49.36	38.17
16	54.99	43.53
17	59.8	48.16
18	64.01	52.26
19	67.76	55.92
20	71.13	59.23
21	74.2	62.25
22	77.02	65.03
23	79.62	67.61
24	82.04	70
25	84.3	72.25
26	86.43	74.35
27	88.43	76.34
28	90.31	78.22
29	92.11	80

****

DO YOU WANT ANOTHER GO ?
ENTER 1 IF YES
?0

**** END OF PROGRAM ****

Often a designer needs to be able to select the number of elements that the filter should have so that a specified attenuation can be achieved at some spot frequency in the filter stop band, i.e. find n in equations (5.4) and (5.5). Rearranging these equations gives for the Butterworth filter

$$n = \frac{\log_{10}(10^{[L(\omega)/10]} - 1)}{2 \log_{10}(\omega/\omega_c)} \tag{5.6}$$

and for the Chebyshev filter, $\omega > \omega_c$,

$$n = \frac{\cosh^{-1}((10^{[L(\omega)/10]} - 1)/(10^{G_r/10} - 1))^{\frac{1}{2}}}{\cosh^{-1}(\omega/\omega_c)} \tag{5.7}$$

Here the identity

$$\cosh^{-1}(x) = \log_e[x + (x^2 - 1)^{\frac{1}{2}}]$$

enables rapid evaluation of equation (5.7).

These expressions have been programmed to provide program 5.2 NOSEC. The output from program NOSEC gives the next largest integer value to the number

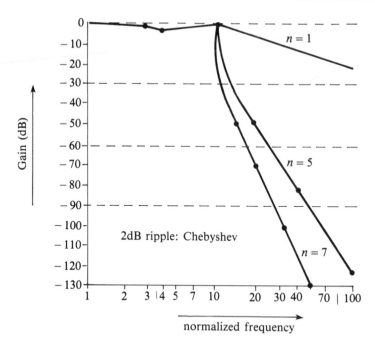

Figure 5.2   Chebyshev Attenuation Curves for 2.0 dB Ripple in the Passband

```
]{ FORMATTED LISTING
FILE: PROGRAM 5.2 NOSEC
PAGE-1


       10    REM
       20    REM   **** NOSEC ****
       30    REM
       40    REM   THIS PROGRAM FINDS
       50    REM   THE NO. OF SECTIONS
       60    REM   REQUIRED TO ACHIEVE
       70    REM   A DESIRED ATTN. ON
       80    REM   CHEBY. OR BUTTERWORTH
       90    REM   FILTER SKIRTS
      100    REM
      110    REM   DATA ENTRY
      120    HOME
      130    PRINT "ENTER DESIRED ATTENUATION IN DB"
      140    INPUT L
      150    PRINT "ENTER FREQ. NORMALIZED TO CUTOFF FREQ."
      160    PRINT "AT WHICH SPECIFIED ATTN. IS DESIRED"
      170    PRINT "THIS SHOULD BE GREATER THAN UNITY"
      180    INPUT W
      190    PRINT "FOR CHEBY. SECTION"
      200    PRINT "ENTER PASSBAND RIPPLE IN DB"
      210    PRINT "IF ZERO IS ENTERED THEN "
      220    PRINT "BUTTERWORTH RESPONSE ONLY"
      230    PRINT "WILL BE COMPUTED"
      240    INPUT G
      250    REM   BUTTERWORTH
      260    LET N = 0.5 *  LOG (10 ^ (L / 10) - 1) /  LOG (W)
      270    LET N =  INT (N + 1)
      280    PRINT
      290    PRINT "*******************"
      300    PRINT
```

```
310    PRINT "BUTTERWORTH RESPONSE NEEDS "N" SECTIONS"
320    PRINT
330    PRINT "FOR "L" DB ATTENUATION"
340    PRINT "AT "W" TIMES THE NORMALIZED CUTOFF FREQ."
350    PRINT
360    IF G = 0 THEN
              490
370    REM   CHEBYSHEV
380    LET A =   LOG (W +   SQR (W ^ 2 - 1))
390    LET B = 10 ^ (G / 10) - 1
400    LET C = 10 ^ (L / 10) - 1
410    LET D =   SQR (C / B)
420    LET N =   LOG (D +   SQR (D * D - 1)) / A
430    LET N =   INT (N + 1)
440    PRINT
450    PRINT "CHEBYSHEV RESPONSE NEEDS "N" SECTIONS"
460    PRINT "FOR "G" DB PASSBAND RIPPLE"
470    PRINT "AND "L" DB ATTENUATION"
480    PRINT "AT "W" TIMES THE NORMALIZED CUTOFF FREQ."
490    PRINT
500    PRINT "********************"
510    PRINT
520    PRINT "DO YOU REQUIRE ANOTHER GO ?"
530    PRINT "ENTER 1 IF YES"
540    INPUT A
550    IF A = 1 THEN
              120
560    PRINT
570    PRINT "**** END OF PROGRAM ****"
580    END
```

END-OF-LISTING


```
'RUN
ENTER DESIRED ATTENUATION IN DB
?18
ENTER FREQ. NORMALIZED TO CUTOFF FREQ.
AT WHICH SPECIFIED ATTN. IS DESIRED
THIS SHOULD BE GREATER THAN UNITY
?1.3
FOR CHEBY. SECTION
ENTER PASSBAND RIPPLE IN DB
IF ZERO IS ENTERED THEN
BUTTERWORTH RESPONSE ONLY
WILL BE COMPUTED
?0.2

******************

BUTTERWORTH RESPONSE NEEDS 8 SECTIONS

FOR 18 DB ATTENUATION
AT 1.3 TIMES THE NORMALIZED CUTOFF FREQ.


CHEBYSHEV RESPONSE NEEDS 6 SECTIONS
FOR .2 DB PASSBAND RIPPLE
AND 18 DB ATTENUATION
AT 1.3 TIMES THE NORMALIZED CUTOFF FREQ.

********************

DO YOU REQUIRE ANOTHER GO ?
ENTER 1 IF YES
?1
ENTER DESIRED ATTENUATION IN DB
?65
ENTER FREQ. NORMALIZED TO CUTOFF FREQ.
AT WHICH SPECIFIED ATTN. IS DESIRED
THIS SHOULD BE GREATER THAN UNITY
?4
FOR CHEBY. SECTION
```

```
ENTER PASSBAND RIPPLE IN DB
IF ZERO IS ENTERED THEN
BUTTERWORTH RESPONSE ONLY
WILL BE COMPUTED
?0

******************

BUTTERWORTH RESPONSE NEEDS 6 SECTIONS

FOR 65 DB ATTENUATION
AT 4 TIMES THE NORMALIZED CUTOFF FREQ.

********************

DO YOU REQUIRE ANOTHER GO ?
ENTER 1 IF YES
?0

**** END OF PROGRAM ****
```

calculated from equations 5.6 and 5.7. An integer number of elements has to occur
since real filters are to be designed.

The synthesis of passive filters has been discussed thoroughly by numerous
authors, one good example being Weinberg [1]. This or a similar book should be
consulted for a more detailed description of what is to follow.

Most filter circuit applications require the filter network to be driven from a
generator with known load impedance and to be terminated in a known load
impedance (figure 5.3). In most cases the generator and load impedances can be
arranged to be real quantities.

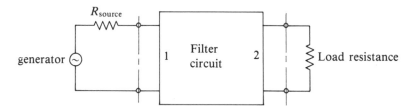

Figure 5.3   Two-Port Filter Representation

This is very convenient since networks of this type, known in network analysis
jargon as doubly-terminated filters, are a favorite topic in books on network
analysis and synthesis. One such filter is shown in figure 5.4, together with the g
notation usual in this type of problem. Here g represents the roots of the governing
filter transfer function for an nth order filter. These g values can be scaled to represent actual electrical circuit elements in order to synthesize a desired filter response.

Figure 5.4 shows a low pass filter constructed from a series of inductors and
capacitors in a TEE configuration. Alternatively, a PI configuration could have been
employed so that g_1 becomes a shunt inductance and g_2 a series capacitance. For
the g notation in figure 5.4, g_0, g_{n+1} represent the source and load resistances if g_1

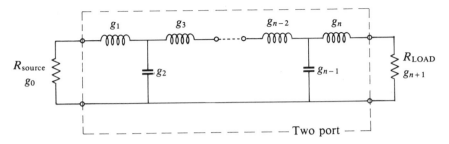

Figure 5.4   Doubly Terminated Low Pass Filter

and g_n are capacitors, otherwise g_0, g_{n+1} represent conductances when g_1, g_n are inductances. Simple expressions are available for the evaluation of the g values necessary for a lumped filter to be designed to a given specification. The expressions themselves are fairly straightforward but their derivation is a complicated business (see, for example, Orchard [2]). Most designers ignore these expressions and instead interpolate tables of calculated data such as those given in chapter 13 of Weinberg [1].

For Butterworth filters, the normalized design equations are

$$\omega_c = 1$$

$$g_0 = g_{n+1} = 1$$

$$g_k = 2 \sin \left[\frac{(2k-1)\pi}{2n} \right] \qquad k = 1, 2, \ldots, n$$

Here, the argument of the sine function is in radians. If n is made either even or odd, symmetry of the component values will result. This means that the filter can be placed between two equal impedance levels without introducing mismatch.

For Chebyshev filters with G_r decibels of ripple in the passband

$$\omega_c = 1$$

$$g_0 = 1$$

$$g_1 = \frac{2a_1}{\psi}$$

$$g_k = \frac{4a_{k-1}\,a_k}{b_{k-1}\,g_{k-1}} \qquad k = 2, 3, \ldots, n$$

$$g_{n+1} = 1 \qquad\qquad n \text{ odd}$$

$$= \coth^2 \left(\frac{\beta}{4} \right) \qquad n \text{ even}$$

where

$$\beta = \log_e \left[\coth \left(\frac{G_r}{17.37} \right) \right]$$

$$\psi = \sinh \left(\frac{\beta}{2n} \right)$$

$$a_k = \sin \left[\frac{(2k-1)\pi}{2n} \right]$$

$$b_k = \psi^2 + \sin^2 \left(\frac{k\pi}{n} \right)$$

If n is odd then the filter components will display symmetry. For even n the component values generated will lead to an asymmetric design. This may be useful especialy when unequal impedance matching is required.

* * *

Example 5.1

Design a third order low pass filter having a maximally flat response. The filter is to operate with its cut-off frequency at 10 MHz from a generator having a 50 ohm source impedance and is to be terminated in a 50 ohm load.

Solution

$$\omega_c = 1, \ \omega = 2\pi \times 10^7 \text{ rad/s}, \ R = 50 \ \Omega$$

Butterworth normalized to 50 Ω

$n = 3$

$g_0 = 1$

$g_4 = 1$

$$g_1 = 2 \sin \left[\frac{(2-1)\pi}{6} \right] = 1$$

$$g_2 = 2 \sin \left[\frac{(4-1)}{6} \pi \right] = 2$$

$$g_3 = 2 \sin \left[\frac{(6-1)}{6} \pi \right] = 1$$

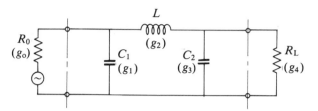

Figure 5.5   Selected Circuit Topology for Example 5.1

Having found the g values for the filter prototype, it is now necessary to transform the prototype according to the design specification.

First select a circuit topology, say that given in figure 5.5. Remember that the g values used represent normalized component values. Therefore it is necessary to scale the prototype values both in magnitude and frequency

$$\frac{\omega_c}{\omega} = \frac{1}{2\pi \times 10^7} = 1.59 \times 10^{-8}$$

$$R_o = Rg_0 = 50 \times 1 = 50 \text{ ohm}$$

$$C_1 = \frac{1}{R}\left(\frac{\omega_c}{\omega}\right) g_1 = 318 \text{ pF}$$

$$L = R\left(\frac{\omega_c}{\omega}\right) g_2 = 1590 \text{ nH}$$

$$C_2 = \frac{1}{R}\left(\frac{\omega_c}{\omega}\right) g_3 = 318 \text{ pF}$$

$$R_L = Rg_4 = 50 \text{ ohm}$$

$$* \quad * \quad *$$

Note from the results above:

(1) How the component values are symmetrical about inductor L.
(2) How the scaling was achieved.

Since the design equations for the g values are based on a normalized breakpoint $\omega_c = 1$ and since they yield normalized values, then to denormalize it is necessary to perform the following scaling transformations.

1 Scale according to frequency. This means divide every normalized g value representing a capacitor or inductor by the desired cut-off frequency expressed in radians per second. Resistors are not included in this operation.

2 To obtain a load resistance of m ohms, multiply all g values representing resistors and inductors by m, similarly divide all g values representing capacitors by m.

* * *

Example 5.2

Design a passive lumped element filter section that exhibits a Chebyshev response with 0.01 dB ripple in the passband. The filter should attenuate the signal by at least 5 dB at frequencies four times above the cut-off frequency. The cut-off frequency is to be 1 GHz and the filter is to operate in a 75 ohm system.

Solution

First, it is necessary to find the number of reactive elements required to meet the specification

$$G_r = 0.01 \text{ dB}$$

$$L(4\omega_c) = 5 \text{ dB}$$

$4\omega_c > \omega_c$ therefore use equation (5.7) appropriate for the Chebyshev response under these conditions

$$n = \frac{\cosh^{-1}[(10^{0.5} - 1)/(10^{0.001} - 1)]^{\frac{1}{2}}}{\cosh^{-1}(4)} = \frac{4.11}{2.06} = 2.0$$

To err on the safe side choose $n = 3$. An odd value for n has been selected, this means that $g_0 = g_4$ and $g_1 = g_3$ for the Chebyshev response.

Next, calculate the necessary g values

$$\beta = \log_e \left[\coth \left(\frac{0.01}{17.37} \right) \right] = 7.5$$

$$\psi = \sinh \left(\frac{7.5}{6} \right) = 1.5893$$

$$a_1 = 0.5$$

$$b_1 = 2.566 + \left[\sin \left(\frac{\pi}{3} \right) \right]^2 = 3.428$$

$$a_2 = 1.0$$

$$b_2 = 3.428$$

$$a_3 = 0.5$$

$$b_3 = 0$$

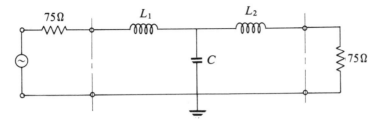

Figure 5.6   Selected Circuit Topology for Example 5.2

hence

$$g_0 = 1$$

$$g_1 = \frac{2 \times 0.5}{1.5893} = 0.6292$$

$$g_2 = \frac{4 \times 0.5 \times 1}{3.428 \times 0.6292} = 0.9274$$

$$g_3 = 0.6291$$

$$g_4 = 1$$

Now select an appropriate topology, e.g. that given in figure 5.6.

$$\frac{\omega_c}{\omega} = \frac{1}{2\pi \times 10^{10}} = 1.592 \times 10^{-11}$$

$$\therefore \quad L_1 = L_2 = 75 \, (1.592 \times 10^{-11})(0.6294) = 75 \text{ nH}$$

$$C = \frac{1}{75} \, (1.592 \times 10^{-11})(0.9274) = 20 \text{ pF}$$

* * *

From the simple examples 5.1 and 5.2, it can be seen that the number of calculations necessary for filter sections containing large numbers of components become large. This is very tedious and error prone especially if Chebyshev filters are being designed. To speed up the process, program 5.3 GVAL calculates the g values necessary to realize prototype Butterworth and Chebyshev low pass filters for any given number order of filter, and in the case of the Chebyshev filter any desired passband ripple in decibels.

```
]| FORMATTED LISTING
FILE: PROGRAM 5.3 GVAL
PAGE-1


10   REM
20   REM   **** GVAL ****
30   REM
```

```
40    REM    THIS PROGRAM COMPUTES
50    REM    NORMALIZED COMPONENT
60    REM    VALUES FOR LUMPED
70    REM    BUTTERWORTH AND CHEBY.
80    REM    FILTER SECTIONS.
90    REM    THE CUTOFF FREQ. HAS
100   REM    NORMALIZED TO UNITY
110   REM
120   DIM  G(100),A(100),B(100)
130   LET  Q = 0
140   HOME
150   PRINT "IF YOU REQUIRE VALUES FOR "
160   PRINT "BUTTERWORTH FILTER ENTER 1"
170   PRINT "IF YOU WANT A CHEBY. FILTER ENTER 0"
180   INPUT  A
190   IF  A = 0 THEN
           440
200   IF  A = 1 THEN
           210
210   REM   BUTTERWORTH RESPONSE
220   PRINT "I/P NO. OF FILTER SECTIONS"
230   INPUT  N
240   FOR  I = 1 TO N
250       LET G(I) = 2 *  SIN (((2 * I - 1) * 3.141592653) / 2 / N)
260   NEXT  I
270   LET  G(0) = 1
280   LET  G(N + 1) = 1
290   PRINT
300   PRINT "***********************"
310   PRINT
320   PRINT "NORMALIZED G-VALUES BUTTERWORTH RESPONSE"
330   PRINT
340   PRINT "FOR "N" SECTIONS"
350   PRINT
360   PRINT "NUMBER          G-VALUE"
370   FOR  I = 0 TO N + 1
380       PRINT I, INT (G(I) * 10000 + .5) / 10000
390   NEXT  I
400   PRINT
410   PRINT "***********************"
420   PRINT
430   GOTO 800
440   REM   CHEBYSHEV RESPONSE
450   PRINT "IF NO. OF FILTER SECTIONS IS"
460   PRINT "ODD ENTER A 1"
470   INPUT  Q
480   PRINT "ENTER NUMBER OF FILTER SECTIONS"
490   INPUT  N
500   PRINT "ENTER IN BAND RIPPLE IN DB"
510   INPUT  G
520   LET  D = G / 17.37
530   LET  E = ( EXP (D) +  EXP ( - D))
540   LET  E = E / ( EXP (D) -  EXP ( - D))
550   LET  E =  LOG (E)
560   LET  C = 0.5 * ( EXP (E / 2 / N) -  EXP ( - E / 2 / N))
570   FOR  I = 1 TO N
580       LET  A(I) =  SIN ((2 * I - 1) * 3.141592653 / 2 / N)
590       LET  B(I) = C * C + ( SIN (I * 3.141592653 / N)) ^ 2
600       IF  I = 1 THEN
              LET  G(1) = 2 * A(1) / C:
              GOTO 620
610       LET  G(I) = 4 * A(I - 1) * A(I) / B(I - 1) / G(I - 1)
620   NEXT  I
630   LET  G(0) = 1
640   IF  Q = 1 THEN
          LET  G(N + 1) = 1
650   IF  Q = 0 THEN
          LET  G(N + 1) = (( EXP (E / 4) +  EXP ( - E / 4)) / ( EXP (E / 4)
          EXP ( - E / 4))) ^ 2
660   PRINT
670   PRINT "*****************"
680   PRINT
690   PRINT "NORMALIZED G-VALUE CHEBYSHEV RESPONSE"
700   PRINT
```

```
710   PRINT "FOR "N" SECTIONS AND "G" DB RIPPLE"
720   PRINT "IN THE PASSBAND"
730   PRINT
740   PRINT "NUMBER          G-VALUE"
750   FOR I = 0 TO N + 1
760       PRINT I, INT (G(I) * 10000 + .5) / 10000
770   NEXT I
780   PRINT
790   PRINT "*********************"
800   PRINT
810   PRINT "DO YOU WANT ANOTHER GO ?"
820   PRINT " INPUT 1 IF YES"
830   INPUT A
840   IF A = 1 THEN
          140
850   PRINT
860   PRINT "**** END OF PROGRAM ****"
870   PRINT
```

END-OF-LISTING

```
]RUN
IF YOU REQUIRE VALUES FOR
BUTTERWORTH FILTER ENTER 1
IF YOU WANT A CHEBY. FILTER ENTER 0
?1
I/P NO. OF FILTER SECTIONS
?4
```

****

NORMALIZED G-VALUES BUTTERWORTH RESPONSE

FOR 4 SECTIONS

NUMBER	G-VALUE
0	1
1	.7654
2	1.8478
3	1.8478
4	.7654
5	1

****

```
DO YOU WANT ANOTHER GO ?
 INPUT 1 IF YES
?1
IF YOU REQUIRE VALUES FOR
BUTTERWORTH FILTER ENTER 1
IF YOU WANT A CHEBY. FILTER ENTER 0
?0
IF NO. OF FILTER SECTIONS IS
ODD ENTER A 1
?0
ENTER NUMBER OF FILTER SECTIONS
?6
ENTER IN BAND RIPPLE IN DB
?1.0
```

****

NORMALIZED G-VALUE CHEBYSHEV RESPONSE

FOR 6 SECTIONS AND 1 DB RIPPLE
IN THE PASSBAND

NUMBER	G-VALUE
0	1
1	2.1547
2	1.1041

```
3                    3.0635
4                    1.1518
5                    2.9368
6                     .8101
7                    2.6599

*********************

DO YOU WANT ANOTHER GO ?
  INPUT 1 IF YES
?0

**** END OF PROGRAM ****
```

5.1.1 Lumped High Pass Filters

After a lumped low pass filter prototype has been designed to meet a certain specification, it is possible to transform the low pass prototype to a high pass prototype with minimum effort. The first thing is to establish the requirements of the frequency transformation to map a low pass filter response to a high pass response. The situation is illustrated in figure 5.7.

From figure 5.7 it can be appreciated that, to go from the idealized low pass response to the idealized high pass response, two criteria have to be obeyed:

1     $\omega^1 = 0$ must become $\omega = \infty$
2     $\omega^1 = 1$ must become $\omega = \omega_c$

Expressing these criteria in the form of a simple mathematical equation yields the low to high pass frequency translation formula

$$\omega^1 = -\frac{\omega_c}{\omega}$$

The amplitude of the response is unaffected by this transformation.

In order to realize the high pass response, the prototype circuit shown in figure 5.5 could be used, on replacing each inductive element in the low pass circuit by a capacitive element. The converse applies to the low pass capacitive elements. In this way a suitable high pass circuit topology can quickly be found.

Thus an inductance in the low pass case becomes a capacitance

$$C\big|_{\mathrm{HP}} = \frac{1}{(\omega_c/\omega)L\big|_{\mathrm{LP}}}$$

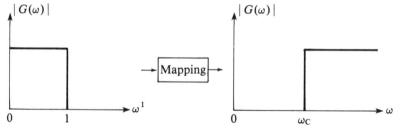

Figure 5.7   Ideal Low Pass to High Pass Frequency Translation

and a capacitance becomes an inductance according to

$$L \big|_{HP} = \frac{1}{(\omega_c/\omega)C \big|_{LP}}$$

Resistance values are unaffected by the transformation. To complete the low pass to high pass mapping it is only necessary to scale all the impedances according to the termination resistances.

* * *

Example 5.3

Repeat example 5.1 but this time produce a high pass Butterworth response.

Solution

From example 5.1 the low pass prototype g values were

$$g_0 = g_1 = g_3 = g_4 = 1$$

$$g_2 = 2$$

$$\omega_c = 10^7 \text{ Hz}$$

From the discussion on low pass to high pass filter mappings, a suitable topology can be derived from figure 5.5, this is given as figure 5.8. Scale values

$$\frac{\omega_c}{\omega} = \frac{1}{2\pi \times 10^7}$$

Now scale according to frequency, remember that a capacitor in the low pass case becomes an inductor and vice versa

$$L_1 = L_2 = \frac{1}{2\pi \times 10^7 \times g_1} = 1.59 \times 10^{-8} \text{ H}$$

$$C_1 = \frac{1}{2\pi \times 10^7 \times g_2} = 7.95 \times 10^{-9} \text{ F}$$

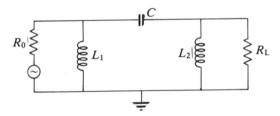

Figure 5.8   Selected Circuit Topology for Example 5.3

Next, scale the design to work at the desired impedance level, 50 ohms in this case

$$L_1 = L_2 = 1.59 \times 10^{-8} \times 50 = 80 \text{ nH}$$

$$C = \frac{7.95 \times 10^{-9}}{50} = 159 \text{ pF}$$

This completes the design.

*   *   *

If the number of reactive elements required to give a desired attenuation at some point in the out-of-band response of a high pass filter (below ω_c) is to be found then the technique used to find n for the low pass filter is still valid since the frequency transformation leaves the filter skirt response unaffected.

5.1.2 Low Pass to Band Pass Transformation

Low pass to band pass transformation is shown in figure 5.9. Here, the transformation required is one that will map

$\omega_1 = 1$ to ω_u

$\omega_1 = 0$ to ω_o

and

$\omega^1 = -1$ to ω_L

A suitable mapping that exhibits these properties is given as

$$\omega^1 = \frac{1}{(\omega_u - \omega_L)} \left(\frac{\omega^2 - \omega_o{}^2}{\omega} \right)$$

where ω^1 refers to the low pass response. Here, ω_o, the center frequency, is taken to be the geometric mean of ω_L and ω_u, i.e.,

$$\omega_o = (\omega_L \omega_u)^{\frac{1}{2}}$$

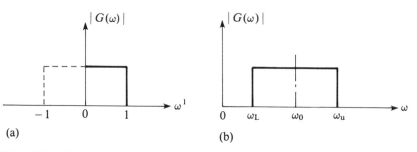

Figure 5.9   Ideal Low Pass to Band Pass Frequency Translation (a) Ideal Low Pass (b) Ideal Bandpass (bandwidth = $\omega_u - \omega_L$)

It should also be noted that the Q factor for a bandpass filter having upper and lower 3 dB breakpoints ω_u, ω_L is simply

$$Q = \frac{\omega_0}{\omega_u - \omega_L}$$

*   *   *

Example 5.4

Design a sixth order bandpass filter exhibiting a Butterworth response. The filter should have a center frequency of 50 MHz and a Q factor of 10. The circuit is to operate in a 50 ohm system.

Solution

Design a third order $n = 3$ Butterworth low pass prototype, using the topology in figure 5.10. Butterworth low pass filter

$$n = 3$$

from program GVAL or example 5.1

$$g_0 = g_1 = g_3 = g_4 = 1$$
$$\therefore \quad L_1 = L_2 : R_S = R_L$$

and

$$g_2 = 2$$

Note that all g values are normalized to $\omega_c = 1$ rad/s and 1 ohm impedance. Now bandwidth

$$\omega_u - \omega_L = \frac{\omega_0}{Q} = \frac{2\pi \times 50 \times 10^6}{10} = \pi \times 10^7 \text{ rad/s}$$

To complete the transformation set series inductors equal to series resonant LC circuits and set the shunt capacitors equal to parallel resonant LC circuits (see, for example, Weinberg [1] (p. 542)). This is shown in figure 5.10.

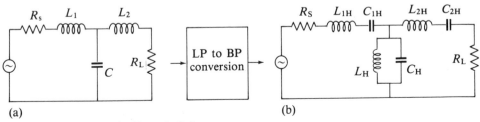

(a)   (b)

Figure 5.10   Topology for Example 5.4

Replace $L_1 = L_2$ in the low pass structure with

$$L_{1H} = L_{2H} = \frac{L_1}{(\omega_u - \omega_L)} = \frac{1}{\pi \times 10^7} = 32 \text{ nH}$$

Similarly

$$C_{1H} = C_{2H} = \frac{\omega_u - \omega_L}{\omega_o^2 L_1} = \frac{\pi \times 10^7}{1 \times (50 \times 2 \times \pi \times 10^6)^2} = 318 \ \text{pF}$$

Next, replace capacitance C in the low pass structure with an inductance

$$L_H = \frac{\omega_u - \omega_L}{\omega_o^2 C} = 0.16 \text{ nH}$$

in parallel with a capacitance

$$C_H = \frac{C}{\omega_u - \omega_L} = 64 \text{ nF}$$

Finally, all the components must be scaled to obtain the desired impedance levels. These rules are the impedance scaling rules already established for low pass and high pass filter networks.

$$R_S = R_o = 50 \text{ ohms}$$

$$L_{1H} = L_{2H} = 1.6 \ \mu\text{H}$$

$$C_{1H} = C_{2H} = 6.4 \text{ pF}$$

$$C_H = 1.3 \text{ nF}$$

$$L_H = 8 \text{ nH}$$

This completes the design.

* * *

In example 5.2 the number of reactive elements required to meet a particular out-of-band attenuation specification is calculated. The procedure outlined in example 5.2 for achieving this can be extended to the band pass network with ease.

* * *

Example 5.5

Suppose a filter displaying a Chebyshev response with 0.1 dB ripple in the passband is required to have upper and lower 3 dB points at 1.8 GHz and 2.2 GHz and to exhibit at least 40 dB attenuation at 1.4 GHz. Find how many reactive elements will be required so that the filter can meet the specification.

Solution

Let $\omega_o = (1.8 \times 2.2)^{\frac{1}{2}}$ GHz $= 1.99$ GHz. From the low pass to band pass mapping

$$\omega^1 = \frac{1}{0.4}\left(\frac{1.4^2 - 1.99^2}{1.4}\right) = -3.57$$

The negative sign here can be ignored since it represents the mapping of ω in the band pass filter to ω^1 in the low pass filter prototype.

For 0.1 dB passband ripple and $\omega^1 = 3.57$ and 40 dB attenuation at 1.4 GHz, $\varepsilon = 1.0233$, so that

$$n = \frac{\cosh^{-1}(3.999/0.0233)}{\cosh^{-1}(3.57)} = \frac{5.84}{1.95} = \text{approximately } 3$$

$\therefore$ choose $n = 4$.

To summarize the transformations used, table 5.2 has been drawn up, this acts as a speedy reference.

Table 5.2   Lumped Filter Transformations

	Filter type	
Low Pass	High Pass	Band Pass
L	$\dfrac{1}{(\omega_c L)}$	$\dfrac{L}{(\omega_u - \omega_L)}$ $\dfrac{(\omega_u - \omega_L)}{\omega_0^2 L}$
		$\dfrac{(\omega_u - \omega_L)}{\omega^2{}_0}$
C	$\dfrac{1}{(\omega_c C)}$	$\dfrac{C}{(\omega_u - \omega_L)}$

* * *

A sinusoidal waveform applied to the input of a filter will experience both an amplitude change and a phase shift when it emerges at the output of the filter. The amplitude variation has been discussed already. If the input signal is considered (for the purpose of this discussion) to be the reference signal, there is a finite time delay associated with the filter, since the input signal cannot appear instantaneously at the filter output. Weinberg [1] gives a complete step-by-step development of the governing expression for the time delay and phase shift of signals passing through

Butterworth and Chebyshev filters of order *n*. For a complete description of the derivation of the relevant formula Weinberg [1] (pp. 497 and 517) should be consulted. The expressions derived in the text cited above are reproduced in appendix C and have been programmed to provide a useful utility routine called FILRESP, program 5.4. Computer program FILRESP can quickly calculate the phase shift and time delay of signals introduced into a filter on passing through *n* order Chebyshev or Butterworth band pass filter networks for a number of spot frequencies up to a selected frequency point on one of the band edges of the filter. The filters are assumed to have symmetrical band pass characteristics, which is approximately true for filters having *Q* factors of three or higher. In some communications applications the phase response of the filter may be as critical, or even more so, than the amplitude response, and in some applications a filter circuit may be deliberately introduced in the signal path to effect a known time delay. Program FILRESP enables a priori calculations to be carried out with minimum effort.

```
][ FORMATTED LISTING
FILE: PROGRAM 5.4 FILRESP
PAGE-1


    10   REM
    20   REM   **** FILRESP ****
    30   REM
    40   REM   THIS PROGRAM CALCULATES
    50   REM   THE AMPLITUDE, PHASE
    60   REM   AND TIME DELAY OF
    70   REM   BUTTERWORTH AND CHEBY.
    80   REM   BANDPASS FILTERS.
    90   REM
   100   REM   FO=CENTER FREQ (GHZ)
   110   REM   Q=BANDWITH (GHZ)
   120   REM   N=NUMBER OF SECTIONS
   130   REM   R=PASSBAND RIPPLE IN
   140   REM   DB, CHEBY. ONLY
   150   REM   S=NO. OF EQUALLY
   160   REM   SPACED FREQ INTERVALS
   170   REM
   180   DIM T(105),L(105),P(105)
   190   HOME
   200   LET PI = 3.14159265
   210   PRINT "I/P CENTER FREQ. (GHZ)"
   220   INPUT FO
   230   PRINT "I/P BANDWIDTH (GHZ)"
   240   INPUT Q
   250   PRINT "I/P NO. OF FILTER SECTIONS"
   260   INPUT N
   270   PRINT "I/P PASSBAND RIPPLE IN DB"
   280   PRINT "IF BUTTERWORTH FILTER ENTER 0"
   290   INPUT R
   300   PRINT "I/P NO. OF FREQ INTERVALS"
   310   PRINT "REQUIRED, LESS THAN 100"
   320   INPUT S
   330   HOME
   340   PRINT
   350   PRINT "WORKING:----"
   360   IF R = 0 THEN
             1050
   370   REM   CHEBY. RESPONSE
   380   LET E = SQR ( EXP (R / 4.3429) - 1)
   390   LET FI = 1 / N * LOG (1 / E + SQR (1 / E / E + 1))
   400   LET K = LOG (1 / E + SQR (1 / E / E - 1))
   410   LET H = 0.5 * ( EXP (K / N) + EXP ( - K / N))
```

```
420    LET W = H / S
430    LET W1 = 0
440    LET S2 = S + 1
450    FOR I = 1 TO S2
460        LET M1 = 0
470        LET S1 = 0
480        IF (W1 - 1) < 0 THEN
               510
490        IF (W1 - 1) = 0 THEN
               550
500        IF (W1 - 1) > 0 THEN
               590
510        LET O =  -  ATN (W1 /  SQR ( - W1 * W1 + 1)) + PI / 2
520        LET U =  SIN ((2 * M1 + 1) * O) /  SIN (O)
530        LET T1 =  COS (N * O)
540        GOTO 640
550        LET O = 0
560        LET U = 1
570        LET T1 = 1
580        GOTO 640
590        LET O =  LOG (W1 +  SQR (W1 * W1 - 1))
600        LET U =  EXP ((2 * M1 + 1) * O) -  EXP ( - (2 * M1 + 1) * O)
610        LET U = U / ( EXP (O) -  EXP ( - O))
620        LET Y = W1 +  SQR (W1 * W1 - 1)
630        LET T1 = (Y ^ N + Y ^ ( - N)) / 2
640        LET K = U /  SIN ((2 * M1 + 1) * PI / 2 / N)
650        LET K = K * 0.5 * ( EXP ((2 * N - 2 * M1 - 1) * FI) -  EXP ((2 * N -
           2 * M1 - 1) *  - 1 * FI))
660        LET S1 = S1 + K
670        LET M1 = M1 + 1
680        IF (M1 - N + 1) <  = 0 THEN
               480
690        LET T(I) = S1 / PI / Q / (1 + E * E * T1 * T1) * E * E
700        IF ( ABS (W1) - 1) < 0 THEN
               730
710        IF ( ABS (W1) - 1) = 0 THEN
               760
720        IF ( ABS (W1) - 1) > 0 THEN
               790
730        LET O =  -  ATN (W1 /  SQR ( - W1 * W1 + 1)) + PI / 2
740        LET K =  COS (N * O)
750        GOTO 810
760        LET O = 0
770        LET K = 1
780        GOTO 810
790        LET O =  LOG ( ABS (W1) +  SQR ( ABS (W1) ^ 2 - 1))
800        LET K = 0.5 * ( EXP (N * O) +  EXP ( - N * O))
810        LET L(I) = 4.34294 *  LOG (1 + (E * K) ^ 2)
820        LET G = PI / 2 / N
830        LET S1 = 0
840        LET J = 0
850        FOR M2 = 1 TO 100 STEP 2
860            LET K =  EXP ( - M2 * FI) / (M2 *  SIN (M2 * G))
870            IF (W1 - 1) <  = 0 THEN
                   910
880            LET O =  LOG ( ABS (W1) +  SQR ( ABS (W1) ^ 2 - 1))
890            LET K = K * 0.5 * ( EXP (M2 * O) +  EXP ( - M2 * O))
900            GOTO 930
910            LET O =  -  ATN (W1 /  SQR ( - W1 * W1 + 1)) + PI / 2
920            LET K = K *  COS (M2 * O)
930            LET S1 = S1 + K
940            IF (K - .1) <  = 0 THEN
                   960
950            GOTO 990
960            LET J = J + 1
970            IF (J - 4) < 0 THEN
                   1000
980            IF (J - 4) >  = 0 THEN
                   1010
990            LET J = 0
1000       NEXT M2
1010       LET P(I) = S1 * 360 / PI
1020       LET W1 = W1 + W
```

```
1030    NEXT I
1040    GOTO 1380
1050    REM   BUTTERWORTH RESPONSE
1060    LET W1 = 0
1070    LET H =   EXP ( LOG (.99526) / 2 / N)
1080    LET W = H / S
1090    LET S2 = S + 1
1100    FOR I = 1 TO S2
1110        LET M1 = 0
1120        LET S1 = 0
1130        LET K = W1 ^ (2 * M1)
1140        LET K = K /   SIN ((2 * M1 + 1) * PI / 2 / N)
1150        LET S1 = S1 + K
1160        LET M1 = M1 + 1
1170        IF (M1 - N + 1) <  = 0 THEN
                1130
1180        LET T(I) = S1 / (1 + W1 ^ (2 * N))
1190        LET L(I) = 10 *   LOG (1 + W1 ^ (2 * N)) /   LOG (10)
1200        LET G = PI / 2 / N
1210        LET S1 = 0
1220        LET J = 0
1230        FOR M2 = 1 TO 100 STEP 2
1240            LET K = W1 ^ (2 * (M2 - 1))
1250            LET K = K / (M2 *   SIN (M2 * G))
1260            LET S1 = S1 + K
1270            IF (K - .001) <  = 0 THEN
                    1290
1280            GOTO 1320
1290            LET J = J + 1
1300            IF (J - 4) < 0 THEN
                    1330
1310            IF (J - 4) >  = 0 THEN
                    1340
1320            LET J = 0
1330        NEXT M2
1340        LET P(I) = S1 * 360 / PI
1350        LET W1 = W1 + W
1360    NEXT I
1370    GOTO 1380
1380    REM   PRINT ROUTINE
1390    PRINT
1400    PRINT "*********************"
1410    PRINT
1420    IF R = 0 THEN
            PRINT "BUTTERWORTH FILTER RESPONSE:----"
1430    IF R <  > 0 THEN
            PRINT "CHEBYSHEV FILTER RESPONSE:----"
1440    PRINT
1450    PRINT "CENTER FREQUENCY "FO" GHZ"
1460    PRINT "BANDWIDTH "Q" GHZ"
1470    PRINT "NO. OF SECTIONS "N
1480    IF R <  > 0 THEN
            PRINT "PASSBAND RIPPLE "R" DB"
1490    PRINT
1500    PRINT   TAB( 1);"FREQUENCY"; TAB( 14);"ATTN."; TAB( 23);"PHASE"; TAB( 33);
        "DELAY"
1510    PRINT   TAB( 3);"(GHZ)"; TAB( 14);"(DB)"; TAB( 22);"DEGREES"; TAB( 33);"nS
        EC."
1520    PRINT
1530    LET H = H * Q / 2 / S
1540    FOR I = 1 TO S + 1
1550        LET F = FO + (I - 1) * H
1560        PRINT   TAB( 3); INT (F * 10000 + .5) / 10000; TAB( 14); INT (L(I) * 1
            000 + .5) / 1000; TAB( 23); -   INT (P(I) * 100 + .5) / 100; TAB( 33);
            INT (T(I) * 1000 + .5) / 1000
1570    NEXT I
1580    PRINT
1590    PRINT "*********************"
1600    PRINT
1610    PRINT "DO YOU WANT ANOTHER GO ?"
1620    PRINT "IF YES ENTER 1"
1630    INPUT R7
1640    IF R7 = 1 THEN
            190
```

```
1650   PRINT
1660   PRINT "**** END OF PROGRAM ****"
1670   END
```

END-OF-LISTING

```
]RUN
I/P CENTER FREQ. (GHZ)
?1
I/P BANDWIDTH (GHZ)
?0.09
I/P NO. OF FILTER SECTIONS
?3
I/P PASSBAND RIPPLE IN DB
IF BUTTERWORTH FILTER ENTER 0
?0.1
I/P NO. OF FREQ INTERVALS
REQUIRED, LESS THAN 100
?10

WORKING:----

********************

CHEBYSHEV FILTER RESPONSE:----

CENTER FREQUENCY 1 GHZ
BANDWIDTH .09 GHZ
NO. OF SECTIONS 3
PASSBAND RIPPLE .1 DB
```

FREQUENCY (GHZ)	ATTN. (DB)	PHASE DEGREES	DELAY nSEC.
1	0	0	5.677
1.0063	.017	−12.76	5.661
1.0125	.056	−25.47	5.638
1.0188	.092	−38.18	5.675
1.025	.096	−51.12	5.864
1.0313	.056	−64.76	6.308
1.0375	3E−03	−79.76	7.073
1.0438	.058	−96.78	8.082
1.05	.446	−116.07	8.91
1.0563	1.414	−134.85	8.881
1.0625	3.01	−152.56	7.829

```
********************

DO YOU WANT ANOTHER GO ?
IF YES ENTER 1
?0

**** END OF PROGRAM ****
```

5.2 LOW PASS DISTRIBUTED FILTER CIRCUITS

At frequencies much above several hundred MHz, lumped components tend to become physically small and therefore difficult to handle. Also, as frequency is increased the lumped components depart from their ideal characteristics due to radiation and loss mechanisms present within the devices. At frequencies high enough to make distributed lines attractive (remember high frequencies imply small line dimensions) then sections of transmission line are used, whose length and

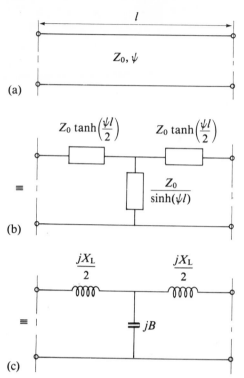

Figure 5.11 Lumped Line Representation of Short Line Section (a) Transmission Line
Section (b) Symmetrical TEE Representation of Circuit (c) Lumped
Representation for Lossless Line

characteristic impedance can be carefully selected to simulate the behavior of a
lumped prototype filter. This type of direct synthesis approach is rather approximate
since a number of important factors such as discontinuity capacitance and dispersion together with the periodic response of the distributed circuit are neglected.
However, filter circuits produced by the direct method as outlined in this section are
sufficiently realistic to allow a preliminary attempt at a circuit design.

In order to gain an understanding of how sections of transmission line can be
used to simulate lumped reactive elements the TEE equivalent circuit for a section
of transmission line discussed in section 1.3 is reproduced in figure 5.11 together
with the relevant design equations. The line sections to be approximated by the TEE
circuit are assumed to be physically short so that line loss can be neglected in the
first approximation. This allows the complex hyperbolic functions shown in figure
5.11 to be reduced to simple trigonometric functions, thus allowing a more convenient method of solution.

From figure 5.11 it is evident that

$$X_L = 2Z_o \tan\left(\frac{\beta l}{2}\right)$$ (5.8)

and

$$X_C = \frac{1}{B} = \frac{Z_o}{\sin{(\beta l)}} \tag{5.9}$$

Here it has been assumed that line attenuation $\alpha = 0$.

If a PI equivalent circuit had been used then a dual set of equations would have been forthcoming. These are

$$X_L = Z_o \sin{(\beta l)} \tag{5.10}$$

$$X_C = \frac{1}{B} = \frac{Z_o}{2 \tan{(\beta l/2)}} \tag{5.11}$$

Equations (5.9) and (5.10) enable X_C and X_L to be evaluated without having to evaluate a tangent function. This is useful when a computer is used since most microcomputers will return the tangent function only in the first and third quadrants so that additional checking is required to establish whether the argument of the function actually lies in the second or fourth quadrants.

By substituting $\beta = \omega/V_p$ into equations (5.8)–(5.11) and by invoking the small angle approximations for the tangent and sine functions, i.e.

$$\tan{\theta} \approx \sin{\theta} \approx \theta \text{ radians}$$

(this is valid since l, the line length, is assumed to be small), then equations (5.12)–(5.15) result

$$X_L = \omega L = 2Z_o \tan{\left(\frac{\omega l}{2V_p}\right)} = \left. \frac{Z_o \omega l}{V_p} \right|_{l < \lambda/8} \tag{5.12}$$

$$B = \omega C = Y_o \sin{\left(\frac{\omega l}{V_p}\right)} = \left. Y_o \frac{\omega l}{V_p} \right|_{l < \lambda/8} \tag{5.13}$$

for the TEE equivalent circuit, and

$$X_L = \omega L = Z_o \sin{\left(\frac{\omega l}{V_p}\right)} = \left. \frac{Z_o \omega l}{V_p} \right|_{l < \lambda/8} \tag{5.14}$$

$$B = \omega C = 2 Y_o \tan{\left(\frac{\omega l}{2V_p}\right)} = \left. \frac{Y_o \omega l}{V_p} \right|_{l < \lambda/8} \tag{5.15}$$

for the PI equivalent circuit.

Equations (5.12) and (5.14) are identical, as are equations (5.13) and (5.15). This establishes the duality of the PI and TEE equivalent circuits. Equations (5.12)–(5.15) show that lumped reactive components can be realized from a distributed transmission line according to equations (5.12)–(5.15). These are summarized below as equations (5.16) and (5.17)

$$L = \frac{Z_o l}{V_p} \tag{5.16}$$

$$C = \frac{l}{Z_o V_p} \tag{5.17}$$

Table 5.3 Distributed Circuit Representations of Lumped Elements

Type	Lumped circuit	Distributed Equivalent	Transformation $(l < \lambda_g/8)$
A		Z_0, length l	$L = \dfrac{Z_0 l}{f\lambda_g}$
B		Z_0, length l	$C = \dfrac{l}{Z_0 f\lambda_g}$
C		Z_0, l ; Z_0, l, $\dfrac{\lambda_g}{4}$	$L = \dfrac{Z_0 l}{f\lambda_g}$
D†		Z_{01}, l_1 ; Z_{02}, l_2, $\dfrac{\lambda_g}{4}$	$C = \dfrac{l_1}{Z_{01} f\lambda_g}$
E		Z_{01}, l_1 ; Z_{02}, l_2	$L = \dfrac{l_2 Z_{02}}{f\lambda_g}$; $Z_{01} \ll Z_{02}$
F†		Z_{01}, Z_0, Z_{01}, $\dfrac{\lambda_g}{2}$	$L = \dfrac{1}{\omega^2 C}$; $Z_{01} \gg Z_0$

† Equivalent representations for parallel LC circuits

Examination of equations (5.16) and (5.17) reveals that if a short line section with a high characteristic impedance is selected and the line is terminated at both ends by a low impedance line, the capacitance in equation (5.17) will tend to zero while equation (5.16) remains valid. Under these circumstances the TEE equivalent circuit reduces to a series inductance L. Similarly when a short line section having low characteristic impedance is terminated with high impedance sections then equation (5.17) becomes dominant and the line section becomes equivalent to a shunt capacitance C.

Variations on the basic theme of deriving distributed analogs of lumped circuits are available. Circuits such as shunt inductances and shunt series resonant circuits together with shunt parallel resonant circuits are all readily derived. The approximations used are compiled in table 5.3 and are depicted as they would physically appear if laid out on microstrip or stripline circuit material.

It should be noted from table 5.3 that in stripline or microstrip material the series circuit equivalents of element types B, D and E are not immediately forth-coming without recourse to special techniques such as manufacturing series gaps in line segments to simulate lumped capacitance. This presents a problem whenever distributed analogs of lumped high pass and band pass prototypes are to be synthesized purely from simple transmission line segments. This problem will be addressed in the next section.

* * *

Example 5.6

Construct a low pass filter from uniform transmission line sections that will operate with a cut-off frequency of 1 GHz and will display a Butterworth characteristic. Overall systems considerations suggest that a choice of a fifth order filter would be appropriate and that the filter should be designed to operate when inserted in a 25 ohm line section. The filter is to be constructed from stripline with a line thickness to ground plane spacing ratio of 0.05 and a relative dielectric constant of four. Calculate the insertion loss of the filter at 2.0 GHz.

Solution

(a) Generate the g values for a fifth order maximally flat filter from program GVAL.

$$g_0 = g_6 = 1.0$$

$$g_1 = g_5 = 0.618$$

$$g_2 = g_4 = 1.618$$

$$g_3 = 2.0$$

(b) Select a topology so that the resulting stripline circuit will be as simple as possible (i.e. no series capacitors). Figure 5.12 shows the topology selected for this problem.

(c) Calculate the values of L and C required for a cut-off frequency of 1 GHz or $2\pi \times 10^9$ rad/s

$$\frac{\omega_c}{\omega} = \frac{1}{2\pi \times 10^9} = 1.59 \times 10^{-10}$$

$$R_1 = R_2 = 25 \text{ ohm}$$

$$C_1 = C_3 = \frac{1}{25} \times 1.59 \times 10^{-10} \times 0.618 = 4 \text{ pF}$$

$$L_1 = L_2 = 6.3 \text{ nH}$$

$$C_2 = 12.7 \text{ pF}$$

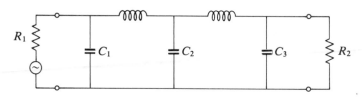

Figure 5.12   Topology for Example 5.6

(d) The insertion loss of the filter at 2 GHz can be found from equation (5.4) as

$$L(2 \text{ GHz}) = 10 \log_{10}\left[1 + \left(\frac{2}{1}\right)^{10}\right] \text{dB}$$

$$= 30 \text{ dB}$$

(e) Draw the distributed circuit that is equivalent to the lumped circuit shown in figure 5.12. This is shown in figure 5.13.

In the general case, when designing with transmission lines, two degrees of freedom are available, these are;

(i) characteristic impedance;
(ii) line length.

Usually, in filter design, the values used for the high and low impedance sections that simulate lumped L and C values are constrained, leaving line lengths as the parameter that is varied to achieve the desired design.

The constraints placed on the maximum and minimum characteristic impedance of the line segments in use in a particular circuit depend on the material out of which the circuit is to be constructed. For the stripline material specified in this problem, program 2.5 given in chapter 2 allows the width of the 25 ohm feeder line to be calculated as 7 mm. Examine now the criteria that must be applied for the selection of line width W_2 shown in figure 5.13, i.e. the low

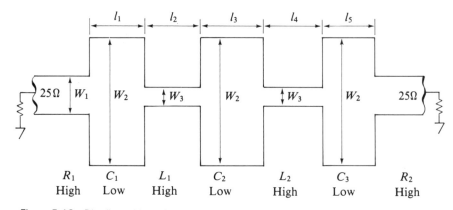

Figure 5.13   Distributed Low Pass Filter

impedance line section representing shunt capacitance C_{In} the lumped prototype. The widest that line W_2 should be made is governed by the width that will support a transverse resonance. For this reason it is wise to make W_2 less than one-quarter of a guide wavelength at the highest frequency to be expected (in this case say 1.5 cm). This should serve to suppress low order transverse modes. For the stripline material specified in this example, 1.5 cm corresponds to about 12.5 ohms. The narrowest line width, width W_3 is governed by fabrication techniques and should be greater than 1 mm or about 70 ohms in this case. Before proceeding any further, it must be emphasized once again that no account of discontinuities that exist between high and low impedance sections of line has been allowed for in this simplified design approach. The discontinuity that exists at the abrupt junction between high and low impedance sections will be discussed further at the end of this section. The further reading section at the end of this chapter gives indication of the various work that has gone into the closed form evaluation of discontinuities for various transmission line types.

(f) Having selected W_2 and W_3, the next step is to find the values of line length that enable the reactive components in the low pass prototype to be simulated. Equations (5.9) and (5.10) provide the necessary tools for this conversion, on substituting $\beta = 2\pi/\lambda_g$ where λ_g is the guide wavelength of the individual line section under consideration.

   The length of the inductive element is

$$l_L = \frac{\lambda_{gc}}{2\pi} \sin^{-1}\left(\frac{\omega L}{Z_{oL}}\right)$$

where the subscript L indicates an inductance.

   The length of the capacitive element is

$$l_C = \frac{\lambda_{gL}}{\pi} \sin^{-1}(\omega C Z_{oC})$$

where the subscript C indicates capacitance.

   Note that for both these expressions the argument of the sine functions is in radians. It should also be noted that in general $\lambda_{gL} \neq \lambda_{gc}$ since guide wavelength can be a function of characteristic impedance, e.g. in microstrip line. Now

$$\lambda_g = \frac{3 \times 10^{10}}{10^9 \sqrt{4}} = 15 \text{ cm}$$

Since $L_1 = L_2 = 6.3$ nH, then

$$l_1 = l_4 = \frac{15}{2\pi} \sin\left(\frac{2\pi \times 10^9 \times 6.3 \times 10^{-9}}{70}\right) = 1.44 \text{ cm}$$

If $\sin^{-1}$ is returned in degrees, it will be necessary to convert the result to radians using the formula

$$\text{radians} = \frac{2\pi}{360} \times \text{number of degrees}$$

Now calculate the remaining lengths required for the capacitive elements C_1, C_2, C_3

$$C_1 = C_3 = 4.0 \text{ pF}$$

$$\therefore \quad l_1 = l_5 = \frac{15}{2\pi} \sin^{-1}(0.025 \times 12.5) = 0.75 \text{ cm}$$

and for

$$C_2 = 12.7 \text{ pF}$$

$$l_3 = 3.6 \text{ cm}$$

This completes the first attempt at the design.

*   *   *

The value of 3.6 cm obtained for l_3 is rather large. To compensate for this, a lower impedance line could have been used. With the line lengths derived above, the overall length of the structure is about 8 cm or just over one-half guide wavelength at the frequency of interest.

In the initial attempt at this design, the end capacitances of the PI line representation of a high impedance, inductive line, were neglected. These capacitances are given as

$$X_C = \frac{Z_{oL}}{\tan(\beta l/2)}$$

or for short lengths

$$C_{end} = \frac{l}{Z_{oL}2f\lambda_g} \tag{5.18}$$

Similarly the end inductances of the TEE line representation of a low impedance, mainly capacitive line, were also neglected. These end inductances for a mainly capacitive line are given as

$$L_{end} = \frac{lZ_{oC}}{2f\lambda_g} \tag{5.19}$$

In order to assess more fully the physical situation, these parameters should be included in the initial attempt. Drawing the lumped equivalent circuit for figure 5.13 and including end capacitances and inductances, the equivalent circuit, more closely approximating figure 5.13 than figure 5.12 the original lumped prototype circuit, is shown as figure 5.14. Figure 5.14 results by substituting full TEE and PI representations for each line segment of the distributed filter.

Continuing on from the initial filter design where these excess end values were neglected, the next step would be to calculate the excess component values according to equations (5.18) and (5.19). An approximate second trial design would proceed as follows. The capacitance values originally specified in the low pass prototype are

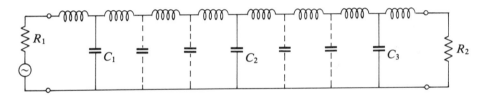

Figure 5.14 Equivalent Circuit for Example 5.6 using Full TEE and PI Representations

corrected. At this point excess inductance can be neglected. This assumption allows the excess capacitance components of mainly inductive line segments to be included in the design. The correction to the original low pass prototype capacitances can now be executed, since all that is required is to subtract the computed excess capacitance from the original lumped prototype capacitance, on a component to component basis. This gives new values for capacitance that can be recomputed to give new line lengths in the distributed filter for each capacitive element. Next, the excess inductance associated with each low impedance line section could be calculated and the lumped prototype inductance values reduced by the excess amount calculated. New line lengths for each inductive element could then be found. This process of calculating the excess inductance and capacitance correcting the original lumped low pass prototype and generating new line lengths is repeated a number of times until either the capacitance or inductance values converge to some fixed value. In this way, the line lengths originally calculated for the lumped prototype filter in the initial design become foreshortened. Computer program 5.5 LPF executes the iterative procedure outlined above for the approximate synthesis of distributed low pass filter circuits.

```
]{ FORMATTED LISTING
FILE: PROGRAM 5.5 LPF
PAGE-1


  10   REM
  20   REM    **** LPF ****
  30   REM
  40   REM    THIS PROGRAM COMPUTES
  50   REM    THE LENGTHS OF LINE
  60   REM    REQUIRED TO SYNTHESISE
  70   REM    MAX. FLAT OR
  80   REM    EQUIRIPPLE FILTERS
  90   REM    FORESHORTING DUE TO
 100   REM    DUE TO TEE AND PI
 110   REM    END IMPS. ARE INCLUDED
 120   REM
 130   REM    Z0=CHARAC. IMP.(OHMS)
 140   REM    Z1=IND. LINE IMP.(OHMS)
 150   REM    Z2=CAP. LINE IMP.(OHMS)
 160   REM    E0=DIE. CONST. FEED
 170   REM    E1=DIE. CONST. IND.
 180   REM    E2=DIE. CONST. CAP.
 190   REM    W=CUTOFF FREQ.(GHZ)
 200   REM    R=RIPPLE (DB)
 210   REM    N=NO. OF SECTIONS
 220   REM
 240   DIM G(100),A(100),B(100),F(100),L(100)
 245   HOME
 250   LET PI = 3.141592654
```

```
260  PRINT
270  PRINT "I/P IMP. OF FEED LINE (OHM)"
280  INPUT Z0
290  PRINT "I/P IMP. OF CAPACITIVE LINE (OHM)"
300  INPUT Z2
310  PRINT "I/P IMP. OF INDUCTIVE LINE (OHM)"
320  INPUT Z1
330  PRINT "I/P EFF. DIE. CONST. FEED"
340  INPUT E0
345  PRINT "I/P EFF. DIE CONST. CAPACITIVE LINE"
347  INPUT E2
350  PRINT "I/P EFF. DIE CONST. INDUCTIVE LINE"
360  INPUT E1
370  PRINT "I/P RIPPLE REQUIRED IN DB"
380  PRINT "FOR BUTTERWORTH FILTER ENTER 0"
390  INPUT R
400  PRINT "I/P THE NUMBER OF FILTER"
410  PRINT "SECTIONS REQUIRED"
415  PRINT "TO MAINTAIN SYMMETRY"
416  PRINT "THIS SHOULD BE AN ODD INTEGER"
417  PRINT "OF VALUE GREATER THAN ONE"
420  INPUT N
424  PRINT "I/P CUTOFF FREQ.(GHZ)"
425  INPUT W
430  PRINT
440  PRINT "*************************"
450  PRINT
460  IF R = 0 THEN
         PRINT "BUTTERWORTH FILTER DESIGN:---"
470  IF R < > 0 THEN
         PRINT "CHEBYSHEV FILTER DESIGN:---"
480  PRINT
490  PRINT "DESIGN PARAMETERS SPECIFIED AT ONSET"
500  PRINT
510  PRINT "LINE IMPEDANCES"
520  PRINT "            (1)FEED"Z0" OHMS"
530  PRINT "            (2)IND. "Z1" OHM"
540  PRINT "            (3)CAP. "Z2" OHM"
545  PRINT
550  PRINT "EFFECTIVE DIE. CONSTS."
560  PRINT "            (1)FEED "E0""
570  PRINT "            (2)IND. "E1""
580  PRINT "            (3)CAP. "E2""
590  PRINT
600  IF R < > 0 THEN
         PRINT "RIPPLE IS "R" DB"
610  PRINT "NUMBER OF FILTER SECTIONS "N
615  PRINT "FILTER CUTOFF FREQ. "W" (GHZ)"
620  LET W1 = 2 * PI * W
630  IF R < > 0 THEN
         700
640  REM   GVALUES BUTTERWORTH
650  LET G(1) = 1
660  LET G(N + 2) = 1
670  FOR I = 1 TO N
680      LET G(I + 1) = 2 *  SIN ((2 * I - 1) * PI / (2 * N))
690  NEXT I
695  GOTO 840
700  REM   GVALUES CHEBY.
710  LET D = R / 17.37
720  LET E = ( EXP (D) +  EXP ( - D))
730  LET E = E / ( EXP (D) -  EXP ( - D))
740  LET E =  LOG (E)
750  LET C = 0.5 * ( EXP (E / 2 / N) -  EXP ( - E / 2 / N))
760  FOR I = 1 TO N
770      LET A(I) =  SIN ((2 * I - 1) * PI / 2 / N)
780      LET B(I) = C * C + ( SIN (I * PI / N)) ^ 2
790      IF I = 1 THEN
             LET F(1) = 2 * A(1) / C:
             GOTO 810
800      LET F(I) = 4 * A(I - 1) * A(I) / B(I - 1) / F(I - 1)
810  NEXT I
812  FOR I = 1 TO N
814      LET G(I + 1) = F(I)
816  NEXT I
```

```
 820   LET  G(1) = 1
 830   LET  G(N + 2) = 1
 840   REM    COMPUTE LENGTHS
 850   LET  MID = (N + 1) / 2 + 1
 860   LET  K = 0
 870   LET  I1 = W1 *  SQR (E1) / 30
 880   LET  I2 = W1 *  SQR (E2) / 30
 890   LET  I3 = 0
 900   FOR  I = 2 TO MID STEP 2
 910        LET  I4 = I
 920        LET  K1 = Z2 * ( TAN (I1 * L(I - 1) / 2) +  TAN (I1 * L(I + 1) / 2))
 930        LET  K1 = (G(I) * Z0 - K1) / Z1
 940        LET  K2 = 1 /  SQR (1 / K1 / K1 - 1)
 950        LET  K3 =   ATN (K2) / I2
 960        IF ( ABS (L(I) - K3) - K3 / 1000) < = 0 THEN
                  980
 970        LET  I3 = 1
 980        LET  L(I) = K3
 990   NEXT I
1000   LET  K = K + 1
1010   IF (I4 - MID) = 0 THEN
                  1030
1020   LET  L(I4 + 2) = L(I4)
1030   FOR  I = 3 TO MID STEP 2
1040        LET  I1 = I
1050        LET  K1 = ( TAN (I2 * L(I - 1) / 2) +  TAN (I2 * L(I + 1) / 2)) / Z1
1060        LET  K1 = (G(I) / Z0 - K1) * Z2
1070        LET  K2 = 1 /  SQR (1 / K1 / K1 - 1)
1080        LET  L(I) =   ATN (K2) / I1
1090   NEXT I
1100   IF (I4 - MID) = 0 THEN
                  1110
1105   LET  L(I4 + 2) = L(I4)
1110   IF (I3 * (K - 10)) < 0 THEN
                  890
1120   LET  K1 = Z0 * Z0 * ( TAN (I2 * L(2) / 2) / Z1) / Z1
1130   LET  K2 = 1 /  SQR (1 / (K1 * K1) - 1)
1140   LET  L(1) =   ATN (K2) / I2
1150   PRINT
1160   REM    PRINT CALCULATED VALUES
1170   PRINT
1180   PRINT   TAB( 1);"LINE"; TAB( 6);"G-VALUE"; TAB( 14);"LENGTH"; TAB( 22);"IM
       PEDANCE"
1190   PRINT   TAB( 15);"(CM)"; TAB( 24);"(OHMS)"
1200   PRINT
1205   LET  I = 0
1210   PRINT   TAB( 1);I; TAB( 7); INT (G(1) * 1000 + .5) / 1000; TAB( 15); INT (
       L(1) * 1000 + .5) / 1000; TAB( 25); INT (Z1 * 100 + .5) / 100
1220   LET  K4 = 2 * (MID - 1)
1230   FOR  I = 2 TO K4
1240        LET  I4 = I - 1
1250        LET  I5 = I
1260        IF (I - MID) < = 0 THEN
                  1280
1270        LET  I5 = K4 - I + 2
1280        IF (2 *  INT (I / 2) - I) < > 0 THEN
                  1310
1290        PRINT   TAB( 1);I4; TAB( 7); INT (G(I5) * 1000 + .5) / 1000; TAB( 15);
            INT (L(I5) * 1000 + .5) / 1000; TAB( 25); INT (Z1 * 100 + .5) / 100
1300        GOTO 1314
1310        PRINT   TAB( 1);I4; TAB( 7); INT (G(I5) * 1000 + .5) / 1000; TAB( 15);
            INT (L(I5) * 1000 + .5) / 1000; TAB( 25); INT (Z2 * 100 + .5) / 100
1314   NEXT I
1316   PRINT   TAB( 1);K4; TAB( 7); INT (G(1) * 1000 + .5) / 1000; TAB( 15); INT
       (L(1) * 1000 + .5) / 1000; TAB( 25); INT (Z1 * 100 + .5) / 100
1318   PRINT
1320   PRINT
1330   PRINT  "***********************"
1340   PRINT
1350   PRINT  "DO YOU WANT ANOTHER GO ?"
1360   PRINT  "IF YES ENTER 1"
1370   INPUT  L
1380   IF  I = 1 THEN
                  245
1390   PRINT
```

```
1400   PRINT "**** END OF PROGRAM ****"
1410   END
```

END-OF-LISTING

]RUN

```
I/P IMP. OF FEED LINE (OHM)
?50
I/P IMP. OF CAPACITIVE LINE (OHM)
?20
I/P IMP. OF INDUCTIVE LINE (OHM)
?130
I/P EFF. DIE. CONST. FEED
?1
I/P EFF. DIE CONST. CAPACITIVE LINE
?1
I/P EFF. DIE CONST. INDUCTIVE LINE
?1
I/P RIPPLE REQUIRED IN DB
FOR BUTTERWORTH FILTER ENTER 0
?0.01
I/P THE NUMBER OF FILTER
SECTIONS REQUIRED
TO MAINTAIN SYMMETRY
THIS SHOULD BE AN ODD INTEGER
OF VALUE GREATER THAN ONE
?5
I/P CUTOFF FREQ.(GHZ)
?1.0
```

****

CHEBYSHEV FILTER DESIGN:---

DESIGN PARAMETERS SPECIFIED AT ONSET

```
LINE IMPEDANCES
              (1)FEED50 OHMS
              (2)IND. 130 OHM
              (3)CAP. 20 OHM

EFFECTIVE DIE. CONSTS.
              (1)FEED 1
              (2)IND. 1
              (3)CAP. 1
```

RIPPLE IS .01 DB
NUMBER OF FILTER SECTIONS 5
FILTER CUTOFF FREQ. 1 (GHZ)

LINE	G-VALUE	LENGTH (CM)	IMPEDANCE (OHMS)
0	1	.091	130
1	.756	1.226	130
2	1.305	.157	20
3	1.577	2.894	130
4	1.305	.157	20
5	.756	1.226	130
6	1	.091	130

****

DO YOU WANT ANOTHER GO ?
IF YES ENTER 1

?0

**** END OF PROGRAM ****

The performance of the final design can be appraised by program CASCADE given in chapter 4.

5.2.1 Discontinuity Effects

At no point in the discussion on low pass filter design for distributed circuits have the discontinuity effects, already mentioned for the high to low impedance transformation between series line segments, been accounted for. The discontinuity associated with this type of abrupt irregularity in a transmission line is mainly capacitive and for most cases can be represented at a fixed frequency by a shunt capacitance or more generally by a TEE equivalent circuit. Figure 5.15 illustrates the electric field behavior and lumped TEE equivalent for this discontinuity. To take the discontinuity into account, the lumped component values in the TEE representation of the discontinuity have to be evaluated. Unfortunately these component values are dependent on the type of line and on the exact nature of the discontinuity under investigation. Large scale computer simulations are normally employed to model the electric and magnetic field distributions that exist at discontinuities and empirical equations are curve fitted to the theoretical data obtained. The further reading section at the end of this chapter should be consulted for more detail.

If the simple low to high or high to low impedance transformation of figure 5.15 were to be modeled by a lumped circuit then figure 5.16 (a) would result. If a lumped TEE representation for the discontinuity in this circuit was also included then figure 5.16 (b) becomes valid. Having included excess end reactances and lumped discontinuity effects in the lumped analog of the distributed low pass filter, it would be necessary to repeat the second attempt design technique outlined in the previous section. From the discussion given it is seen that if more physically realistic parameters are taken into account the more the distributed circuit will depart from the lumped prototype circuit it was derived from initially. In practice, post manufacture trimming of circuits often cannot be avoided.

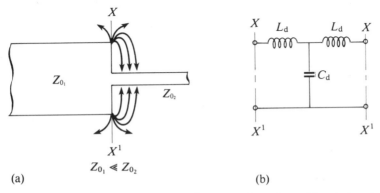

(a)   (b)

Figure 5.15 Lumped Discontinuity Representation of an Abrupt Junction in Uniform Transmission Line (a) Junction of Low Impedance and High Impedance Line (b) Discontinuity Equivalent Circuit

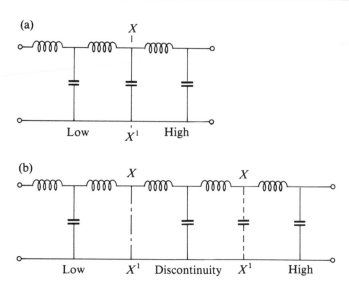

Figure 5.16   Refined Lumped Equivalent Circuits for Example 5.6

5.3   DISTRIBUTED HIGH PASS FILTER DESIGN

Up to now the main concern has been the synthesis of low pass filter circuits utilizing distributed elements. Logical progression leads now to the development of distributed high pass filter circuitry from lumped prototypes. A problem immediately becomes evident on examination of the lumped prototype high pass filter shown, for example, in figure 5.8. How can series capacitors be realized in transmission line structures? The simplest solution is to produce a physical break in the center conductor of a coaxial cable or in the current carrying conductor of microstrip or stripline transmission line. At first sight this appears to be an attractive proposition but in practice it is not recommended for two main reasons. First, the physical gap needed to produce the desired capacitance value suggested by the lumped prototype circuit is normally so small that it presents problems in fabrication due to gap repeatability. Second, a physical break in the current carrying conductor of a transmission line does not manifest itself as a simple lumped capacitance but appears as a more complicated PI circuit comprised of series and shunt capacitance values. For these reasons the use of series breaks are discouraged unless they can be carefully controlled and modeled. Distributed high pass filter circuits therefore assume a hybrid form where lumped components and distributed elements are combined to give the desired circuit performance. At frequencies up to 20 GHz, chip capacitors formed by thick or thin film techniques can be successfully used. Example 5.7 demonstrates the design and construction of a coaxial high pass filter from a hybrid combination of lumped and distributed elements to simulate the behavior of a lumped element prototype.

* * *

Example 5.7

Design a high pass filter that uses coaxial line and lumped capacitors. The filter should exhibit the following electrical characteristics:

1   Third order Butterworth response.
2   Cut-off frequency 1 GHz.
3   To be embedded in a 50 ohm system.

Assume the coaxial line outer dimension to be given as 0.762 cm.

Solution

$n = 3$, Butterworth

$R_L = R_o = 50$ ohm
$\omega_c = 2\pi \times 10^9$ rad/s

(a) Select the topology in figure 5.17 and find g values for equivalent low pass prototype (program GVAL)

$g_0 = g_1 = g_3 = g_4 = 1$

$g_2 = 2$

Find C_1, C_2, L for high pass response

$$C_1 = C_2 = \frac{1}{2\pi \times 10^9 \times 1} \frac{1}{50} = 3.2 \text{ pF}$$

$L = 3.98$ nH

(b) Select the coaxial line impedance of the line section required to simulate inductance L, say 100 ohms.

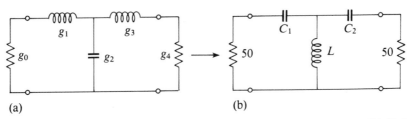

(a)   (b)

Figure 5.17   Circuit Topology for Example 5.7 (a) Low Pass Prototype (b) High Pass Equivalent

Reference to Table 5.3 shows a suitable method for realizing a shunt inductance is by means of a short circuit stub of length l that

$$l = \frac{Lf\lambda_g}{Z_o} = \frac{LV_p}{Z_o}$$

for air-spaced coaxial line $V_p = 3 \times 10^{10}$ cm/s

$$\therefore \quad l = \frac{3 \times 10^{10} \times 3.98 \times 10^{-9}}{100} = 1.194 \text{ cm}$$

(c) The filter is to operate in a 50 ohm system. Program COAX in chapter 2 yields, for a 50 ohm characteristic impedance air-spaced coaxial line with an outer conductor having an internal diameter of 0.762 cm, a center conductor diameter of 0.33 cm. Now the impedance of the inductive line section is 100 ohms, which corresponds to a center conductor diameter of 0.14 cm for a coaxial line having the same outer diameter as the 50 ohm cable section.

(d) Next, synthesize the lumped series capacitors needed to complete the design. Since coaxial line has cylindrical geometry, circular parallel plate capacitors could be used for C_1 and C_2 in the lumped prototype. If the diameter of the capacitor plate is made large with respect to the plate spacing then the effects of fringing fields can be neglected (see section 3.7). For a parallel plate capacitor the capacity is given as

$$C = \frac{\varepsilon_0 \varepsilon_r A}{d}$$

where

   A = plate area = πr^2 for a circular plate capacitor
   d = plate separation
   r = plate radius

Now plate diameter d should lie within $0.33 < (d = 2r) < 0.762$ cm so that the capacitor will fit the line geometry. Select a value for d, say 0.6 cm

   $\therefore \quad r = 0.3$ cm

Choosing a polyolefin material, with $\varepsilon_r = 2.3$ as the capacitor dielectric, yields a capacitor plate spacing of

$$d = \frac{8.854 \times 10^{-14} \times 2.3 \times \pi \times (0.3)^2}{3.2 \times 10^{-12}} = 0.018 \text{ cm}$$

This completes the design. A dimensioned diagram for the filter is shown in figure 5.18.

   Figure 5.19 shows the computed response for the distributed high pass filter designed in this example. This is compared with that defined by the high pass lumped prototype in figure 5.17.

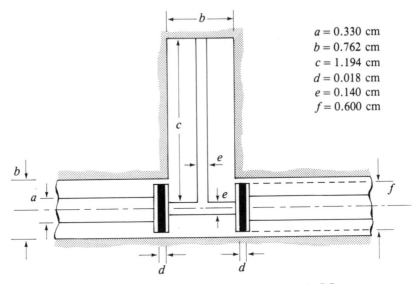

$$a = 0.330 \text{ cm}$$
$$b = 0.762 \text{ cm}$$
$$c = 1.194 \text{ cm}$$
$$d = 0.018 \text{ cm}$$
$$e = 0.140 \text{ cm}$$
$$f = 0.600 \text{ cm}$$

Figure 5.18   Coaxial High Pass Filter Construction for Example 5.7

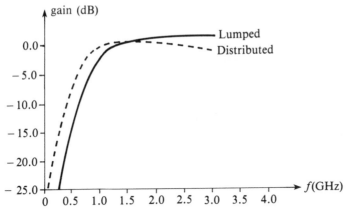

Figure 5.19   Computed Responses for Lumped and Distributed High Pass Filter, Example 5.7

* * *

5.4   DISTRIBUTED BAND PASS FILTER DESIGN

In example 5.4 it was found that when a lumped low pass prototype circuit is converted to a lumped band pass circuit a number of series and parallel resonant LC combinations must be used. The necessity of having to employ parallel and series resonant circuits at first sight appears daunting. Parallel resonant circuits can be constructed fairly readily in a distributed medium, see, for example, table 5.3 types D and F. For the type D element shown in table 5.3, stub lines simulate lumped L and C components placed in parallel across the line. For the type F element, a lightly loaded section of series line with uniform characteristic impedance Z_o behaves in the same manner as a parallel resonant circuit. These elements then indicate that parallel shunt resonant circuits are available. What about series resonant circuits in series with the transmission line? The easiest way to simulate series resonant LC circuits in series with a line is to use a high impedance line section, inductance, in cascade with a lumped capacitance, see, for example, example 5.7. This type of solution is really only feasible where the frequency of operation of the circuit is low enough to allow the use of lumped elements. As frequency is increased an alternative solution is required. One way around the use of lumped elements is to replace the series resonant series LC circuit with an alternative arrangement.

Examination of figure 5.20 shows that the terminal characteristics of a series resonant LC circuit can be made to behave like those of a parallel resonant LC circuit on performing an impedance inversion. Such an inversion can be simply achieved at a single frequency by the use of a quarter wavelength section of uniform transmission line. The application of impedance inverters is fully explained by Matthaei *et al.* [4] (sect. 4.12) which should be consulted for the development of the design equations used in this section.

To illustrate the facility afforded by impedance inverters in the design of distributed band pass circuitry derived from a lumped prototype, an explanation based on the series of diagrams given as figure 5.21 will be used.

Figure 5.21(a) shows a lumped third order band pass prototype circuit whose elements are assumed to have undergone impedance scaling and frequency transformation from a normalized lumped low pass circuit. Once this stage is reached, the next step is to insert impedance inverter sections designated as K sections. These convert series resonant circuits to their parallel equivalents, figure 5.21(b).

In order to maintain constant the internal impedance of the filter Z after

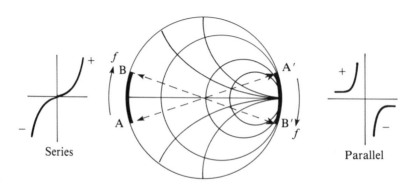

Figure 5.20   Impedance Inversion caused by Quarter Guide Wavelength Line Section

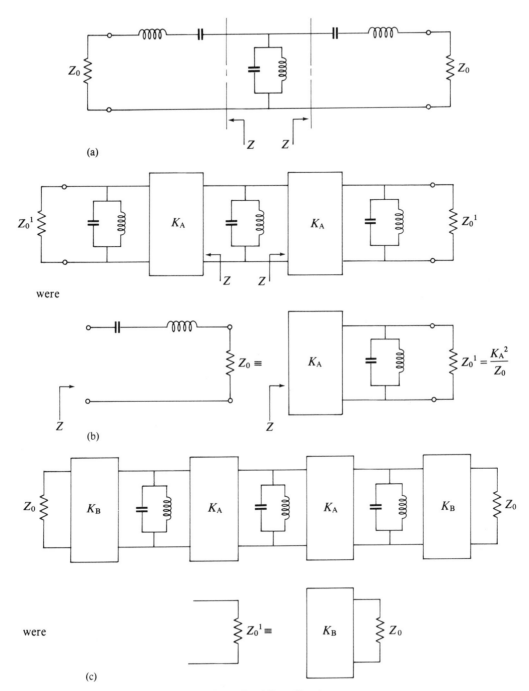

were

Figure 5.21   (a) Lumped Third Order Band Pass Circuit
(b) Prototype after Impedance Inversion
(c) Prototype Circuit with Corrected End Impedances

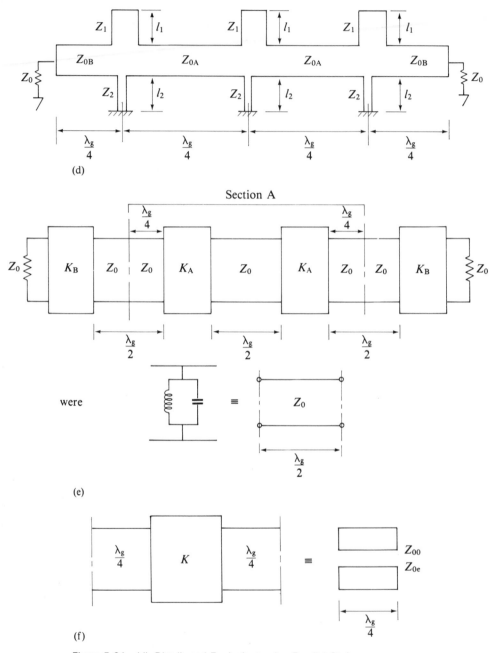

(d)

Section A

(e)

were

(f)

Figure 5.21 (d) Distributed Equivalent using Parallel Stubs
(e) Semi-Distributed Prototype Equivalent
(f) Equivalence between Edge Coupled Lines and Composite Inverter Section
(g) Interdigitated Filter Section with Modified End Sections
(h) Conventional Interdigital Band Pass Configuration

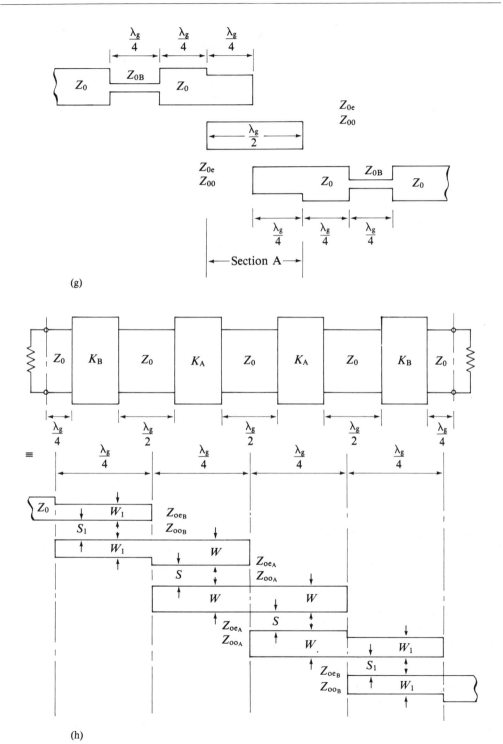

(g)

(h)

impedance inverting sections have been inserted in the circuit, the terminating impedances of the filter have been modified from their original value Z_0 to their new value $Z_0{}^1$. The modification to Z_0 allows the same L and C values to be used throughout the filter thereby simplifying the synthesis. The modification made to Z_0 can then be compensated for by the introduction of two additional impedance inverting sections, designated as K_B in figure 5.21(c) at both ends of the filter. The lumped prototype filter of figure 5.21(a) is now ready for synthesis.

Synthesis of distributed band pass circuitry can be carried out in a variety of ways. One way suitable for microstrip realization could use open circuit and short circuit stubs to generate the parallel resonant circuits of figure 5.21(c). The impedance inverting sections are simulated by series quarter wavelength sections (figure 5.21(d)). In this circuit all the open circuit shunt stubs have a fixed characteristic impedance Z_1 and length l_1 and all the short circuit stubs are identical having fixed length l_2 and characteristic impedance Z_2. In this design it is possible to add an additional open circuit quarter wavelength section to each short circuit stub thereby simulating at a fixed frequency the behavior of a short circuit. This has the added advantage of allowing the filter to transfer a d.c. bias from generator to load. The design equations for the stubs can be found from element D in table 5.3, remember to make Z_1 much smaller than Z_2. All that remains to be done to complete the stub design is to establish the characteristic impedance of the impedance inverting sections, Z_{0A}, Z_{0B} [4]

$$Z_{0A} = \frac{2Z_0}{\pi B} (g_1 g_2)^{1/2} \tag{5.20}$$

$$Z_{0B} = Z_0 \left(\frac{2g_1}{\pi B}\right)^{1/2} \tag{5.21}$$

Here Z_0 is the impedance of the system in which the filter is to operate. The g values g_1, g_2 occur from the original low pass filter prototype (see example 5.1 for instance). Finally B is the fractional bandwidth of the filter given as

$$B = \frac{\omega_u - \omega_L}{\omega_0}$$

Another method for simulating the parallel LC circuit is to replace the LC lumped components by a lightly loaded transmission line that has the same characteristic impedance as the network into which the filter is to be embedded. These transmission line segments should be one-half guide wavelength at the center frequency of the filter, figure 5.21(e) shows the situation.

Seymour Cohn [5] in one of his now many classical papers showed that an impedance inverter embedded between two one-quarter wavelength line segments (figure 5.21(f)) has the same terminal characteristics as a pair of parallel edge coupled lines, provided the coupled lines have even and odd mode impedances constrained to be

$$Z_{oe} = Z_0 \left[1 + \frac{Z_0}{K} + \left(\frac{Z_0}{K}\right)^2 \right] \tag{5.22}$$

$$Z_{00} = Z_0\left[1 - \frac{Z_0}{K} + \left(\frac{Z_0}{K}\right)^2\right]$$
(5.23)

Equations (5.22) and (5.23) are applied for each embedded K section in the filter. Once Z_{oe} and Z_{oo} are known, program CMIC or CSTRIP can give the line spacing and line width needed for final construction of the filter. Exploiting the equivalence between a set of parallel edge coupled lines and an embedded K section, as shown in figure 5.21(f), leads to a very useful filter topology. This type of filter, shown in figure 5.21(g) for a third order filter, is called an interdigitated filter. Notice how a d.c. block is inherent in the filter design and how the end sections of the filter are realized by a one-quarter wavelength line followed by a one-quarter wavelength impedance inverter section.

A more traditional form of the interdigital filter, shown in figure 5.21(g), is evolved by adding additional quarter wavelength sections at the input and output side of the filter. This maintains symmetry so that the circuit in figure 5.21(h) results.

The topology in figure 5.21(h) is commonly used and will provide 15 to 20 percent bandwidth. The filter bandwidth is limited mainly by the difference in even and odd mode propagation velocities that exist at frequencies removed from the design frequency. When bandwidths greater than about 20 percent are required then the spacing between the lines forming the end sections of the filter become very narrow leading to fabrication and repeatability problems. To avoid this problem the topology given in figure 5.21(g) should be used. A further advantage gained by this topology is that the number of coupled line pairs required to construct a third order filter is reduced from four in figure 5.21(h) to two for figure 5.21(g).

When filters of a high order are needed, large substrate areas are required. Various packing schemes are available for interdigitated filters that allow a reduction in the amount of substrate area required for a filter of given order. Two suggestions for the third order filter are given in figure 5.22 as being representative of circuit layouts currently in use. Other layouts are available and the final type selected will depend on the individual requirements of each application.

One practical point that has not been mentioned so far is the choice for the length of the edge coupled resonators. Throughout this section it has been suggested that resonator lengths should be selected to be one-half guide wavelength long. If

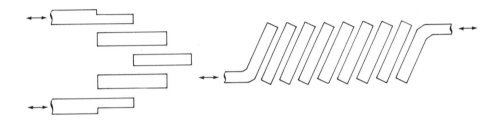

Figure 5.22 Equivalent Folded Interdigital Filters

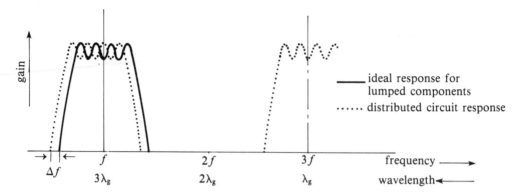

Figure 5.23   Deviation of Interdigital Bandpass Filter Response from Prototype Response

this length is selected then, neglecting all other deleterious effects, a response such as that shown in figure 5.23 can result.

From figure 5.23 two problems can be immediately identified. The first is the existence of a spurious response at some frequency other than the design frequency. This is due to the periodic nature of transmission line, and should be borne in mind from a systems point of view. The lumped prototype of the band pass circuit will not exhibit this spurious response. The second feature exhibited by figure 5.23 is that the measured response is shifted some frequency Δf to the lower side of the center frequency. This shift implies that the resonator lengths have been uniformly selected to be too long, i.e. greater than one-half guide wavelength at the filter center frequency. The physical length of the resonators when measured is one-half guide wavelength. What then could be causing the apparent lengthening of the resonator lines? The major effect contributing to this apparent lengthening is the end capacitance associated with each open circuit length of line forming the interdigital fingers of the filter. Figure 5.24 shows a typical resonator section together with its electrical equivalent when the capacitance due to fringing fields at each end of the section is taken into account.

The effect of these end capacitances formed by the fringing fields is to make the resonator section appear electrically longer than its physical length would indicate. This apparent lengthening, designated Δl, is easily related to end capacitance. In section 1.5 it can be seen that for an open circuit stub having negligible loss the input impedance is given as

$$X_{oc} = -jZ_o \cot (\beta l)$$

Referring back to figure 5.24, a section of perfectly terminated open circuit length Δl added to the original resonator can be made to appear like the resonator with fringing capacitance. This equivalence follows if the reactance due to the end capacitance representing the fringing capacitance is equated to the open circuit stub equation

$$X_{oC} = \frac{1}{j\omega C_{end}} = -jZ_o \cot (\Delta l\beta)$$

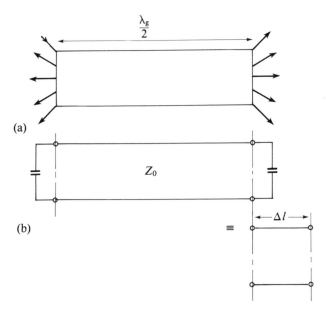

$$\frac{\lambda_g}{2}$$

(a)

Z_0

(b)

$\equiv$

Δl

Figure 5.24   Effect of End Capacitance on Transmission Line Length (a) Physical Resonator
(b) Electrical Equivalent including End Capacitance

Here, a small angle approximation can be made since $\Delta l \rightarrow 0$, hence

$$\frac{1}{j\omega C_{end}} = \frac{-jZ_o}{\Delta l\beta}$$

with

$$\beta = \frac{2\pi}{\lambda_g}$$

then

$$\Delta l = Z_o \lambda_g f C_{end}$$

or

$$\Delta l = Z_o C_{end} v_p \tag{5.24}$$

where

$$v_p = \text{phase velocity} = \frac{c}{(\varepsilon_{eff})^{1/2}} \text{ for microstrip line}$$

$$= \frac{c}{(\varepsilon_r)^{1/2}} \text{ for TEM line}$$

In equation (5.24) Z_o and v_p will be known or can be closely estimated. This leaves
the end capacitance to be found so that the additional line length attributed to the

fringing fields can be compensated for by modifying the lengths of the resonator sections according to equation (5.25)

$$\frac{\lambda_g^{\ 1}}{2} = \frac{\lambda_g}{2} - 2\Delta l \tag{5.25}$$

Foreshortening the original resonator sections by $2\Delta l$ should cause the measured filter response in figure 5.23 to move closer to the desired filter response. What is needed to finally establish Δl is a mechanism by which the fringing capacitance associated with open circuited sections of transmission line can be evaluated.

Now since interdigital filters are most commonly constructed on microstrip or stripline substrates, approximate expressions for the fringing capacitance, or better still the equivalent length Δl, will now be given for both microstrip and stripline. The first of these applies to microstrip and is due to Hammerstadt and Bekkadal [6].

$$\Delta l = 0.412h \left[\frac{\varepsilon_{eff} + 0.3}{\varepsilon_{eff} - 0.258} \right] \left[\frac{W/h + 0.262}{W/h + 0.813} \right]$$

This expression should be used cautiously since it has been derived from curve fitted results for a particular class of substrate material and line aspect ratios. A better approach is to use the calculated values for end capacitance according to the approach suggested by Silvester and Benedek [7]. Equations (5.24) and (5.25) can then be invoked to give the corrected resonator length $\lambda_g^{\ 1}$. The calculation of C_{end} by this approach is rather cumbersome since it involves the application of weighting terms selected from a look-up table designed to give usable results over a wide range of materials and frequencies. Their work should be consulted for a comprehensive detailed explanation. One final method, possibly the best one, is to compute the fringing capacitance as and when required from first principles. The method of moments solution given in chapter 3, after some modification to account for the introduction of a dielectric material, is particularly suitable for this type of calculation.

A correction Δl for stripline given as

$$\Delta l = 0.165h$$

where h is the ground plane spacing, is often used.

* * *

Example 5.8

Construct a series of band pass filters based on the topologies outlined in this section. The filters constructed should exhibit Chebyshev responses with 1.0 dB ripple in the passband and be derived from a third order lumped element low pass filter section prototype. The frequency of operation for the filters is centered on 4 GHz and the filters should be compatible with a 50 ohm characteristic impedance system. The filters should have a bandwidth of 1 GHz and be constructed on a microstrip substrate.

Solution

$f_o = 4$ GHz, $f_u - f_L = 1$ GHz

$Z_o = Z_L = 50$ ohms, $n = 3$, ripple in passband 1 dB

Program GVAL for a Chebyshev response yields

$g_1 = g_4 = 1$

$g_1 = g_3 = 1.0315$

$g_2 = 1.1474$

Execute low pass to band pass conversion for the circuit configuration given in figure 5.21(a). The requirements of this problem necessitate a knowledge of L_H and C_H only.

$$L_H = \frac{2\pi \times 10^9}{(8\pi \times 10^9)^2 1.1474} = 8.67 \times 10^{-12} \text{ H}$$

$$C_H = \frac{1.1474}{2\pi \times 10^9} = 0.183 \times 10^{-9} \text{ F}$$

Scale these components to the desired impedance level, then

$L_H = 0.434$ nH

$C_H = 3.7$ pF

Select first the low pass prototype, figure 5.21(d). Let $Z_1 = 25$ ohm and $Z_2 = 100$ ohm

$$B\frac{\omega_u - \omega_L}{\omega_0} = \frac{2\pi \times 10^9}{8\pi \times 10^9} = 0.25$$

$$l = \frac{\lambda_{ga}}{4}$$

$$Z_{oA} = \frac{2 \times 50}{\pi(0.25)}(1.0315 \times 1.1474)^{\frac{1}{2}} = 138 \text{ ohms}$$

$$l = \frac{\lambda_{gb}}{4}$$

$$Z_{oB} = 50\left(\frac{2 \times 1.0315}{\pi(0.25)}\right)^{\frac{1}{2}} = 81 \text{ ohms}$$

From table 5.3

$$l_1 = Z_1 C_H f \lambda_{g_1} = 25 \times 3.66 \times 10^{-12} \times 4 \times 10^9 \times \lambda_{g_1}$$
$$= 0.366\lambda_{g_1}$$

$$l_2 = \frac{L_H f \lambda_{g_2}}{Z_2} = \frac{0.434 \times 10^{-9} \times 4 \times 10^9}{100}\lambda_{g_2} = 0.0174\lambda_{g_2}$$

Here the guide wavelength corresponding to each line of different characteristic can be found from program MIC since the filter is to be constructed on a microstrip medium. Next choose the layout given in figure 5.21(g).

In the second prototype filter Z_{0B} is equivalent to Z_{0B} in the first prototype, i.e. 81 ohms

$$Z_0 = 50 \text{ ohms}$$

Let K in equations (5.22) and (5.23) be Z_{0A} of the first prototype

$$Z_{\text{oe}} = 50\left[1 + \frac{50}{138} + \left(\frac{50}{138}\right)^2\right] = 75 \text{ ohms}$$

$$Z_{\text{oo}} = 50\left[1 - \frac{50}{138} + \left(\frac{50}{138}\right)^2\right] = 39 \text{ ohms}$$

The quarter wavelength sections comprising each resonator section can be found from program CMIC as can the line widths and spacings of the parallel edge coupled lines for the even and odd mode characteristic impedances derived above. Program MIC can be used to find the widths and guide wavelengths for the remaining transmission lines in the circuit.

Finally, consider the third prototype filter section given in figure 5.21(h)

$$\left.\begin{array}{l} Z_{\text{oeA}} = 75 \text{ ohms} \\ Z_{\text{ooA}} = 39 \text{ ohms} \end{array}\right\} \text{as for the second prototype}$$

Now, calculate Z_{oeB} and Z_{ooB}, this time $K = 81$ ohms obtained from the low pass prototype in figure 5.21(f)

$$Z_{\text{oeB}} = 50\left[1 + \frac{50}{81} + \left(\frac{50}{81}\right)^2\right] = 100 \text{ ohms}$$

$$Z_{\text{ooB}} = 50\left[1 - \frac{50}{81} + \left(\frac{50}{81}\right)^2\right] = 38 \text{ ohms}$$

Once again program CMIC will provide the additional line spacings, widths and guide wavelengths necessary to complete the design.

∗   ∗   ∗

5.5   QUARTER WAVE TRANSFORMERS WITH EXTENDED BANDWIDTH

Often it is necessary to transform from one impedance level to another. One reason perhaps could be that it is necessary to interface a component with a very large or very small impedance to a standard 50 ohm generator. In this case, it is necessary to use impedance transformers to provide a feasible range of impedances over which transmission lines can be successfully fabricated. The simple transformer sections discussed in the last chapter have limited bandwidth so they are virtually useless when wide band performance is required. To obtain wide band performance it is necessary to employ multisection transformers. Transformers having two or three sections can give bandwidths of up to 150 percent. The actual bandwidth achieved for a particular number of sections comprising the transformer is dependent on the ratio R of the final and initial load impedances, considered real, to be matched.

Collin [8] has investigated transformers displaying Chebyshev, equiripple responses and Butterworth, maximally flat responses. For most practical purposes transformers with more than three one-quarter wavelength sections are seldom used since they become physically long for frequencies below 10 GHz. This section will be confined in its scope to the design of two and three section transformers for the reason cited above. The notation used for the design of stepped impedance transformers is illustrated in figure 5.25. Here Z_0 and Z_3 represent the resistances that require matching.

The nominal length of each section in the transformer is one-quarter guide wavelength selected at the center frequency of the transformer. For stepped impedance transformers constructed from dispersive line sections, i.e. lines where the effective permittivity hence guide wavelength is a function of frequency, the approximation for the physical length of the line as given by equation (5.26) should be employed

$$l = \frac{\lambda_{gH}\lambda_{gL}}{2(\lambda_{gH} + \lambda_{gL})} \tag{5.26}$$

Here, the subscripts L and H indicate the lowest and highest frequencies at which the transformer is to be operated.

In either case, at the midband frequency, the electrical length of the line should be 90 degrees. The fractional bandwidth B is defined for the purposes of this discussion by equation (5.27)

$$B = 2\left(\frac{\lambda_{gH} - \lambda_{gL}}{\lambda_{gH} + \lambda_{gL}}\right) \tag{5.27}$$

Collin showed for two and three section transformers that the equations given

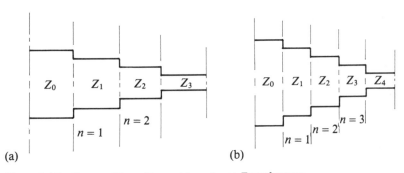

Figure 5.25 Quarter Wave Stepped Impedance Transformers
(a) Two Section (b) Three Section

Table 5.4 Stepped Impedance Transformer Design Equations (see ref. 4)

	$n = 2$	$n = 3$
Butterworth	$Z_1 = Z_o R^{\frac{1}{4}}$	$V_1{}^2 + 2R^{\frac{1}{2}}V_1 - \dfrac{2R^{\frac{1}{2}}}{V_1} - \dfrac{R}{V_1{}^2} = 0$
	$Z_2 = Z_o R^{\frac{1}{2}}$	$V_2{}^2 = \dfrac{R^{\frac{1}{2}}}{V_1}$
		$Z_1 = Z_o V_1;\ Z_2 = Z_o R^{\frac{1}{2}}$
		$Z_3 = \dfrac{Z_o R}{V_1}$
Chebyshev	$V_1{}^2 = (D^2 + R)^{\frac{1}{2}} + D$	$V_1{}^2 + 2R^{\frac{1}{2}}V_1 - \dfrac{2R^{\frac{1}{2}}}{V_1} - \dfrac{R}{V_1{}^2} = \dfrac{3k^2(R-1)}{4 - 3k^2}$
	$Z_1 = Z_o V_1$	$V_2 = R^{\frac{1}{2}}V_1$
	$Z_2 = \dfrac{Z_o R}{V_1}$	$Z_1 = Z_o V_1$
	where	$Z_2 = Z_o R^{\frac{1}{2}}$
	$D = \dfrac{(R-1)k^2}{2(2-k^2)}$	$Z_3 = \dfrac{Z_o R}{V_1}$
	and	For all equations
	$k = \sin\left(\dfrac{\pi}{4} B\right)$	$R = \dfrac{Z_o}{Z_3}$
		or $\dfrac{Z_3}{Z_o}$ whichever > 1

in table 5.4 are valid. Inspection of this table reveals that the design equations for two section Chebyshev and Butterworth filters are simple enough to allow rapid hand calculation. However, when three section transformers are to be designed the situation is rather more complex since it involves the solution of a nonlinear equation. This solution is best found by a straightforward iterative process on a digital computer. Once the nonlinear expression for V_1 has been solved, values for Z_1, Z_2 and Z_3 can be found. A computer program, program 5.6 STRANS, has been written which allows two and three section transformers having Butterworth and Chebyshev responses to be designed. Once designed, one of the cascade analysis programs can assess overall circuit performance.

Since discontinuities have not been taken into account in the design, some empirical adjustment of the final circuit will be required in order to trim the response of the circuit after construction. Analysis of transformer circuits shows that the final bandwidth of an ideal multisection transformer is a slightly sensitive function of the number of sections in the circuit. The degree of sensitivity is dependent on the ratio

```
][ FORMATTED LISTING
FILE: PROGRAM 5.6 STRANS
PAGE-1

    10   REM
    20   REM    **** STRANS ****
    30   REM
    40   REM    THIS PROGRAM COMPUTES
    50   REM    THE CHARAC. IMPS.
    60   REM    NECESSARY FOR STEPPED
    70   REM    IMPEDANCE TRANSFORMERS
    80   REM    DISPLAYING BUTTERWORTH
    90   REM    AND CHEBYSHEV RESPONSES
   100   REM    TWO AND THREE SECTION
   110   REM    TRANSFORMERS ARE
   120   REM    CATERED FOR.
   130   REM
   140   REM    R=IMPEDANCE RATIO
   150   REM    Z0,Z4=IMPS. TO BE
   160   REM    MATCHED
   170   REM    BW=BANDWIDTH AS A
   180   REM    PERCENTAGE (E.G. 20)
   190   REM
   200   HOME
   210   PRINT "INPUT LOWER OF THE TWO"
   220   PRINT "IMPEDANCES TO BE MATCHED"
   230   INPUT Z0
   240   PRINT "INPUT HIGHER OF THE TWO "
   250   PRINT "IMPEDANCES TO BE MATCHED"
   260   INPUT Z4
   270   LET R = Z4 / Z0
   280   PRINT "INPUT PERCENTAGE BANDWIDTH"
   290   INPUT Q
   300   PRINT
   310   PRINT "*********************"
   320   PRINT
   330   PRINT "LOW IMPEDANCE "Z0" OHMS"
   340   PRINT "HIGH IMPEDANCE "Z4" OHMS"
   350   PRINT "GIVING IMPEDANCE RATIO "R
   360   PRINT "FOR A "Q" PERCENT BANDWIDTH"
   370   REM   TWO SECTION CHEBYSHEV
   380   LET Q =  SIN (3.141592653 / 4 * Q / 100)
   390   LET Y = Q * Q * (R - 1) / (2 - Q * Q) / 2
   400   LET V1 =  SQR ( SQR (Y * Y + R) + Y)
   410   LET Z1 = Z0 * V1
   420   LET Z2 = Z0 * R / V1
   430   PRINT
   440   PRINT "TWO SECTION CHEBYSHEV RESULTS"
   450   PRINT "IMP. OF FIRST SECTION " INT (Z1 * 1000 + .5) / 1000" OHMS"
   460   PRINT "IMP. OF SECOND SECTION " INT (Z2 * 1000 + .5) / 1000" OHMS"
   470   PRINT
   480   REM   TWO SECTION BUTTERWORTH
   490   LET Z1 = Z0 *  SQR ( SQR (R))
   500   LET Z2 = Z0 *  SQR (R)
   510   PRINT "TWO SECTION BUTTERWORTH RESULTS"
   520   PRINT "IMPEDANCE OF FIRST SECTION " INT (Z1 * 1000 + .5) / 1000" OHMS"
   530   PRINT "IMPEDANCE OF SECOND SECTION " INT (Z2 * 1000 + .5) / 1000" OHMS"
   540   PRINT
   550   REM   THREE SECTION CHEBYSHEV
   560   LET V1 = 0
   570   LET Y = 3 * Q * Q * (R - 1) / (4 - 3 * Q * Q)
   580   FOR I = 1000 TO 10000
   590       LET V1 = I / 1000
   600       LET X = V1 * V1 + 2 *  SQR (R) * V1
   610       LET X = X - 2 *  SQR (R) / V1 - R / V1 / V1
   620       LET Z = (X - Y) * (X - Y)
   630       IF Z < 0.0001 THEN
                 650
   640   NEXT I
   650   PRINT "THREE SECTION CHEBYSHEV RESULTS"
   660   PRINT "IMPEDANCE OF FIRST SECTION " INT (Z0 * V1 * 1000 + .5) / 1000" OHM
         S"
   670   PRINT "IMPEDANCE OF SECOND SECTION " INT (Z0 *  SQR (R) * 1000 + .5) / 10
         00" OHMS"
```

```
680   PRINT "IMPEDANCE OF THIRD SECTION " INT (Z0 * R / V1 * 1000 + .5) / 1000"
         OHMS"
690   PRINT
700   REM   THREE SECTION BUTTERWORTH
710   LET V1 = 0
720   FOR I = 1000 TO 10000
730       LET V1 = I / 1000
740       LET X = V1 * V1 + 2 *  SQR (R) * V1
750       LET X = X - 2 *  SQR (R) / V1 - R / V1 / V1
760       LET Z = X * X
770       IF Z < 0.0001 THEN
              790
780   NEXT I
790   PRINT "THREE SECTION BUTTERWORTH RESULTS"
800   PRINT "IMPEDANCE OF FIRST SECTION " INT (Z0 * V1 * 1000 + .5) / 1000" OHM
         S"
810   PRINT "IMPEDANCE OF SECOND SECTION " INT (Z0 *  SQR (R) * 1000 + .5) / 10
         00" OHMS"
820   PRINT "IMPEDANCE OF THIRD SECTION " INT (Z0 * R / V1 * 1000 + .5) / 1000"
         OHMS"
830   PRINT
840   PRINT "*********************"
850   PRINT
860   PRINT "DO YOU REQUIRE ANOTHER GO ?"
870   PRINT "IF YES ENTER 1"
880   INPUT T
890   IF T = 1 THEN
              200
900   PRINT
910   PRINT "**** END OF PROGRAM ****"
920   END
```

END-OF-LISTING

```
]RUN
INPUT LOWER OF THE TWO
IMPEDANCES TO BE MATCHED
?25
INPUT HIGHER OF THE TWO
IMPEDANCES TO BE MATCHED
?50
INPUT PERCENTAGE BANDWIDTH
?20
*********************

LOW IMPEDANCE 25 OHMS
HIGH IMPEDANCE 50 OHMS
GIVING IMPEDANCE RATIO 2
FOR A 20 PERCENT BANDWIDTH

TWO SECTION CHEBYSHEV RESULTS
IMP. OF FIRST SECTION 29.795 OHMS
IMP. OF SECOND SECTION 41.953 OHMS

TWO SECTION BUTTERWORTH RESULTS
IMPEDANCE OF FIRST SECTION 29.73 OHMS
IMPEDANCE OF SECOND SECTION 35.355 OHMS

THREE SECTION CHEBYSHEV RESULTS
IMPEDANCE OF FIRST SECTION 27.3 OHMS
IMPEDANCE OF SECOND SECTION 35.355 OHMS
IMPEDANCE OF THIRD SECTION 45.788 OHMS

THREE SECTION BUTTERWORTH RESULTS
IMPEDANCE OF FIRST SECTION 27.25 OHMS
IMPEDANCE OF SECOND SECTION 35.355 OHMS
IMPEDANCE OF THIRD SECTION 45.872 OHMS
*********************
DO YOU REQUIRE ANOTHER GO ?
IF YES ENTER 1
?0

**** END OF PROGRAM ****
```

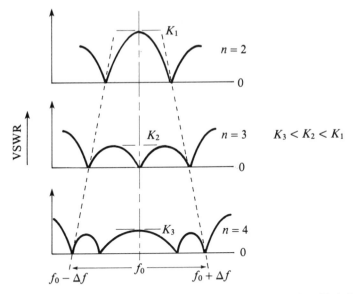

Figure 5.26   Bandwidth Spreading and Ripple Attenuation Variation for Chebyshev
             Transformers

of the impedances to be matched. Post-design analysis also shows that for Chebyshev transformers an increased number of sections leads to reduced ripple in the passband of the circuit. Figure 5.26 illustrates in a qualitative manner the effects described above.

The actual VSWR in the transformer passband depends on the ratio of the impedances to be matched and on the fractional bandwidth required. For a two section transformer displaying a Chebyshev response that performs an impedance matching function between impedances in the ratio $3:1$ with $B = 0.2$, the maximum VSWR will be 1.01, while for $B = 1.0$ the maximum VSWR rises to 1.47. For a three section Chebyshev transformer operating with the same impedance matching ratio with $B = 0.2$, VSWR $= 1.0$ and when $B = 1.0$, VSWR is 1.18. From these sample results it is obvious that when more sections are used to construct a transformer for a given impedance ratio the maximum VSWR is reduced provided the fractional bandwidth is held constant.

* * *

Example 5.9

Design a two section stepped impedance transformer that will operate over a 40 percent bandwidth between two unequal resistances of magnitude 50 ohms and 100 ohms. The transformer should display a Chebyshev frequency response.

Solution

First find the impedance ratio $R = 100/50 = 2$, fractional bandwidth $B = 40$ percent or 0.4

$$R = \sin\left(\frac{\pi}{4}0.4\right) = 0.31$$
$$\underbrace{\phantom{\left(\frac{\pi}{4}0.4\right)}}_{\text{radians}}$$

therefore

$$D = \frac{(2-1)(0.31)^2}{2(2-0.31^2)} = 0.0251$$

next

$$V_1^2 = (0.0252^2 + 2)^{1/2} + 0.0252 = 1.44$$

$$\therefore \quad V_1 = 1.2$$

denormalize to 50 ohms

$$Z_1 = 50(1.2) = 60.0 \text{ ohms}$$

$$Z_2 = \frac{50(2.0)}{1.2} = 83.3 \text{ ohms}$$

This completes the design.

* * *

In this example D is small and the design becomes approximately equivalent to a Butterworth design in which case the characteristic impedances needed to complete the design become

$$Z_1 = 59.5 \text{ ohms}$$

$$Z_2 = 84.1 \text{ ohms}$$

For large impedance ratios, $R \gg 1$ the correspondence between Chebyshev and Butterworth designs becomes less marked.

5.6   IMPEDANCE MATCHING TAPERS

An alternative to cascaded sections of uniform transmission line for impedance transformation is the use of a length of transmission line with non-uniform cross-section. Generally speaking, lines with large cross-section have a lower impedance than those with small cross-sections. By carefully selecting impedance change as a function of line length, smooth variations of the line impedance can be made to

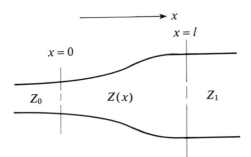

Figure 5.27   Taper Matching Circuit

occur. This, in turn, reduces the abrupt change between adjacent line sections thereby minimizing discontinuities, hence reflections.

To demonstrate how a transmission line having nonuniform cross-section can transform one impedance level to another different impedance level, both assumed to be real, consider the discussion presented below.

Consider a taper profile that can be represented by some well known mathematical function, say an exponential function. Figure 5.27 shows how two sections of transmission line each with different characteristic impedance are matched with a tapered transmission line. The characteristic impedance of the line is selected to be a function of distance measured from the end of the first section of transmission line characteristic impedance Z_o to the beginning of the second section of lower characteristic impedance Z_1.

The boundary conditions imposed on the circuit are that the impedance at $x = 0$ should be Z_o while the impedance at distance $x = l$ should be Z_1. If the electrical profile of the taper is assumed to obey an exponential law then

$$\frac{Z(x)}{Z_o} = \exp\left(\frac{\eta x}{l}\right)$$

where η represents the taper rate.

The taper rate can be weighted in various ways in order to provide improved matching at frequencies away from the design frequency. One suitable choice for η is $\log_e(Z_1/Z_o)$[8]. This then gives

$$\frac{Z(x)}{Z_o} = \exp\left[\frac{\log_e(Z_1/Z_o)x}{l}\right] \tag{5.28}$$

This expression obeys the boundary conditions imposed on the structure

at $x = 0$

$$Z(0) = Z_o \exp(0) = Z_o$$

at $x = l$

$$Z(l) = Z_o \exp\left[\log_e\left(\frac{Z_1}{Z_o}\right)\right] = Z_1$$

The normalized reflection coefficient corresponding to equation (5.28) is plotted as a function of physical taper length normalized to guide wavelength (figure 5.28). This shows the reflection coefficient follows a $\sin X/X$ curve.

From figure 5.28 the tapered line response displays a series of side lobes whose magnitude diminishes with increasing frequency. These lobes have minima at $l/\lambda_g = 0.5$, 1.0, 1.5, etc., indicating that the best possible match between the terminating impedances will occur at these points. The graph also indicates that the first side lobe maximum has a value of approximately 0.2 while the second maximum has magnitude 0.075. Subsequent side lobe maxima have even lower values. From this behavior it can be deduced that the exponential taper will give excellent broadband performance. Say, for example, the taper length is selected to be one guide wavelength (point A, figure 5.28). Then, as the excitation frequency of the signal feeding the taper is increased, the fixed length of tapered line will appear electrically longer, i.e. more wavelengths per unit length. The reflection coefficient seen looking into the tapered section of line will execute a series of sinusoidal variations of diminishing amplitude according to the $\sin X/X$ relationship pertinent to exponential taper profiles. In terms of figure 5.28, the measured reflection coefficient curve will appear to move left relative to a stationary viewing point. By proper choice of design frequency, normally one that is several times lower than the frequency at which the circuit is to be operated, it is possible to ensure both broad bandwidth and low loss. In some applications the side lobe ratios obtained for adjacent lobe in the exponential taper may be too large or perhaps the sharp nulls obtained may be undesirable. For these applications tapers having different mathematical formulae governing the taper profile should be used. Other taper profiles have been suggested in the literature, some of these are listed in the further reading section at the end of this chapter.

From a constructional viewpoint, it is quite difficult to build a section of transmission line whose cross-section varies continuously according to some strictly defined mathematical equation. For circular or rectangular waveguides, the diameter

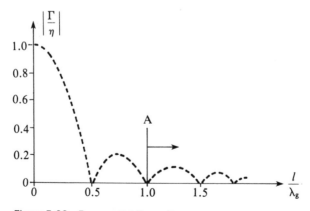

Figure 5.28   Exponential Taper Response

or height of the guide would have to be accurately machined to provide a continuous taper. This is possible if expensive computer controlled machinery is available. This type of machinery would allow automatic interpolation between defined sample positions on the taper. In this way smooth taper profiles could be machined, or more likely the taper profile would be machined in a piecewise linear fashion. For the inner conductor of a coaxial system this will result in a series of truncated cones. If manual machining is the only option, then a step approximation, piecewise constant, will be necessary. This approximation to the smooth taper will produce a more pronounced departure to the theoretical response of the taper than would a smooth taper due to the effects of discontinuities. For a coaxial inner conductor, the conductor will appear as a series of stepped rings. When tapers are to be constructed on stripline or microstrip substrates, piecewise constant approximations may be necessary if the circuit pattern is to be laid out manually.

In practice, if a piecewise constant approximation is to be used to represent a taper profile then about twenty equally spaced sample points along the length of the taper will usually yield good performance. As the number of points by which the profile section is defined increases, the taper behavior will approach more closely its ideal response. Before synthesizing taper sections according to equation (5.28), one should be aware of a pitfall.

For a TEM transmission line the expressions governing guide wavelength and characteristic impedance are independent. This makes the design of a tapered matching circuit in a coaxial line a relatively easy task. However, when tapered sections are to be designed on microstrip or stripline material the synthesis problem is somewhat more complex. For stripline or microstrip line, guide wavelength depends primarily on the effective dielectric constant of the substrate material and to a lesser extent on the aspect ratio of the line. The dependence of guide wavelength on line aspect ratio canot be neglected, so that in a tapered section of transmission line where the line aspect ratio is varying as a function of length, the guide wavelength along the taper will also vary. One way of overcoming this problem is to take the geometric mean of the terminating impedances. The guide wavelength corresponding to this value of impedance can then be subdivided into the number of segments required to construct the taper. A better solution would be to calculate the length of each segment on a per segment basis, i.e. at each position required along the taper calculate the required characteristic impedance then find the guide wavelength for this impedance. The length of each segment selected initially as a fraction of a guide wavelength would then be scaled so that the total number of segments, when added together, would have the same length as that desired for the complete taper.

Program 5.7 EXTAP, generates the characteristic impedance of a series of line segments at a selected number of equally spaced points along an exponential taper profile. The actual physical dimensions corresponding to these values of characteristic impedance can be found for a particular type of transmission line from one of the synthesis programs given in chapter 2. Due to its simplicity, the synthesis of coaxial exponential taper profiles has been included in program EXTAP for the purposes of demonstration.

```
]| FORMATTED LISTING
FILE: PROGRAM 5.7 EXTAP
PAGE-1


   10   REM
   20   REM   **** EXTAP ****
   30   REM
   40   REM   EEF=EFF/REL DIE. CONST.
   50   REM   F1=OPERATING FREQ (GHZ)
   60   REM   N=NO. OF PTS. ON PROFILE
   70   REM   ZO=INITAL LINE IMP.
   80   REM   Z1=FINAL LINE IMP.
   90   REM   Z=TAPER IMPEDANCE
  100   HOME
  110   REM
  120   REM   INPUT DATA
  130   PRINT "I/P REL. OR EFF. PERM."
  140   INPUT EEF
  150   PRINT "I/P FREQUENCY (GHZ)"
  160   INPUT F1
  170   PRINT "I/P REQUIRED NO. OF POINTS"
  180   INPUT N
  190   PRINT "I/P INITAL LINE IMPEDANCE (OHMS)"
  200   INPUT ZO
  210   PRINT "I/P FINAL LINE IMPEDANCE (OHMS)"
  220   INPUT Z1
  230   PRINT "TAPER LTH. IN FRACTIONS OF A WAVELTH."
  240   INPUT L
  250   LET D = 0.8
  260   LET T = 30 / F1 /  SQR (EEF)
  270   LET T1 = L * T
  280   PRINT
  290   REM   COAX DEMO.
  300   PRINT "IF YOU REQUIRE COAX"
  310   PRINT "SYNTHESIS DEMO. I/P 1"
  320   INPUT K
  330   PRINT
  340   PRINT "****************"
  350   PRINT
  360   PRINT "EXPONENTIAL TAPER RESULTS"
  370   PRINT
  380   IF K <  > 1 THEN
             420
  390   PRINT "COAX. DEMO."
  400   PRINT "INT. DIA. OF OUTER COND. IS 0.8 CM"
  410   PRINT
  420   PRINT "REL/EFF DIE. CONST."EEF
  430   PRINT "FREQUENCY OF OPERATION "F1" GHZ"
  440   PRINT "NO. OF POINTS ON TAPER PROFILE "N
  450   PRINT "INITAL LINE IMP. = "ZO" OHMS"
  460   PRINT "FINAL LINE IMP. = "Z1" OHMS"
  470   PRINT "TAPER LTH. FRACTIONS OF A WAVE LTH. "L
  480   PRINT "GUIDE WAVELTH. = "T" CM"
  490   PRINT "TAPER LTH. = "T1" CM"
  500   LET DN = N + 1
  510   PRINT
  520   PRINT "CMS              IMP.          DIA. CMS."
  530   PRINT
  540   LET D1 = 0
  550   FOR F = 1 TO DN
  560       LET G = (F - 1) / N
  570       REM   TAPER CALCULATIONS
  580       LET Z = ZO *  EXP (G *  LOG (Z1 / ZO))
  590       IF K <  > 1 THEN
               610
  600       LET D1 = D /  EXP (Z *  SQR (EEF) / 60)
  610       LET H = G * T1
  620       PRINT  INT (H * 100 + .5) / 100,  INT (Z * 100 + .5) / 100,  INT (D1 *
            1000 + .5) / 1000
  630   NEXT F
  640   PRINT
  650   PRINT "******************"
```

```
660    PRINT
670    PRINT "DO YOU REQUIRE ANOTHER GO ?"
680    PRINT "ENTER 1 IF YES"
690    INPUT R
700    IF R = 1 THEN
           100ELSE600
710    PRINT
720    PRINT "**** END OF PROGRAM ****"
730    END
```

END-OF-LISTING

```
]RUN
I/P REL. OR EFF. PERM.
?1
I/P FREQUENCY (GHZ)
?10
I/P REQUIRED NO. OF POINTS
?10
I/P INITAL LINE IMPEDANCE (OHMS)
?50
I/P FINAL LINE IMPEDANCE (OHMS)
?75
TAPER LTH. IN FRACTIONS OF A WAVELTH.
?3

IF YOU REQUIRE COAX
SYNTHESIS DEMO. I/P 1
?1
```

****

EXPONENTIAL TAPER RESULTS

COAX. DEMO.
INT. DIA. OF OUTER COND. IS 0.8 CM

REL/EFF DIE. CONST.1
FREQUENCY OF OPERATION 10 GHZ
NO. OF POINTS ON TAPER PROFILE 10
INITAL LINE IMP. = 50 OHMS
FINAL LINE IMP. = 75 OHMS
TAPER LTH. FRACTIONS OF A WAVE LTH. 3
GUIDE WAVELTH. = 3 CM
TAPER LTH. = 9 CM

CMS	IMP.	DIA. CMS.
0	50	.348
.9	52.07	.336
1.8	54.22	.324
2.7	56.47	.312
3.6	58.8	.3
4.5	61.24	.288
5.4	63.77	.276
6.3	66.41	.264
7.2	69.16	.253
8.1	72.02	.241
9	75	.229

****

DO YOU REQUIRE ANOTHER GO ?
ENTER 1 IF YES
?0

**** END OF PROGRAM ****

5.7   BRANCHLINE COUPLER

It has already been stated in previous sections that parallel edge coupled lines tend to be difficult if not impossible to manufacture in a repeatable way especially if the coupling ratios are tight. Voltage couplings of greater than 3 dB result in parallel edge coupled lines with very narrow line spacings so that there exists a high probability of a short circuit occurring across the gap due to defects in the fabrication process. Fortunately another type of coupler exists which can produce couplings as tight as 0 dB. This is the so-called branch line coupler and is illustrated in figure 5.29. The branch line coupler is particularly suitable for fabrication using microstrip or stripline transmission lines. Branch line couplers can also be constructed with a little ingenuity from coaxial or waveguide transmission lines.

  Figure 5.29 shows two different representations of branch line structures, one uses a rectangular format normally associated with power combiners or splitters and the other shows the circular format associated with mixer design, both are of course equivalent. The length of each branch is selected to be one-quarter guide wavelength (note that guide wavelength may be a function of line characteristic impedance, this is the case for microstrip). In the ring form, the branch line coupler has a mean diameter of one guide wavelength and comprises four ports arranged at right angles. This sets the distance between ports at the design frequency of the coupler to one-quarter of a guide wavelength. This type of coupler will provide a 90 degree phase shift between the signal at port two relative to port three. Hence, the alternative name for the branch line coupler is the 90 degree hybrid coupler.

  The operation of the branch line coupler can be described in terms of the phase shift associated with each quarter wavelength line section. The branch line coupler is shown schematically in figure 5.30 in its ring form. Here, the signal applied to port one of the coupler is taken as the phase reference. The signal at port one will split equally and will experience a 90 degree phase lag relative to the input signal when it arrives at ports two and four. The portions of the signal present at these ports will

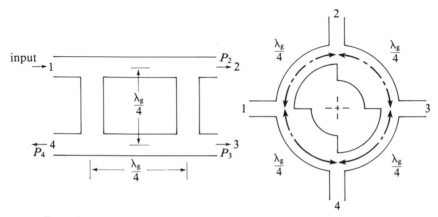

Figure 5.29   Two-Section Branch Line Couplers

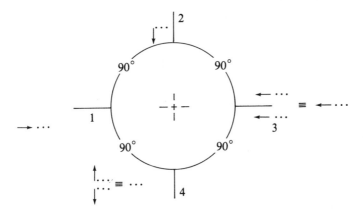

Figure 5.30 Branch Line Coupler Scheme

experience a further 90 degree phase lag relative to the original input signal. The signals arriving from ports two and four will arrive in phase and will reinforce each other at port three. The relative phase shift between port two and port three is therefore 90 degrees. Cancellation of the signals arriving at port four occurs, leaving this as the isolated port.

Apart from the obvious attraction of using a branch line directional coupler as a power splitter or combiner, it has an additional feature as was pointed out by Chen [9]. Chen showed the branch line coupler to be capable of providing, in addition to a power split, an impedance matching facility between unequal impedances at the input and output ports of the coupler. This is a very attractive feature that is often overlooked. Exploited correctly, intermediate impedance transforming circuitry can be minimized, thereby reducing circuit loss and complexity, this results in a saving of space.

The notation used for a general two branch coupler, one capable of matching equal or unequal input and output impedances and equal and unequal power splits, is shown in figure 5.31.

When dealing with waveguide circuitry the lines required to realize figure 5.31 will be series lines so that impedance relationships are the most appropriate. While for coaxial, microstrip and stripline circuits, shunt lines are used so admittance is

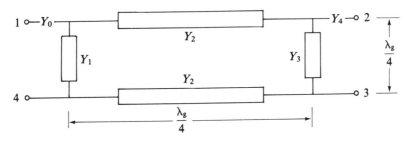

Figure 5.31 Two Branch Coupler Schematic

the best quantity to work in. In figure 5.31 shunt admittances Y_1 and Y_3 together with series admittance Y_2 have been normalized to the input port admittance. Similarly Y_4, the output port admittance, is normalized to the input port admittance Y_o.

For a perfect match to occur for the coupler

$$Y_1 = Y_3 Y_4$$

and for perfect directivity, i.e. none of the output signal reflected back into the coupler reaching the input ports,

$$Y_2^2 = Y_4 + Y_1 Y_3$$

These conditions imply that all the input power will reach ports two and three.

If K represents the ratio of power split between ports two and three

$$K = \frac{P_2}{P_3}$$

then

$$Y_3 = \frac{Y_4}{K^{1/2}}$$

$$Y_2 = \left[\frac{(K+1)Y_4}{K} \right]^{1/2}$$

$$Y_1 = \frac{1}{K^{1/2}}$$

To test these simple design equations consider the next example.

* * *

Example 5.10

Design a two branch coupler having a power coupling ratio of 3 dB and an input and output impedance of 50 ohms.

Solution

Normalize the admittance at the output port

$$Y_o = \frac{1}{50} \qquad Y_4 = \frac{1}{50} \bigg/ \frac{1}{50} = 1$$

now

$$K = \frac{P_2}{P_3}$$

$$\therefore \quad \text{since } P_3 = 10^{-(3/10)} P_{in} = 0.5 P_{in} = P_2$$

then $K = 1$. Applying the design equations gives

$$Y_1 = 1$$

$$Y_2 = \left[(1 + 1) \frac{1}{1} \right]^{\frac{1}{2}} = \sqrt{2}$$

$$Y_3 = 1$$

Denormalizing gives

$$Y_1 = \frac{1}{50} \text{ S} \quad \text{or} \quad Z_1 = 50 \text{ ohms} \rightarrow \text{shunt arm}$$

$$Y_2 = \frac{\sqrt{2}}{50} \text{ S} \quad \text{or} \quad Z_2 = 35.4 \text{ ohms} \rightarrow \text{series arm}$$

$$Y_3 = \frac{1}{50} \text{ S} \quad \text{or} \quad Z_3 = 50 \text{ ohms} \rightarrow \text{shunt arm}$$

where $Z_o = Z_4 = 50$ ohms.

The dimensions corresponding to these values of impedance for a particular transmission line type can be found from one of the synthesis programs given in chapter 2.

* * *

The results obtained in example 5.10 are often quoted for a two arm power splitter circuit providing an equal power split. Couplers of this type are normally useful over a 10 percent bandwidth and are operated with a matched termination on port four.

* * *

Example 5.11

Repeat the above example but this time for 6 dB power coupling.

Solution

$$6 \text{ dB} \rightarrow 10^{(-6/10)} = 0.25$$

$$\therefore \quad P_2 = 0.75 \times \text{input power} \quad \text{and} \quad P_3 = 0.25 \times \text{input power}$$

Hence

$$K = \frac{0.75}{0.25} = 3$$

From the design equations

$$Y_1 = Y_3 = \frac{1}{\sqrt{3}}$$

$$Y_2 = \left(\frac{4}{3}\right)^{\frac{1}{2}}$$

giving

$$Z_1 = Z_3 = 86.6 \text{ ohms}$$

$$Z_2 = 43.3 \text{ ohms}$$

This completes the design.

* * *

For couplings of less than -8 dB, the series arm impedance becomes quite high causing fabrication problems.

To illustrate the impedance matching facility of the branch line coupler, consider an equal power split between unequal terminating impedances.

* * *

Example 5.12

Use a two arm branch line coupler to match a 75 ohm source to a 50 ohm load impedance so that the power flowing in each of the output ports is equal.

Solution

$P_2 = P_3$ for equal power flow in both output ports

$$Y_o = 1$$

$$Y_4 = \frac{1}{50}\frac{75}{1} = 1.5$$

(see Table 5.5). This completes the design.

* * *

As a final example, consider a two to one power split between 100 ohm and 50 impedances giving the values shown in Table 5.6. The value of 141 ohms obtained for the shunt arm between ports one and four is quite high and may lead to fabrication problems.

Table 5.5

Normalized Admittance	Denormalized Admittance (S)	Denormalized Impedance (ohms)
$Y_3 = \dfrac{1.5}{1}$	0.02	50
$Y_2 = 1.5$	0.02	50
$Y_1 = 1$	0.0133	75
$Y_0 = 1$	0.0133	75
$Y_4 = 1.5$	0.02	50

Table 5.6

Normalized Admitance	Denormalized Impedance (ohms)
$Y_4 = 2$	$Z_4 = 50$
$Y_3 = 1.444$	$Z_3 = 70.7$
$Y_2 = 1.7$	$Z_2 = 58.8$
$Y_1 = 0.707$	$Z_1 = 141$
$K = 2$	$Z_o = 100$ ohm

For couplings tighter than 3 dB the line widths required to synthesize the necessary arm impedances of a branch line coupler become quite large so that the circuit soon becomes impractical. When couplings of greater than 3 dB are needed, it is possible to cascade a number of branch line couplers each having reasonably loose coupling so that tight overall coupling is achieved. The situation is shown schematically in figure 5.32.

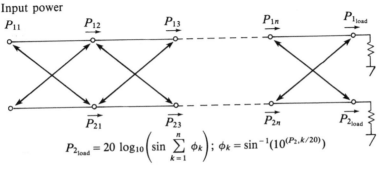

Input power

P_{11} P_{12} P_{13} P_{1n} P_{1load}

P_{21} P_{23} P_{2n} P_{2load}

$$P_{2load} = 20 \log_{10}\left(\sin \sum_{k=1}^{n} \phi_k\right); \quad \phi_k = \sin^{-1}(10^{(P_2,k/20)})$$

Figure 5.32 Cascaded Coupler Schematic

From figure 5.32 it can be calculated that for a -12 dB voltage coupling factor

$$\phi = \sin^{-1}(10^{(-12/20)}) = 14.5°$$

so that three -12 dB couplers in cascade will produce

$$P_{2load} = 20 \log_{10}[\sin(3 \times 14.5)] = -3.24 \text{ dB}$$

four -12 dB couplers in cascade will produce -1.4 dB, etc.

5.8   RAT RACE OR 180 DEGREE HYBRID

A derivative of the 90 degree hybrid coupler is the rat race circuit or 180 degree hybrid ring. This type of circuit is frequently employed in waveguide circuitry where the arms connecting into the circuit form series elements (figure 5.33). Figure 5.33 shows the physical construction of a waveguide rat race circuit together with an electrical equivalent. The total mean circumference of the ring is 1.5 guide wavelengths.

An alternative form of folding is shown in figure 5.33 such that rectangular construction can be used when the circuit is to be fabricated on stripline or microstrip substrates. In this form the three-quarter guide wavelength long section is bent to accommodate the rectangular geometry.

To illustrate the operation of the circuit a phasor representation is shown in figure 5.34.

A signal introduced at port one will split between ports two and four. The signal at port two will lag the signal at port one by 90 degrees. However, the signal arriving at port four will lag behind that at port one by 270 degrees since it has three times further to travel. The signal arriving at port four from port one in the

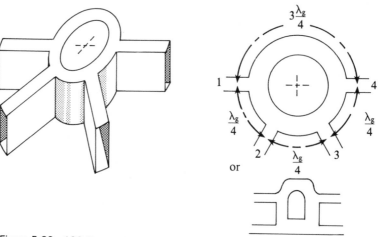

Figure 5.33   180 Degree Hybrid Ring

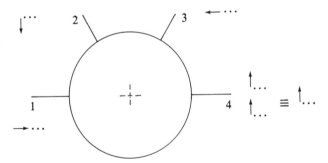

Figure 5.34   180 Degree Hybrid Operation

clockwise direction will reinforce the signal at port four that arrived along the direct anticlockwise route between ports one and four. Both these signals add to give a resultant at port four that is 180 degrees out of phase with respect to the signal at port two. This is a very useful property and is often exploited in balanced mixer or image rejection mixer design which some form of signal cancellation is necessary. Total cancellation occurs at port three, the isolated port.

To illustrate more fully the use of a rat race circuit in a mixer configuration consider figure 5.35. This time signals are applied to ports one and three respectively, these could be the local oscillator and the input signal of a mixer circuit. For convenience these signals are assumed initially to be in phase.

The signal entering port one splits between ports two and four. These arrive at ports two and four, phase shifted by 90 degrees and 270 degrees respectively relative to port one. A signal injected at port three will split between ports two and four, here the phase shift relative to port three will be 90 degrees. The signals arriving at port two from port one and port three are in phase and add together. The converse is true for port four. If the signals injected at ports one and three are not in phase, the circuit operation is the same except that vector arithmetic must be used to determine the actual signal magnitudes at ports two and four.

The operation of this circuit for a simple equal power split is conceptually very

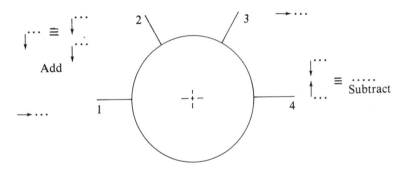

Figure 5.35   180 Degree Hybrid used as a Mixer

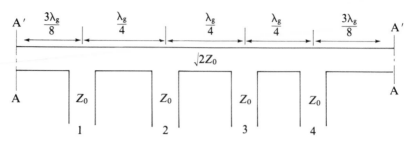

Figure 5.36   Unfolded 180 Degree Hybrid

easy to visualize. If the waveguide circuit in figure 5.33 is examined, then the lines connected to the hybrid ring appear in series with it, for microstrip or stripline they appear as shunt arms. To help visualize what is happening within the structure, the ring has been unfolded. A series arm configuration is selected for discussion (figure 5.36).

Relative to port one, port two and port four are spaced at odd multiples of one-quarter wavelength. Port three is separated from port one by one-half guide wavelength. At the design frequency of the ring a halfwave transformer is effectively interposed between ports one and three. This means that the load presented to port three will appear at port one. Now, since the branches are in series, the total impedance seen by the ring at port one will be $2Z_0$. In order to maintain a match in the system the impedance inverting action between port one and ports two and four is noted. From this, it is evident that the impedance of the line forming the ring should be made equal to

$$Z_{\text{ring}} = (2Z_0 Z_0)^{\frac{1}{2}} = \sqrt{2}Z_0$$

for a coupler with a nominal 50 ohm characteristic impedance then the ring impedance will be 70.7 ohms. This simple design is the basis of a 3 dB rat race circuit. The total mean circumference of the circuit should be 1.5 guide wavelengths. Consider now a modified 3 dB 180 degree hybrid where the characteristic impedance of adjacent line segments are made unequal and where alternate sections have equal characteristic impedance. This type of configuration enables unequal power splitting between ports, while maintaining the 180 degree phase shift between ports as discussed earlier. The modified 180 degree hybrid is shown in figure 5.37.

A signal entering port one will be split between port three and port four. The amount of power actually delivered to each port will depend on the choice that has been made for the characteristic impedances of the line segments forming the ring. The power entering port four P_4 and the power into port three P_3 are related to the characteristic admittances Y_1 and Y_2 at the design frequency by

$$\frac{Y_1}{Y_0} = \left(\frac{P_4}{P_1}\right)^{\frac{1}{2}} \quad \text{and} \quad \frac{Y_2}{Y_0} = \left(\frac{P_3}{P_1}\right)^{\frac{1}{2}}$$

where Y_0 is the characteristic admittance of the feedlines.

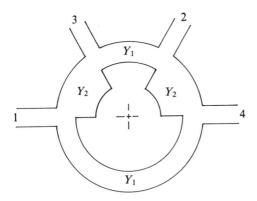

Figure 5.37   Modified 180 Degree Hybrid

At the design frequency, output ports three and four have a 180 degree phase difference.

For a signal applied to port two the relationships

$$\frac{Y_1}{Y_0} = \left(\frac{P_3}{P_2}\right)^{\frac{1}{2}} \quad \text{and} \quad \frac{Y_2}{Y_0} = \left(\frac{P_4}{P_2}\right)^{\frac{1}{2}}$$

are valid. This time output ports three and four have a zero degree phase difference between them.

*   *   *

Example 5.13

Design a 50 ohm, 3 dB power splitter, based on a hybrid coupler that produces outputs from each port that have a 180 degree phase difference between them.

Solution

For a 180 degree phase difference, the signal should be applied to port one and outputs taken from ports three and four

$$3 \text{ dB} \rightarrow 10^{(-3/10)} \rightarrow 0.5$$

$$\therefore \quad \frac{P_3}{P_2} = 0.5$$

hence

$$\frac{Y_1}{Y_0} = 0.707$$

hence $Y_1 = Y_0$ (0.707). For a 50 ohm characteristic impedance

$$Y_0 = 0.02$$

this gives

$$Z_1 = \frac{1}{Y_1} = \frac{1}{0.02 \times 0.707} = 70.7 \text{ ohms}$$

similarly

$$Z_2 = \frac{1}{Y_2} = \frac{1}{0.02(1 - 0.5)^{\frac{1}{2}}} = 70.7 \text{ ohms}$$

This is the same result derived by heuristic reasoning earlier in this section.

Example 5.14

Design a -6 dB coupler with characteristic impedance of 50 ohms such that the output from each port will be in phase.

Solution

This time the signal should be applied to port two. This will ensure zero phase difference between output ports three and four.

$$6 \text{ dB} \rightarrow 10^{(-6/10)} \rightarrow 0.25$$

0.25 of the input power at port two will reach port three.

$$\frac{P_3}{P_2} = 0.25$$

$$\therefore \quad Z_1 = 100 \text{ ohms}$$

the rest of the power $(1 - 0.25)P_{in}$ will reach port four giving

$$Z_2 = 57.7 \text{ ohms}$$

This completes the design.

* * *

5.9 POWER SPLITTER/COMBINER

Many applications require the use of power splitting and power combining circuitry. One example might be in the design of a feed array for a system of transmitting or receiving antennae. Another example could be a balanced amplifier or perhaps in circuits where a single load oscillator signal is required to feed both the transmit and receive sections of a radio.

$$I/P \rightarrow$$
$$P_{in} = P_1 + P_2$$

$$P_{out} = P_1 + P_2 - loss$$
$$O/P \leftarrow$$

1   Three port

2   O/P   I/P
    $P_1 - loss$ P_1
    O/P   I/P
3   $P_2 - loss$ P_2

Figure 5.38 Three-Port Representation of Power Splitter/Combiner Circuit

In all cases, a three port device is required with zero phase shift between output ports (figure 5.38). Such three port devices are reciprocal and can be used in power combining mode. At frequencies up to around 500 MHz, power divider/combiner circuitry can be constructed from lumped components. Successful operation depends on neat compact construction; when adhered to, good results can be obtained. Figure 5.39 shows a power combiner circuit constructed entirely from lumped components.

The design of lumped element power combiner circuits is straightforward if

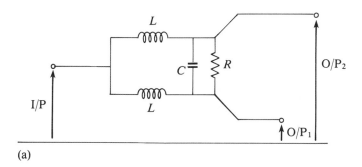

(a)

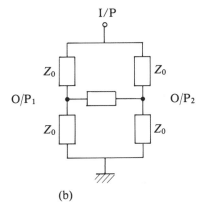

(b)

Figure 5.39 Lumped Power Splitter Circuit (a) Physical Construction
(b) Electrical Equivalent Circuit at Design Frequency

equations (5.29)–(5.31) are employed

$$C = \frac{1}{2\omega Z_0} \tag{5.29}$$

$$L = \frac{Z_0}{\omega} \tag{5.30}$$

and

$$R = 2Z_0 \tag{5.31}$$

where Z_0 represents the impedance of the splitter circuit. It should be noted that the circuit design equations described above are useful only if equal power splits are being considered.

*   *   *

Example 5.15

Design a power divider circuit to be operated in a 75 ohm system at a frequency of 144 MHz.

Solution

$$Z_0 = 75 \text{ ohm}, \quad f = 144 \text{ MHz}$$

from equation (5.31)

$$R_1 = 150 \ \Omega$$

from equation (5.30)

$$L = \frac{75}{2\pi \times 144 \times 10^6} = 83 \text{ nH}$$

finally from equation 5.28

$$C = \frac{1}{4\pi \times 144 \times 10^6 \times 75} = 7.5 \text{ pF}$$

This completes the design.

*   *   *

To understand how the power splitter circuit works, examine equations (5.29)–(5.31). These show

$$|X_C| = \frac{1}{\omega C} = 2Z_0$$

while

$$|X_L| = \omega L = Z_o$$

and

$$R = 2Z_o$$

The parallel combination of R and $|X_C|$ gives at the design frequency of the splitter circuit an impedance of Z_o. When the output terminals of the circuit are terminated with their matched impedance Z_o, the splitter circuit, reduces to a balanced bridge having zero potential difference between output ports. The circuit has an input impedance of Z_o ohms thereby matching the input line impedance (see figure 5.39). The implication of the balanced bridge is that the resistor R, placed between the branches forming port two and port three of the splitter, will dissipate zero power. This condition is valid only when symmetrical components are used and the circuit is operated at exactly the design frequency. Resistor R absorbs reflected power when the circuit is operated in an out-of-balance condition.

As frequencies increase into the UHF and higher frequency region, distributed circuits come into their own. A rather lengthy analysis of the general problem of providing two isolated outputs with arbitrary power division was considered by Parad and Moynihan [10]. Their paper which is an extension of the work originally produced by Wilkinson [11], should be consulted for a full derivation of the results utilized in this section. Figure 5.40 shows the circuit layout for a general splitter/combiner circuit as it would appear if synthesized from a microstrip or stripline material.

In order to achieve an unequal power split, the three port coupler has to be made asymmetric. The balancing resistor R can be made from carbon or metal oxide for frequencies up to 1 GHz and from thin or thick film for frequencies in excess of this. The relevant design equations are cited as

$$\frac{\text{Power at port 2}}{\text{Power at port 3}} = \frac{1}{K^2}$$

$$Z_{o2} = Z_o[K(1 + K^2)]^{\frac{1}{2}}$$

$$Z_{o3} = Z_o\left(\frac{1 + K^2}{K^3}\right)^{\frac{1}{2}}$$

$$Z_{o4} = Z_o(K)^{\frac{1}{2}}$$

$$Z_{o5} = \frac{Z_0}{K^{\frac{1}{2}}}$$

$$R = Z_0\left(\frac{1 + K^2}{K}\right)$$

In the derivation of these equations discontinuity effects have been ignored and should be compensated for empirically. For this type of circuit an increased power split between ports two and three decreases the bandwidth of the three-port.

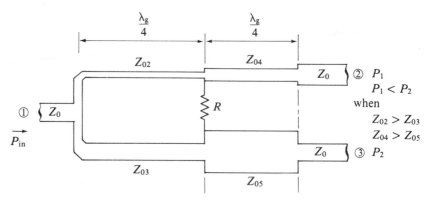

Figure 5.40   Unequal Power Splitter Circuit

However, since the circuit coupling remains almost constant for frequency ranges of up to an octave, it can out-perform 90 degree or 180 degree hybrid circuits whose bandwidths for simple two arm structures rarely exceed 10 or 15 percent. The bandwidth of three-port power splitters can be extended further to provide excellent performance over frequency ranges of up to a decade provided multisection couplers are used. The design of this type of coupler has been thoroughly discussed by Cohn [12] and will not be subject to any further debate here.

* * *

Example 5.16

Design a 3 dB power divider circuit to operate in a 50 ohm system. Draw a suitable coaxial version of the circuit.

Solution

$Z_o = 50$ ohms

$$\frac{P_2}{P_3} = \frac{\frac{1}{2}}{\frac{1}{2}} = 1 \quad \therefore \quad K = 1$$

from the design equations for this structure

$Z_{o2} = Z_{o3} = Z_o\sqrt{2} = 70.7$ ohms

$Z_{o4} = Z_{o5} = Z_o = 50.0$ ohms

$R = 2Z_o = 100$ ohms

* * *

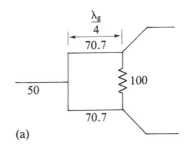

(a)

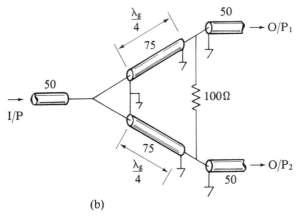

(b)

Figure 5.41   3 dB Power Splitter for Example 5.16 (a) Schematic (b) Coaxial Realization

A practical version of this circuit can be crudely constructed from standard 50 and 75 ohm coaxial cable (figure 5.41). When constructed according to figure 5.41 the splitter will provide less than 0.25 dB difference in signal level between ports.

* * *

Example 5.17

Design a 50 ohm three-port power splitter capable of providing 6 dB power coupling between output ports.

Solution

6 dB → 0.25

∴   power out of port 2 = $0.25 P_{in}$

power out of port 3 = $0.75 P_{in}$

Here losses have been neglected.

$$\frac{P_2}{P_3} = \frac{0.25}{0.75} = \frac{1}{3}$$

from this $K = (3)^{1/2} = 1.732$. Since

$$Z_o = 50 \text{ ohms}$$

$$Z_{o2} = 50[1.732(1 + 3)]^{1/2} = 132 \text{ ohms}$$

$$Z_{o3} = 50 \left(\frac{1+3}{1.732^3} \right)^{1/2} = 44 \text{ ohms}$$

$$Z_{o4} = 50(1.732)^{1/2} = 66 \text{ ohms}$$

$$Z_{o5} = \frac{50}{(1.732)^{1/2}} = 38 \text{ ohms}$$

$$R = \frac{50(1 + 3)}{1.732} = 115 \text{ ohms}$$

* * *

Comparison of the results of example 5.17 with those of example 5.16 shows that the circuit length doubles when unequal power splits are required. When synthesizing the circuit designed in example 5.17, two points are worthy of note. First, characteristic impedance Z_{o2} is rather high and will result in a very thin line, current carrying conductor, which in turn will cause fabrication problems. Therefore, it looks like 5 dB or 6 dB power coupling factors are the most that can be easily realized. The second point is that each section has differing impedance so that the physical length of each section may differ very slightly, this is true if the line does not support pure TEM propagation.

5.10 HYBRID QUADRATURE (LANGE) COUPLERS

In section 2.7 the idea of even and odd mode excitation was discussed with reference to parallel edge coupled lines. Some of the major problems associated with parallel edge coupled lines are stated below.

1   It is almost impossible to obtain coupling coefficients of greater than -3 dB due to the very narrow intergap separations required between lines. These narrow separations required for tight coupling often lead to repeatability problems and to intergap short circuits because of fabrication imperfections.
2   In a simple parallel edge coupler there exists no inherent mechanism for the compensation of even and odd mode propagation velocities of the electromagnetic wave within the structure. In fact, couplers are often designed so that coupler length is based on the odd mode propagation velocity (since under these condi-

tions tighter coupling will occur than if the even mode propagation velocity is used). Lack of compensation between even and odd mode propagation velocities within parallel edge couplers ultimately means their bandwidth is reduced.

The two problems cited for parallel edge couplers were apparently overcome with the introduction of the Lange Coupler in 1969 [13]. Lange suggested that an interdigitated structure be used with its alternate fingers short circuited. The construction of the Lange coupler for an odd number of fingers is shown in figure 5.42 and is best suited for synthesis on microstrip. Notice how the circuit is no longer planar due to the introduction of the shorting links.

For a Lange coupler the circuit pattern is etched on microstrip and short lengths of bonding wire are placed to short together alternate pairs of lines. In this configuration, symmetry is carefully preserved along lines $X - X'$ and $Y - Y'$. Lange suggested that it was partially due to this symmetry that even and odd mode phase velocities were equalized giving the Lange configuration an operational bandwidth in excess of an octave (i.e. a doubling of frequency). The spacing between lines in this configuration is much wider for a given coupling coefficient if the same coefficient was to be realized in a parallel edge coupled system. Ideally, the bond wires affixed between alternate pairs of lines should have zero inductance and should present as small a discontinuity to the electromagnetic fields in the structure as possible. For this reason the bond wires are kept as short as possible without shorting together adjacent lines. Sometimes multiple wires are used in parallel to reduce wire inductance so that the electrical behavior of the coupler is enhanced. A few other practical points relevant to figure 5.42 are worthy of note. The first is that the feed lines to and from the four port coupler have equal characteristic impedance. These feeds are mitered to reduce discontinuity between the coupler and the surrounding circuitry. Second, port C is ideally isolated so that no power flows in

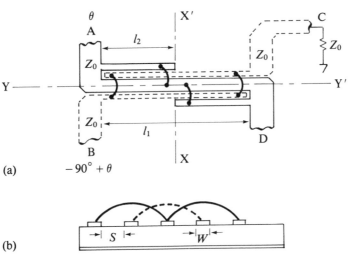

Figure 5.42   Construction Symmetrical Lange Coupler (a) Plan View (b) Cross-Section

it. However, because of fringing fields and other imperfections, a matched load normally terminates this port. Since the amount of power leaking from this port is small, the load does not have to have excellent electrical properties so that a fairly crude termination can be used. Third, all line spacings and line widths within the coupler are assumed equal.

The coupler is of the quadrature type which implies that a signal applied to the input port, port A, will emerge from the coupled output port, port B, with a 90 degree phase shift relative to the input signal. Experimental results on Lange couplers show that over an octave bandwidth worst case deviations in phase from the 90 degree nominal figure of around 2 degrees can be expected. Since the Lange coupler exhibits insensitivity of phase shift between input and output ports over a wide frequency range, it has found ready application areas in broadband mixer and balanced amplifier circuits.

One simple analysis of the Lange coupler has been proposed by Osmani [14] based on the work of Ou [15]. This literature should be consulted for full details, only a qualitative description shall be recounted here. Ou analyzed the even and odd mode admittances of pairs of adjacent lines. This analysis was confined to even numbers of equal length lines and neglected multiple coupling between nonadjacent lines. The effects of discontinuities and bond wire inductances are also neglected. As for the original Lange configuration in figure 5.42, alternate pairs of lines are shorted together. This situation is shown in figure 5.43 for four lines. The port notation for this configuration is that described in figure 5.42. It is noted that, when figure 5.43 is examined, the resulting physical structure is no longer identical to that of figure 5.42. Despite this, all the comments made previously in this section with regard to the electrical excellence of the original structure are still valid.

The results of the analysis produced by Ou are reproduced as

$$Z_0{}^2 = \frac{Z_{oe}Z_{oo}(Z_{oo} + Z_{oe})^2}{[Z_{oe} + (k-1)Z_{oo}][Z_{oo} + (k-1)Z_{oe}]} \tag{5.32}$$

where k is the number of lines which must be an even integer 2, 4, 6, $\ldots$, $n + 2$ and

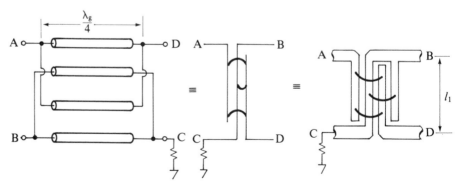

Figure 5.43   Modified Lange Coupler

the desired voltage coupling C is defined in equation (5.33)

$$C = \frac{(\text{voltage coupling in dB})}{20}$$

$$= \frac{(k-1)Z_{oe}^2 - (k-1)Z_{oo}^2}{(k-1)(Z_{oe}^2 + Z_{oo}^2) + 2Z_{oe}Z_{oo}} \tag{5.33}$$

These formulae can be easily applied to the analysis of an available Lange coupler constructed according to figure 5.41.

Osmani multiplied equation (5.32) by the voltage coupling factor given below as

$$\frac{1+C}{1-C}$$

and, after rearranging the resulting expression, obtained a set of simple design equations given as equations (5.34) and (5.35)

$$Z_{oo} = Z_o \left(\frac{1-C}{1+C}\right)^{\frac{1}{2}} \frac{(k-1)(1+q)}{(C+q)+(k-1)(1-C)} \tag{5.34}$$

$$Z_{oe} = Z_{oo} \frac{C+q}{(k-1)(1-C)} \tag{5.35}$$

where

$$q = [C^2 + (1-C^2)(k-1)^2]^{\frac{1}{2}}$$

Equations (5.32)–(5.35) are straightforward enough to enable hand calculations to be performed for the design and subsequent analysis of Lange couplers of the type in figure 5.43. When synthesizing a Lange coupler on microstrip, program CMIC is used in the symmetrical mode with its data entry section appropriately modified to allow direct entry of the even and odd mode characteristic impedances of the coupler. This will generate the line widths and spacings necessary to satisfy the even and odd modes impedance generated by equations (5.34) and (5.35).

* * *

Example 5.18

Given a four line Lange coupler designed according to figure 5.43 and having W/H and S/H values of 0.15 and 0.25 respectively, calculate its characteristic impedance and the degree of voltage coupling in decibels. The microstrip material on which the coupler is to be constructed has a relative dielectric constant of 9.5 and minimal conductor thickness.

Solution

This is an analysis problem, i.e. given physical dimensions find electrical characteristics. From tables, program CMIC, or otherwise for a microstrip material having

$$\frac{W}{H} = 0.15, \quad \frac{S}{H} = 0.25, \quad \varepsilon_r = 9.5, \quad t = 0$$

$$Z_{oe} \approx 176\ \Omega \quad \text{and} \quad Z_{oo} \approx 52\ \Omega$$

The number of lines in the system is four so that

$$k = 4$$

Substituting for Z_{oe}, Z_{oo} and k into equations (5.32) and (5.33) yields

(a) Characteristic impedance for the coupler

$$Z_o{}^2 = \frac{176 \times 52 \times (52 + 176)^2}{[176 + (3 \times 52)]\,[52 + (3 \times 176)]} = 2471$$

$$\therefore \quad Z_o = 49.7 \text{ ohms}$$

and

(b) the voltage coupling coefficient

$$C = \frac{3(176)^2 - 3(52)^2}{3(52^2 + 176^2) + 2(52 \times 176)} = 0.7107 \text{ or } -2.97 \text{ dB}$$

Example 5.19

Design a Lange coupler having four fingers that will have a characteristic impedance of 50 ohms and a coupling coefficient of 10 dB. The coupler is to be built on microstrip material having a substrate with a relative permittivity of 2.23. Assume conductor thickness to be 0.017 mm and the dielectric height to be 0.254 mm. The frequency of operation is 10.0 GHz.

Solution

$$Z_o = 50 \text{ ohms}, \quad C = 10 \text{ dB or } 0.3162, \quad \varepsilon_r = 2.23$$

$$t = 0.017 \text{ mm}, \quad k = 4, \quad h = 0.254 \text{ mm}$$

Use equations (5.34) and (5.35) to find Z_{oo} and Z_{oe}

$$q^2 = [0.3162^2 + (1 - 0.3162^2)(4 - 1)^2]$$

$$\therefore \quad q = 2.864$$

$$Z_{oo} = 50 \left(\frac{1 - 0.3162}{1 + 0.3162}\right)^{\!\frac{1}{2}} \left[\frac{3(3.864)}{(0.3162 + 2.864) + 3(1 - 0.3162)}\right]$$

$$= 80 \text{ ohms}$$

and

$$Z_{oe} = 80 \left[\frac{0.3162 + 2.864}{3(1 - 0.3162)} \right]$$

$$= 124 \text{ ohms}$$

From program CMIC for the material specified above

$$\frac{W}{H} = 1.103 \qquad \frac{S}{H} = 0.39$$

These values correspond to

$$\frac{W}{H} = 2.335 \qquad \frac{S}{H} = 0.095$$

obtained for a 10 dB parallel edge coupled directional coupler. These results show that the Lange coupler configuration has greatly reduced the fabrication tolerances of the interline spacing for the same coupling factor. This completes the design.

* * *

Having obtained the line widths and spacing necessary to give the required impedance and coupling, the length of the coupler l_1 in figure 5.43 must be obtained. This length, selected to be one-quarter of a guide wavelength at the lowest frequency of operation of the coupler, can be chosen in one of three ways.

1   Take the guide wavelength corresponding to the odd mode, this results in the tightest coupling.
2   Select the arithmetic mean of the even and odd mode guide wavelength.
3   Select the geometric mean of the even and odd mode guide wavelength.

Ideally, in a Lange structure the even and odd mode propagation velocities are the same. This means that the three methods suggested above for choosing the length of the coupler should produce the same result. In figure 5.42 the short fingers are made to be one-quarter of the mean value of the even and odd mode guide wavelengths at the highest frequency in the band over which the coupler is to operate. However, in practice, a constant of proportionality is needed as a correction factor in order to obtain operation at the exact center frequency that is required by the design. This constant of proportionality is usually chosen empirically and can be made to include correction for discontinuity and bond wire effects. Once the length of the coupler has been obtained, the bond wires are positioned by hand. Several bond wires may be used in parallel to minimize excess inductance.

So far nothing has been done in the design to account for discontinuities that exist within the coupler structure and which tend to degrade its performance. Correction for these discontinuities is best obtained empirically perhaps in the following manner. In the first cut design, parasitics and discontinuities will normally tend to cause the center frequency of a constructed coupler to lie below the design center frequency. To compensate for this the length of the coupler will have to be reduced a fractional amount. If the coupling provided by the structure is found to be below the design coupling figure, then the spacing between fingers must be reduced. This will tend to raise the characteristic impedance of the coupler so that, to compensate, the width of the coupling fingers must be increased slightly. In this way the total width of the interdigital section comprising overlapped fingers should remain approximately constant. One final point worth mentioning is this; for most applications four fingers will be sufficient to give reasonable electrical performance with low insertion loss. Program 5.8 LANGE, enables the even and odd mode impedance of couplers with given even numbers of fingers and coupling factor to be computed. The program also allows the characteristic impedance and degree of coupling to be calculated given the even and odd mode impedances of the coupler. Program LANGE should be operated in conjunction with program CMIC to allow first cut synthesis of Lange couplers of the type shown in figure 5.43.

```
][ FORMATTED LISTING
FILE: PROGRAM 5.8 LANGE
PAGE-1


  10   REM
  20   REM   **** LANGE ****
  30   REM
  40   REM   THIS PROGRAM ENABLES
  50   REM   SYNTHESIS AND ANALYSIS
  60   REM   OF LANGE TYPE COUPLERS
  70   REM   HAVING EVEN NO.S OF
  80   REM   INTERDIGITAL FINGERS
  90   REM
 100   REM   ZE=EVEN MODE IMP.(OHMS)
 110   REM   ZO=ODD MODE IMP.(OHMS)
 120   REM   Z=CHARAC. IMP.(OHMS)
 130   REM   C=VOLTAGE COUPLING(DB)
 140   REM
 150   HOME
 160   PRINT "IF YOU REQUIRE"
 170   PRINT "SYNTHESIS ENTER 1"
 180   PRINT "FOR ANALYSIS ENTER 0"
 190   INPUT T
 200   IF T = 0 THEN
             500
 210   REM   SYNTHESIS
 220   PRINT "I/P DESIRED VOLTAGE COUPLING (DB)"
 230   INPUT C
 240   PRINT "I/P NO. OF FINGERS"
 250   PRINT "MUST BE AN EVEN NUMBER"
 260   INPUT K
 270   PRINT "I/P CHARAC. IMP REQUIRED"
 280   INPUT Z
 290   LET D = 10 ^ ( - C / 20)
 300   LET Q =  SQR (D * D + (1 - D * D) * (K - 1) ^ 2)
 310   LET ZO = Z *  SQR ((1 - D) / (1 + D))
 320   LET ZO = ZO * (K - 1) * (1 + Q)
 330   LET ZO = ZO / ((D + Q) + (K - 1) * (1 - D))
 340   LET ZE = ZO * (D + Q)
 350   LET ZE = ZE / (K - 1) / (1 - D)
 360   PRINT
 370   PRINT "***********************"
 380   PRINT
 390   PRINT "LANGE COUPLER SYNTHESIS RESULTS"
 400   PRINT
 410   PRINT "VOLTAGE COUPLING FACTOR " INT (D * 100 + .5) / 100" OR " INT (C *
       100 + .5) / 100" DB"
 420   PRINT "AND "K" INTERDIGITAL STRIPS"
```

```
430    PRINT
440    PRINT "EVEN MODE IMP REQUIRED IS " INT (ZE * 100 + .5) / 100" OHMS"
450    PRINT "ODD MODE IMP. REQUIRED IS " INT (ZO * 100 + .5) / 100" OHMS"
460    PRINT "FOR COUPLER CHARAC. IMP. OF "Z" OHMS"
470    PRINT
480    PRINT "*********************"
490    GOTO 790
500    REM  ANALYSIS SECTION
510    PRINT
520    PRINT "I/P EVEN AND ODD MODE IMP."
530    INPUT ZE,ZO
540    PRINT "I/P NO. OF INTERDIGITAL FINGERS"
550    PRINT "MUST BE AN EVEN NUMBER"
560    INPUT K
570    LET A = ZE * ZO * (ZO + ZE) ^ 2
580    LET B = (ZE + (K - 1) * ZO) * (ZO + (K - 1) * ZE)
590    LET Z =  SQR (A / B)
600    LET A = (K - 1) * ZE * ZE - (K - 1) * ZO * ZO
610    LET B = (K - 1) * (ZE * ZE + ZO * ZO) + 2 * ZO * ZE
620    LET C = A / B
630    LET D =  LOG (C) /  LOG (10)
640    LET D = 20 * D
650    PRINT
660    PRINT "*********************"
670    PRINT
680    PRINT "LANGE COUPLER ANALYSIS RESULTS"
690    PRINT
700    PRINT "FOR AN ODD MODE IMP. OF "ZO" OHMS"
710    PRINT "AND AN EVEN MODE IMP. OF "ZE" OHMS"
720    PRINT "GIVEN THAT THERE ARE "K" STRIPS"
730    PRINT
740    PRINT "THEN THE CHARAC. IMP. OF THE COUPLER IS " INT (Z * 100 + .5) / 100
       " DB"
750    PRINT "AND THE COUPLING FACTOR IS " INT (C * 100 + .5) / 100
760    PRINT "OR " INT ( - D * 100 + .5) / 100" DB"
770    PRINT
780    PRINT "*********************"
790    PRINT
800    PRINT "DO YOU WANT ANOTHER GO ?"
810    PRINT "ENTER 1 IF YES"
820    INPUT A
830    IF A = 1 THEN
          150
840    PRINT
850    PRINT "**** END OF PROGRAM ****"
860    END
```

END-OF-LISTING


```
]RUN
IF YOU REQUIRE
SYNTHESIS ENTER 1
FOR ANALYSIS ENTER 0
?1
I/P DESIRED VOLTAGE COUPLING (DB)
?10
I/P NO. OF FINGERS
MUST BE AN EVEN NUMBER
?4
I/P CHARAC. IMP REQUIRED
?50

*********************

LANGE COUPLER SYNTHESIS RESULTS

VOLTAGE COUPLING FACTOR .32 OR 10 DB
AND 4 INTERDIGITAL STRIPS

EVEN MODE IMP REQUIRED IS 123.78 OHMS
ODD MODE IMP. REQUIRED IS 79.85 OHMS
FOR COUPLER CHARAC. IMP. OF 50 OHMS
```

```
*********************

DO YOU WANT ANOTHER GO ?
ENTER 1 IF YES
?1
IF YOU REQUIRE
SYNTHESIS ENTER 1
FOR ANALYSIS ENTER 0
?0


I/P EVEN AND ODD MODE IMP.
?123
??80
I/P NO. OF INTERDIGITAL FINGERS
MUST BE AN EVEN NUMBER
?4

*********************

LANGE COUPLER ANALYSIS RESULTS

FOR AN ODD MODE IMP. OF 80 OHMS
AND AN EVEN MODE IMP. OF 123 OHMS
GIVEN THAT THERE ARE 4 STRIPS

THEN THE CHARAC. IMP. OF THE COUPLER IS 49.88 DB
AND THE COUPLING FACTOR IS .31
OR 10.15 DB

********************

DO YOU WANT ANOTHER GO ?
ENTER 1 IF YES
?0

**** END OF PROGRAM ****
```

5.11   FURTHER READING

1   Weinberg, L., *Network Analysis and Synthesis*, McGraw Hill, 1962, chapter 11 (Revision published by R.E. Krieger, 1975).

Butterworth, S., 'On the Theory of Filter Amplifiers', *Wireless Engineer*, **7**, 536–41, Oct. 1930.

These references when combined give an insight into the relative merits of maximally flat, Chebyshev, and Bessel filter responses.

2   Orchard, N. J., 'Formulae for Ladder Filters', *Wireless Engineer*, **30**, 3–5, Jan. 1953.

This paper gives a list of useful equations that can be applied to ladder filters.

3   Garg, R. and Bahl, I. J. 'Microstrip Discontinuities', *Int. J. Electronics*, **45**(1), 81–7, 1978.

Altschuler, H. M. and Oliner, A. A., 'Discontinuities in the Center Conductor of Symmetric Strip Transmission Line', Microwave Theory and Techniques, **8**, 328–39, May 1960.

Marcuvitz, N., *Waveguide Handbook*, MIT Radiation Laboratory Series, Vol 10, McGraw-Hill, 1951.

Somlo, P. I., 'The Computation of Coaxial Line Step Capacitances', *IEEE Trans. MTT*, **15**(1) 48–53, Jan. 1967.
These references are pertinent to the estimation of discontinuity effects in microstrip, stripline, waveguide and coaxial circuitry.

4   Matthaei, G. L., Young, L. and Jones, E. M. T., *Microwave Filters, Impedance-Matching Networks, and Coupling Structures*, Artech House, 1965.
This reference appears once again since it must be one of the most definitive and complete books of its kind. Stepped impedance transformers are exhaustively discussed.

5   Cohn, S. B., 'Optimum Design of Stepped Transmission-Line Transformers', *IRE Trans.*, **3**, 16–21, April 1955; 'Parallel Coupled Transmission Line Resonator Filters', *IRE Trans.* **6**, 223–3, April 1958.
More of Cohn's classical papers. This time dealing with stepped transmission line transformer circuits and edge coupled filters.

6   Hammerstad, E. O. and Bekkadal, F., *A Microstrip Handbook*, ELAB Report STF 44 A74169, N7034, University of Trondheim-NTH, Norway, 1975.
Includes an equation that allows the effective increase in line length due to the fringe fields associated with a uniform section of microstrip transmission line to be evaluated directly.

7   Silvester, P., and Benedek, P., 'Equivalent Capacitances of Microstrip Open Circuits', *IEEE Trans.*, **20**(8), 511–76, Aug. 1972.
Provides a cumbersome but comprehensive method for the evaluation of open and excess capacitance for microstrip lines.

8   Collin, R. E., 'Theory and Design of Wide-Band Multisection Quarter-Wave Transformers', *Proc. IRE*, **43**, 179–85, Feb. 1955.

Arnold, R. P. and Bailey, W. L., 'Match Impedances with Tapered Lines, *Electronic Design*, **12**, 136–9, June 7 1974.

Wheeler, H. A., 'Transmission Lines with Exponential Tapers,' *Proceeding IRE*, **27**, 65–71, Jan. 1939.

Klopfenstein, R. W., 'A Transmission Taper of Improved Design', *Proc. IRE*, **44**, 31–5, Jan. 1956.
The information contained in these references allow broadband impedance matching sections based on the stepped impedance transformer and non-uniform transmission line to be designed.

9   Ho, Chen, Y., 'Transform Impedance with a Branchline Coupler', *Microwaves*, **15**(5), 47–52, May 1976.
Gives the theory necessary to develop a simple set of design rules that enable branchline couplers to be designed for power splitting and impedance matching applications.

10 Parad, L. I. and Moynihan, R. L., 'Split Tee Power Divider', *Microwave Theory and Techniques*, **13**(1), 91–5, Jan. 1965.
This paper develops a simple set of design rules for the design of three-port power divider/combiner circuits.

11 Wilkinson, E. J., 'An N-Way Hybrid Power Divider', *IRE Trans.*, **8**, 116–18, Jan. 1960.
The source paper for power divider circuits of this type. Here a coaxial configuration was used unlike the more typical planar constructions currently in use.

12 Cohn, S. B., 'A Class of Broadband Three-Port to TEM-mode Hybrids', *IEEE Trans.* **16**(2), 110–16, Feb. 1968.
Once again Cohn turns his expertise to the solution of a challenging problem and comes up with the goods.

13 Lange, J., 'Interdigitated Stripline Quadrature Hybrid', *IEEE Trans.*, **17**(11), 1150–1, Dec. 1969.
One of the original Lange coupler papers to be published. Contains an experimental account of the circuit performance.

14 Osmani, R. M., 'Synthesis of Lange Couplers', *IEEE Trans.*, **29**(2) 168–70, Feb. 1981.
In this paper a simple set of design equations were given for the Lange coupler geometry given in this chapter. Some experimental results are also given as well as an approximate coupled line synthesis technique.

15 Ou, W. P., 'Design Equations for an Interdigitated Directional Coupler', *IEEE Trans.*, **19**(2), 253–5, Feb. 1975.
Here Ou derived a set of easy-to-use Lange coupler analysis equations that were based on a solution of a simplified matrix representation of the coupler.

16 Besser, L., 'Computer Tweaking Yields 3-dB Interdigitated Coupler', *Microwave Systems News*, **9**(9), 114–18, Sept. 1979.
In this review article the design of a Lange coupler including the effects of discontinuities is discussed. The design is made possible by the use of an advanced commercially available computer program.

APPENDIX A

Extrinsic Mathematical Functions

Function	Equivalent
Inverse sine	ARCSIN(X) = ATN(X/SQR(1 − X∗X))
Inverse cosine	ARCCOS(X) = −ATN(X/SQR(1 − X∗X)) + PI/2
Inverse secant	ARCSEC(X) = ATN(X/SQR(X∗X − 1))
Inverse cosecant	ARCCSC(X) = ATN(X/SQR(X∗X − 1)) + (SGN(X) − 1)∗PI/2
Inverse cotangent	ARCCOT(X) = ATN + PI/2

APPENDIX B

Useful Hyperbolic Functions

$$\sinh \theta = \frac{e^\theta - e^{-\theta}}{2}$$

$$\cosh \theta = \frac{e^\theta + e^{-\theta}}{2}$$

$$\tanh \theta = \frac{\sinh \theta}{\cosh \theta} = \frac{e^\theta - e^{-\theta}}{e^\theta + e^{-\theta}}$$

$$e^{j\theta} = \cos \theta + j \sin \theta$$

$$e^{-j\theta} = \cos \theta - j \sin \theta$$

$$\cos \theta = \frac{e^{j\theta} + e^{-j\theta}}{2}$$

$$j \sin \theta = \frac{e^{j\theta} - e^{-j\theta}}{2}$$

$$\sinh j\theta = j \sin \theta$$

$$\cosh \theta = \cos j\theta$$

$$j \sinh \theta = \sin j\theta$$

$$\cosh j\theta = \cos \theta$$

$$\tan j\theta = j \tanh \theta$$

$$\tanh j\theta = j \tan \theta$$

$$\sinh^{-1}(\theta) = \log_e [\theta + (\theta^2 + 1)^{\frac{1}{2}}]$$

$$\cosh^{-1}(\theta) = \log_e [\theta + (\theta^2 - 1)^{\frac{1}{2}}]$$

Here, θ is a real quantity.

APPENDIX C

Phase and Time Delay Response for Chebyshev and Butterworth Filters of Order n

Chebyshev filter

1 Phase $= -\beta = \dfrac{360}{\pi} \displaystyle\sum_{m=0}^{\infty} \dfrac{\varepsilon^{-(2m+1)\phi_2} C_{2m+1}(\omega)}{(2m+1)\sin(2m+1)\psi}$ degrees $\qquad$ (C.1)

where $\psi = \dfrac{\pi}{2n}$

2 Delay $= \dfrac{-d\beta}{d\omega} = \displaystyle\sum_{m=0}^{n-1} \left[\dfrac{U_{2m}(\omega)\sinh(2n-2m-1)\phi_2}{\varepsilon^2 \sin(2m+1)\psi} \right] \Big/ [1 + \varepsilon^2 C_n{}^2(\omega)]$ seconds

$\qquad$ (C.2)

where

$$\phi_2 = \frac{1}{n} \sinh^{-1}\left(\frac{1}{\varepsilon}\right)$$

and

$U_n(\omega) =$ Chebyshev function order n of the second kind

$$= \frac{\sin(n+1)\cos^{-1}\omega}{\sin[\cos^{-1}(\omega)]}$$

Butterworth filter

1 Phase $= -\beta = \dfrac{360}{2\pi} \displaystyle\sum_{m=0}^{\infty} \dfrac{\omega^{2m}}{(2m+1)\sin(2m+1)\psi}$ degrees $\qquad$ (C.3)

where

$$\psi = \frac{\pi}{2n}$$

2   Delay $= \displaystyle\sum_{m=0}^{n-1} \left[\frac{\omega^{2m}}{\sin{(2m + 1)\psi}} \right] \Big/ [1 + \omega^{2n}]$ seconds   (C.4)

APPENDIX D

Functional Computer Program Index

Index

mode, *see also* mode, designation, 53
  highest, 52, 53
open circuit, 50
RECTGUIDE, 56–58
short circuit, 50
stability, mechanical, 55
wavelength, cut-off, 51
  guide, 51
Wavelength, free space, 9, 41
  cut off, 51, 58, 60, 61

dependence, on dielectric constant, 41
effect on microstrip characteristic
    impedance, 313
guide, 41, 51, 59, 61, 70, 72, 80, 89, 90, 92
practical considerations, power splitter, 332
Wave number, 8

Y-parameter, *see* Short circuit

Z-parameter, *see* Open circuit